Managing
Mississippi and Ohio River
Landscapes

Kenneth R. Olson and Lois Wright Morton

Soil and Water Conservation Society
Ankeny, Iowa

© 2016 by the Soil and Water Conservation Society. All rights reserved.

Edited by Annie Binder
Layout by Jody Thompson
Index by Mary Pelletier-Hunyadi
Maps by Mic Greenberg
Photography by Lois Wright Morton and Kenneth R. Olson unless otherwise noted

On the cover: Front—Satellite image of the confluence of the Mississippi and Ohio rivers during the flood of 2011. The sediment in the Mississippi has a much darker color reflecting the soil organic rich sediment. Photo credit: GeoEye and the USDA Farm Service Agency. Back—Map of the confluence of the Mississippi and Ohio rivers during the flood of 2011.

The following chapters are adapted from previous publications and used with permission:

Chapter 3: Olson, K.R., and F. Christensen. 2014. How waterways, glacial melt waters, and earthquakes re-aligned ancient rivers and changed Illinois borders. Journal of Earth Sciences and Engineering 4(7):389-399.

Chapter 8: Olson, K.R., and L.W. Morton. 2016. Flooding and levee breach impacts on protected agricultural land. Encyclopedia of Soil Science. R. Lal, ed. New York: Taylor and Francis.

Chapter 24: Morton, L.W., and K.R. Olson. 2014. Addressing soil degradation and flood risk decision making in levee protected agricultural lands under increasingly variable climate conditions. Journal of Environmental Protection 5:1220-1234, http://dx.doi.org/10.4236/jep.2014.512117.

Printed in the United States of America
10 9 8 7 6 5 4 3 2 1
ISBN 978-0-9856923-1-5

Library of Congress Cataloging-in-Publication Data
Names: Olson, Kenneth R. (Kenneth Ray), 1947- author. | Morton, Lois Wright, 1951- author.
Title: Managing Mississippi and Ohio River landscapes / Kenneth R. Olson, Lois Wright Morton.
Description: Ankeny, Iowa : Soil and Water Conservation Society, [2016] | Includes index.
Identifiers: LCCN 2016023855| ISBN 9780985692315 (hardcover) | ISBN 0985692316 (hardcover)
Subjects: LCSH: Watersheds--United States. | Mississippi River Watershed. | Ohio River Watershed.
Classification: LCC GB990 .O47 2016 | DDC 333.730977--dc23
LC record available at https://lccn.loc.gov/2016023855

The Soil and Water Conservation Society is a nonprofit scientific and professional organization that fosters the science and art of natural resource management to achieve sustainability. The Society's members promote and practice an ethic that recognizes the interdependence of people and their environment.

Soil and Water Conservation Society
945 SW Ankeny Road, Ankeny, IA 50023
www.swcs.org

To the next generation of soil, water, and social
scientists as they seek to understand the complexities
of the ever-changing human-natural system.

CONTENTS

List of Figures — vii

List of Maps — xiii

Foreword — xv

Acknowledgements — xvi

Chapter 1. Management of Mississippi and Ohio River Landscapes — 1

Chapter 2. Geologic and Climatic Impacts on Ancient Mississippi and Ohio River Systems — 8

Chapter 3. How Realigned Ancient Rivers Influenced the Border Locations of Eight Central States — 17

Chapter 4. Multifunctional Bottomlands: Sny Island Levee Drainage District — 25

Chapter 5. Conversion of Missouri's Big Swamp to Fertile Agricultural Land — 32

Chapter 6. Upland Diversions and Bottomland Drainage Systems: Intended and Unintended Consequences — 42

Chapter 7. St. Johns Levee and Drainage District Attempts to Mitigate Internal Flooding — 52

Chapter 8. Flooding and Levee Breach Impacts on Protected Agricultural Lands — 61

Chapter 9. Impacts of 2008 Flooding on Agricultural Lands in Illinois and Indiana — 69

Chapter 10. Impacts of 2011 Induced Levee Breaches on Agricultural Lands of the Mississippi River Valley — 76

Chapter 11. Repair of the 2011 Flood-Damaged Birds Point–New Madrid Floodway — 83

Chapter 12. Settlement and Land Use Changes in the New Madrid Floodway — 90

Chapter 13. Impact of Levee Breaches, Flooding, and Land Scouring on O'Bryan Ridge Soil Productivity — 98

Chapter 14. The 2011 Ohio River Flooding of the Cache River Valley in Southern Illinois — 108

Chapter 15. Impacts of the 2011 Len Small–Fayville Levee Breach on Private and Public Illinois Lands — 114

Chapter 16. The City of Cairo, Illinois, at the Confluence of the Mississippi and Ohio Rivers — 123

Chapter 17. Managing River Pressure from the 2011 Record Flood on Ohio and Mississippi River Levees at Cairo — 132

Chapter 18. Navigation and Flooding on the Ohio River — 141

Chapter 19. Managing the Tennessee River Landscape — 150

Chapter 20. Managing the Cumberland River Landscape — 159

Chapter 21. Managing the Upper Mississippi River to Improve Commercial Navigation — 165

Chapter 22. Dredging of the Fractured Bedrock-Lined Mississippi River Channel at Thebes, Illinois — 176

Chapter 23. The Illinois Waterway Connecting the Mississippi River and Great Lakes — 182

Chapter 24. Soil Degradation and Flooding Risk Decision Making in Leveed Agricultural Landscapes — 193

Chapter 25. Managing Ohio and Mississippi River Landscapes for the Future — 202

About the Authors — 213

Index — 215

FIGURES

Figure 1.1 A Birds Point levee breach crater lake — 6

Figure 1.2 The Cairo, Illinois, river gage on the Ohio River — 6

Figure 1.3 One of two Sny River aqueducts — 7

Figure 2.1 Geologic time scale of Earth — 9

Figure 2.2 The Hennepin Canal — 13

Figure 3.1 Starved Rock locks and dam — 19

Figure 3.2 The arch of St. Louis, Missouri — 21

Figure 4.1 A 640-acre settling basin (Kiser) — 28

Figure 4.2 A 640-acre settling basin (McCraney) — 28

Figure 4.3 The Horton-Dutch Creek settling basin — 29

Figure 4.4 The Kiser Creek Diversion channel — 29

Figure 4.5 The Sny River aqueduct under the Kiser Creek Diversion channel — 30

Figure 4.6 The merged Mississippi River and Kiser Creek Diversion levee — 30

Figure 4.7 The Pigeon Creek settling basin — 31

Figure 4.8 No-till corn planted on Sny River watershed uplands — 31

Figure 5.1 The Mingo National Wildlife Refuge — 33

Figure 5.2 Agricultural lands created by draining Big Swamp — 35

Figure 5.3 Fertile, irrigated soils at the eastern base of Crowley's Ridge — 35

Figure 5.4 The Little River Drainage District diversion channel — 38

Figure 5.5 The Little River Drainage District diversion embankment — 38

Figure 5.6 Total cropland acres from 1880 to 2012 for five crops in counties in southeast Missouri — 40

Figure 5.7 Cropland acres by crop from 1880 to 2012 in seven counties of southeast Missouri — 41

Figure 6.1 Illinois Shawnee Upland mining operation — 44

Figure 6.2 The aqua green headwaters of the Castor River — 44

Figure 6.3 Reductions in woodland acres and increases in land in farm acres in southeast Missouri — 46

Figure 6.4 The murals on the Cape Girardeau floodwall — 47

Figure 6.5 Mississippi flooding at Thebes, Illinois, on January 5, 2016 — 48

Figure 6.6 Flooding after the Len Small levee breached on January 2, 2016 — 49

Figure 7.1 The outlet of Main Ditch in middle Cache River valley — 54

Figure 7.2 The New Madrid Floodway setback levee gate — 54

Figure 7.3 Floodwater heights in 1997, 2008, and 2011, and corresponding flooded areas — 55

Figure 7.4 Repaired levee breach south of Commerce, Missouri — 56

Figure 7.5 A land auction sign on land protected by the Commerce farmer levee — 56

Figure 7.6 The pond at Big Oak Tree State Park — 58

Figure 8.1 A corn field covered by a sand delta created by a levee breach — 62

Figure 8.2 A breached levee on the Embarras River in June of 2008 — 62

Figure 8.3 The Commerce farmer levee located south of Commerce, Missouri — 62

Figure 8.4 A reconstructed levee at Birds Point, Missouri — 63

Figure 8.5 The concrete floodwall on the east side of Cairo, Illinois — 63

Figure 8.6 Anatomy of a sand boil — 64

Figure 8.7 A crater lake extending through a levee breach — 65

Figure 8.8 A water-filled O'Bryan Ridge gully — 65

Figure 8.9 Bulldozers filling wetlands and ponds with soil at O'Bryan Ridge — 66

Figure 8.10 Tillage to incorporate sediment left behind from flooding — 66

Figure 8.11 Trees transported by floodwaters into adjacent fields — 67

Figure 9.1 Aerial view of ponded depressions and potholes on the uplands in central Iowa after heavy rain — 70

Figure 9.2 Rapid runoff from central Iowa flooding in 2008 — 71

Figure 9.3 Temporary storage ponds surrounded by flooding from the Embarras River near Sainte Marie, Illinois — 71

Figure 9.4 The Embarras River cutting through a levee near Sainte Marie, Illinois, in 2008 — 71

Figure 9.5 The Embarras River, Illinois, adjacent to a missing section of the levee and a crater lake — 73

Figure 9.6 Fields covered by deltaic sand deposits, water, and trees following a levee break on the Embarras River — 73

Figure 9.7 Sand piled for transport off the agricultural lands previously protected by a levee — 74

Figure 9.8 Drainageways, waterways, and culverts clogged by corn stalks after flood events — 74

Figure 10.1 The New Madrid Floodway and basin floodwater on May 20, 2011 — 78

Figure 10.2 Excavator cleaning out sediment in road ditches — 80

Figure 10.3 An irrigation system overturned by floodwater and wind and buried by a sand deposit — 81

Figure 10.4 Gullies extending into cropland of O'Bryan Ridge — 81

Figure 10.5 Gullies and channels created from May 4 to May 16, 2011 — 82

Figure 10.6 A crater lake at the Big Oak Tree frontline levee — 82

Figure 11.1 Repairs to the Big Oak Tree levee blast site in October of 2011 — 85

Figure 11.2 The 2011 Birds Point levee patch built to 51 feet — 85

Figure 11.3 The site of the third explosion on the frontline levee near Big Oak Tree State Park — 86

Figure 11.4 Wet, low-lying soils on October 24, 2011 — 86

Figure 11.5 Drowned wheat collecting significant sediment and protecting against soil erosion — 87

Figure 11.6 Organic and clay coating on plants after flooding and floodwater drainage — 87

Figure 11.7 Removal of sediment from a drainage ditch — 88

Figure 11.8 The 2011 soybean crop planted around gullies — 88

Figure 11.9 A home rebuilt on a mound of soils to protect against future flooding — 89

Figure 12.1 A site on US 60 to commemorate sharecroppers' loss of livelihoods — 91

Figure 12.2 One of the 80 homes in the floodway damaged by floodwater in 2011 — 93

Figure 12.3 Farm structures damaged by floodwaters and wind when the floodway was opened in 2011 — 93

Figure 12.4 A new home built on 10 feet of soil materials to prevent future flood damage — 94

Figure 12.5 Abandoned homes in the village of Pinhook in November of 2011 — 94

Figure 12.6 Birds nesting in the low, wet areas adjacent to the Big Oak Tree levee blast site — 96

Figure 13.1 Land scouring of the bottomland located below the gully fields — 99

Figure 13.2 May of 2011 aerial view of O'Bryan Ridge gully fields — 100

Figure 13.3 Gully development in the O'Bryan Ridge soybean field — 101

Figure 13.4 Wetlands and ponds formed from deep gullies in O'Bryan Ridge — 103

Figure 13.5 Soybeans growing on Udifluvents and adjacent land scoured ridgetops — 103

Figure 13.6 Rill and gully erosion after soybean planting in 2013 on regraded slopes and filled land — 105

Figure 14.1 Wetland habitats reestablished at Grassy Slough near Karnak in the middle Cache River watershed — 110

Figure 14.2 Middle Cache River water redirected into the Mississippi River by the diversion embankment — 110

Figure 14.3 The Post Creek Cutoff — 111

Figure 14.4 Main Ditch in Cache River valley — 112

Figure 14.5 The Reevesville levee next to Bay Creek — 113

Figure 14.6 The unrepaired Karnak levee breach on middle Cache River — 113

Figure 15.1 Diagram of levee topping by the Mississippi River above flood stage — 116

Figure 15.2 Bald cypress trees and American lotus at Horseshoe Lake conservation area — 117

Figure 15.3 Vegetation management on the Commerce to Birds Point mainline levee — 117

Figure 15.4 Land scouring, gullies, and erosion north of the Len Small levee breach — 120

Figure 15.5 Small berms built around farmsteads following the 2011 Len Small levee breach — 121

Figure 15.6 A home surrounded by a farmer-built levee — 122

Figure 16.1 The Cairo floodwall built on the Ohio River side — 124

Figure 16.2 An open floodwall gate in Cairo — 124

Figure 16.3 Fort Defiance State Park south of Cairo — 128

Figure 16.4 A raised railroad bed that serves as a levee for the northern boundary of Cairo, Illinois — 128

Figure 16.5 Remnants of the Birds Point fuse plug levee and a crater lake — 129

Figure 16.6 Barges anchored on the Mississippi River bank next to the flooded Fort Defiance State Park — 130

Figure 17.1 Commercial Avenue sinkholes in Cairo — 133

Figure 17.2 Diagram of a sand boil at Cairo on April 28, 2011 — 133

Figure 17.3 The 40th Street mega sand boil in Cairo — 135

Figure 17.4 The exposed Ohio River floodwall during the 2012 drought — 135

Figure 17.5 Relief wells used to pump groundwater into a drainage ditch to relieve substratum pressure — 136

Figure 17.6 Relationship of the Mississippi River, slurry trench, levee, relief wells, ditches, and Route 3 north of Cairo — 138

Figure 17.7 Bentonite bags dumped into ponds and mixed with water to create slurry — 138

Figure 17.8 A slurry trench between the levee and Mississippi River — 139

Figure 17.9 A soybean field located between the upper Mississippi River and the Cairo, Illinois, levee in 2011 — 140

Figure 18.1 The confluence of the Allegheny and Monongahela rivers in Pittsburgh, Pennsylvania — 143

Figure 18.2 The Olmsted Lock and Dam — 143

Figure 18.3 Lock and Dam 52 downstream of Brookport, Illinois — 143

Figure 18.4 Lock and Dam 53 near Olmsted — 145

Figure 18.5 The McAlpine Dam at Louisville, Kentucky, and Clarksville, Indiana — 145

Figure 18.6 The Wheeling Suspension Bridge and the US Interstate 70 Fort Henry Bridge — 146

Figure 18.7 Major floods and flood crest heights on the Ohio River from 1884 to 2011 — 147

Figure 18.8 The Covington to Cincinnati bridge over the Ohio River — 148

Figure 18.9 A giant 5,304-ton lift built on the construction site of the Olmsted Lock and Dam — 149

Figure 19.1 Flood crests at Paducah, Kentucky, marked on a downtown building — 151

Figure 19.2 The river side of the floodwall at Paducah, Kentucky — 151

Figure 19.3 The Kentucky Dam at Gilbertsville, Kentucky — 152

Figure 19.4 The Tennessee River at the confluence of the French Broad and Holston rivers — 153

Figure 19.5 The Tennessee and Ohio river confluence east of Paducah, Kentucky — 153

Figure 19.6 The Tennessee River in Chattanooga as viewed from Lookout Mountain — 154

Figure 19.7 Murals on the Paducah floodwall — 155

Figure 19.8 Watts Bar Dam on the Tennessee River — 155

Figure 19.9 Nuclear cooling towers near Watts Bar and adjacent to the Tennessee River — 157

Figure 20.1 The Cumberland River and Ohio River confluence at Smithland, Kentucky — 161

Figure 20.2 A railroad built above the Barkley Lock and Dam on the Cumberland River — 161

Figure 20.3 The Barkley Dam on the Cumberland River — 163

Figure 20.4 The Barkley Canal between Kentucky and Barkley reservoirs — 163

Figure 20.5 Puddingstone (quartz pebbles) in small streams in the Land Between the Lakes — 164

Figure 21.1 The confluence of the Missouri River and Mississippi River — 167

Figure 21.2 Monks Mound, a Mississippian mound located at Cahokia, Illinois — 169

Figure 21.3 The St. Anthony Lock and Dam and the Falls of St. Anthony, Minneapolis, Minnesota — 170

Figure 21.4 List of pools and locks on the upper Mississippi River — 171

Figure 21.5 Barge traffic on the upper Mississippi River near St. Louis — 172

Figure 21.6 Barge traffic at the Illinois River confluence with the Mississippi River at Grafton, Illinois — 172

Figure 21.7 The Hennepin Canal walking path at Rock Island, Illinois — 172

Figure 21.8 The Little Missouri River in Theodore Roosevelt National Park, North Dakota — 173

Figure 21.9 Lookout Point near Mississippi Palisades State Park — 173

Figure 21.10 The backwaters of the upper Mississippi River south of Lake Pepin — 174

Figure 22.1 Exposed bedrock near the Thebes railroad bridge — 178

Figure 22.2 River bottom bedrock dredged using an excavator — 179

Figure 22.3 The historic Thebes courthouse — 180

Figure 22.4 The reinforced concrete, two track railroad bridge connecting Illinois to Missouri — 180

Figure 22.5 Exposed valley wall bedrock at Thebes, Illinois — 181

Figure 23.1 Barges pushed by a tugboat downstream on the Chicago Sanitary and Ship Canal — 184

Figure 23.2 The old Illinois and Michigan Canal meets the Chicago Sanitary and Ship Canal — 186

Figure 23.3 Flying Asian carp in the Illinois River at the Marseilles Dam — 186

Figure 23.4 Electric fish barriers constructed in the Chicago Sanitary and Ship Canal — 188

Figure 23.5 Seven partially sunken barges at the Marseilles Dam in April of 2013 — 188

Figure 23.6 A partially sunken barge in April of 2013 — 189

Figure 23.7 Stream bank erosion in the spring of 2013 at Illini State Park — 189

Figure 23.8 Removal of damaged items from flooded homes in the Illinois River bottomlands — 190

Figure 23.9 Trucks to haul away damaged household items in the town of Marseilles, Illinois — 191

Figure 23.10 A dike and pond to temporarily manage the water pool behind the dam at Marseilles — 191

Figure 24.1 New Madrid and Mississippi counties' land uses from the USDA Census of Agriculture, 1930 to 2007 — 195

Figure 24.2 New Madrid and Mississippi counties' major crops from the USDA Census of Agriculture, 1930 to 2007 — 196

Figure 24.3 Wetlands and ponds in the gullies of O'Bryan Ridge replace a productive soybean field — 198

Figure 25.1 The confluence of the Mississippi and Ohio rivers south of Cairo, Illinois, during the flood of 2011 — 204

MAPS

Map 1.1 The location of the Mississippi and Ohio river basins — 2

Map 1.2 The confluence of the Mississippi and Ohio rivers during the flood of 2011 — 3

Map 1.3 The 1803 Louisiana Purchase from the French — 4

Map 2.1 The Gulf Coastal Plain and the central Interior Lowlands of the United States — 10

Map 2.2 The southernmost glacial advance in the Northern Hemisphere during the Pleistocene — 11

Map 2.3 Areas of surficial deposits left by the Kansan, Illinoian, and Wisconsinan glaciers — 11

Map 2.4 Ancient Mississippi, Ohio, and Tennessee rivers and their current river channels — 12

Map 2.5 The flow paths of the ancient rivers of the eastern Mississippi and Ohio basin — 14

Map 2.6 The Illinois land bridge between the Southern Appalachians and the Ozark Highlands — 15

Map 2.7 Historic Lake Agassiz in the Northern Hemisphere — 15

Map 3.1 First proposed northern Illinois state boundaries within Illinois Territory — 18

Map 3.2 The ancient Mississippi River flow and state boundaries — 20

Map 3.3 The path of the ancient Ohio River through southern Illinois — 22

Map 3.4 Location of the current borders of the state of Illinois — 23

Map 3.5 The effect of ancient Mississippi and Ohio river realignment on central state boundaries — 24

Map 4.1 Sny River bottomlands and the adjacent uplands watershed — 26

Map 5.1 Five drainage ditches in southern Little River Drainage District in Missouri — 33

Map 5.2 The Ozark Plateau in the Headwaters Diversion watershed — 34

Map 5.3 The St. Francis River watershed — 36

Map 6.1 The Headwaters Diversion watershed — 43

Map 6.2 The Headwaters Diversion channel — 45

Map 7.1 St. Johns Bayou Drainage District and the New Madrid Floodway in Missouri — 53

Map 9.1 Total rainfall between May 30 and June 12, 2008, in the north-central United States — 70

Map 9.2 The Wabash watershed and major tributaries — 72

Map 10.1 The Ohio and Mississippi river confluence at Cairo, Illinois — 77

Map 11.1 Birds Point–New Madrid Floodway — 84

Map 12.1 Proposed changes to the Birds Point–New Madrid Floodway — 93

Map 13.1 The O'Bryan Ridge gully fields in the New Madrid Floodway — 99

Map 13.2a March of 2011 soils prior to levee breaching in the floodway on O'Bryan Ridge — 102

Map 13.2b May of 2011 soils following levee breaching in the floodway on O'Bryan Ridge — 102

Map 13.3a October of 2013 moderately eroded soils of O'Bryan Ridge — 104

Map 13.3b April of 2014 O'Bryan Ridge ponds drained and wetland areas partially filled — 105

Map 14.1 Cache River valley in southern Illinois — 109

Map 15.1 Len Small levee in Alexander County, Illinois — 115

Map 15.2 Thebes Gap and the Illinois and Missouri bottomlands — 118

Map 15.3 Levee-protected river bottomlands in Missouri, Arkansas, Tennessee, and Kentucky — 119

Map 16.1 Cairo, Illinois, at the confluence of Mississippi and Ohio rivers — 125

Map 17.1 Location of sinkholes and sand boils in Cairo, Illinois, in April of 2012 — 134

Map 17.2 Relief wells and slurry trenches installed after the flood of 2011 — 137

Map 18.1 The Ohio River watershed — 142

Map 18.2 Locks and dams on the Ohio River — 144

Map 19.1 The Tennessee River basin — 152

Map 19.2 The Kentucky Dam and lake on the Tennessee River — 156

Map 20.1 The Cumberland River watershed — 160

Map 20.2 The Kentucky Reservoir on the Tennessee River and the Barkley Reservoir on the Cumberland River — 162

Map 21.1 Six major subwatersheds of the Mississippi River basin — 166

Map 21.2 Locks and dams on the upper Mississippi River — 168

Map 22.1 Bedrock controlled stretch of Mississippi River at Thebes, Illinois — 177

Map 23.1 The St. Lawrence Continental Divide prior to the 1900s — 183

Map 23.2 Locks and dams on the Illinois Waterway — 185

Map 23.3 The reversed flow of the Chicago River — 187

Map 23.4 Lock and dam system on the Illinois River at Marseilles — 190

Map 25.1 Major levees breached on the Ohio and Mississippi rivers and their tributaries in 2011 — 203

Foreword

"The face of the water, in time, became a wonderful book — a book that was a dead language to the uneducated passenger, but which told its mind to me without reserve, delivering its most cherished secrets as clearly as if it uttered them with a voice. And it was not a book to be read once and thrown aside, for it had a new story to tell every day."

— Mark Twain, *Life on the Mississippi*

Ken Olson and Lois Wright Morton have written a book on the wonderful stories of the Mississippi and Ohio rivers. They interpret the language of the rivers and their management through the keen eyes of a soil scientist and a sociologist. More importantly, they interpret the two rivers' complex human and natural relationships in terms that both scientists and non-scientists can understand.

The demand for soil and water resources is ever increasing, and wise land use planning involves diverse and often conflicting goals. These issues require both technical expertise and a human touch. Drs. Olson and Morton provide exactly that, with historical background, unique perspective, clear understanding, and sharp insight to confront current problems and discover new opportunities these two great rivers present.

Public and private landowners and managers of the Mississippi and Ohio river landscapes will benefit greatly from the intensive research and presentation of case studies. Other beneficiaries might include soil scientists, sociologists, conservationists, wetland specialists, human and physical geographers, urban planners, public health specialists, economists, geomorphologists, geologists, hydrologists, agronomists, foresters, and river lovers in general.

A book of this scope and detail is not to be read once and thrown aside, but to be reviewed and studied over time. For as the young steamboat cub pilot Mark Twain stated, these rivers indeed have "a new story to tell every day."

Samuel J. Indorante, Ph.D.
January 2016
Certified Professional Soil Scientist/Soil Classifier
Adjunct Professor of Plant, Soil Science and Agricultural Systems
Southern Illinois University-Carbondale, Carbondale, Illinois

ACKNOWLEDGEMENTS

Our thanks to the following scientists who contributed to one or more chapters in the book: Fred Christensen (chapter 3), David Speidel (chapters 5 and 6), Mike Reed (chapter 9), John Sloan (chapter 13), Jeff Matthews (chapter 13), and Birl Lowery (chapter 25).

We also want to acknowledge and thank Sam Indorante, Larry Dowdy, David Speidel, Birl Lowery, Charles Camillo, John Crivello, Jim Lang, and Bill Kruidenier for fact checking two or more chapters. In addition, Jim Lang proofread all 25 chapters. Thanks to Renea Miller for help with the map book cover, tables, and documents for the final read.

Special thanks to Mic Greenberg, our amazing graphic mapmaker, who transformed 60 hand-drafted drawings and maps into professional electronic images. He also rendered the front cover image and the back cover map.

Our thanks to the funders of our great rivers research over the last few years and supporters of this book. Primary and direct support for this book comes from Department of Natural Resources and Environmental Sciences in the College of Agricultural, Consumer, and Environmental Sciences at University of Illinois, Urbana, Illinois; and from the Department of Sociology in the College of Agriculture and Life Sciences at Iowa State University, Ames, Iowa. This work was completed in cooperation with Regional Research Project No. 15-372 and North Central Regional Project No. NCERA-3 Soil Survey; the Director of the Illinois Office of Research; and the Iowa Agriculture and Home Economics Experiment Station, North-Central Regional Project No.1190, Catalysts for Water Resource Protection and Restoration: Applied Social Science Research. Additional support includes Heartland Regional Water Coordination Initiative, USDA National Institute of Food and Agriculture Integrated Water Program under agreement 2008-51130-19526; and the National Great River Research and Education Center in Alton, Illinois.

Management of Mississippi and Ohio River Landscapes

Two powerful rivers, the Ohio and Mississippi, and their tributaries drain more than 41% of the interior continental United States of America (map 1.1). Their shifting paths have shaped and reshaped the landscapes through which they flow and the confluence (map 1.2) where their sediment-laden waters comingle on the voyage to the Gulf of Mexico. Changing climates and extreme weather events over the millennia have carved new channels through river bottomlands, leaving rock-exposed uplands and fertile valleys behind while altering the location where the Ohio and Mississippi rivers meet. These great rivers often became state boundaries, and their historic realignments have added and subtracted land from many states that border them. For much of their history, the lands adjacent to these rivers were low-lying bottomlands that, unconstrained by human structures, flooded with the seasons.

However, in the last century these rivers—highways of trade, settlement, and adventure—have become agricultural economic engines as humans reengineered the rivers and their bottomlands. Locks and dams, levees and floodwalls, aqueducts, and an extensive system of reservoirs have been constructed to manage these rivers for navigation and to protect communities, agriculture, and other high-value land uses. Alongside attempts to control the height and courses of these rivers and their tributaries, diversion ditches and systematic draining of interior swamps and wetlands have transformed hydric but fertile soils into highly productive, intensely managed agricultural lands. Paradoxically, these infrastructure investments, intended to facilitate navigation and reduce direct risks of flooding, have led to unexpected consequences to the larger ecosystem. Recent levee breaching has created unanticipated shocks to the river ecosystem while generating new knowledge about hydrology, soils, and the vegetation of rivers and their bottomlands. The occasional failure of well-engineered structures reminds us that the river landscape is a complex human-natural system. This complex system is dynamic, ever changing, and often managed based on assumptions of steady state—expectations that the past predicts the future. These assumptions do not well prepare communities to deal with diverse and often competing societal goals under an increasingly variable climate, increasing populations, and intensified land uses [1, 2].

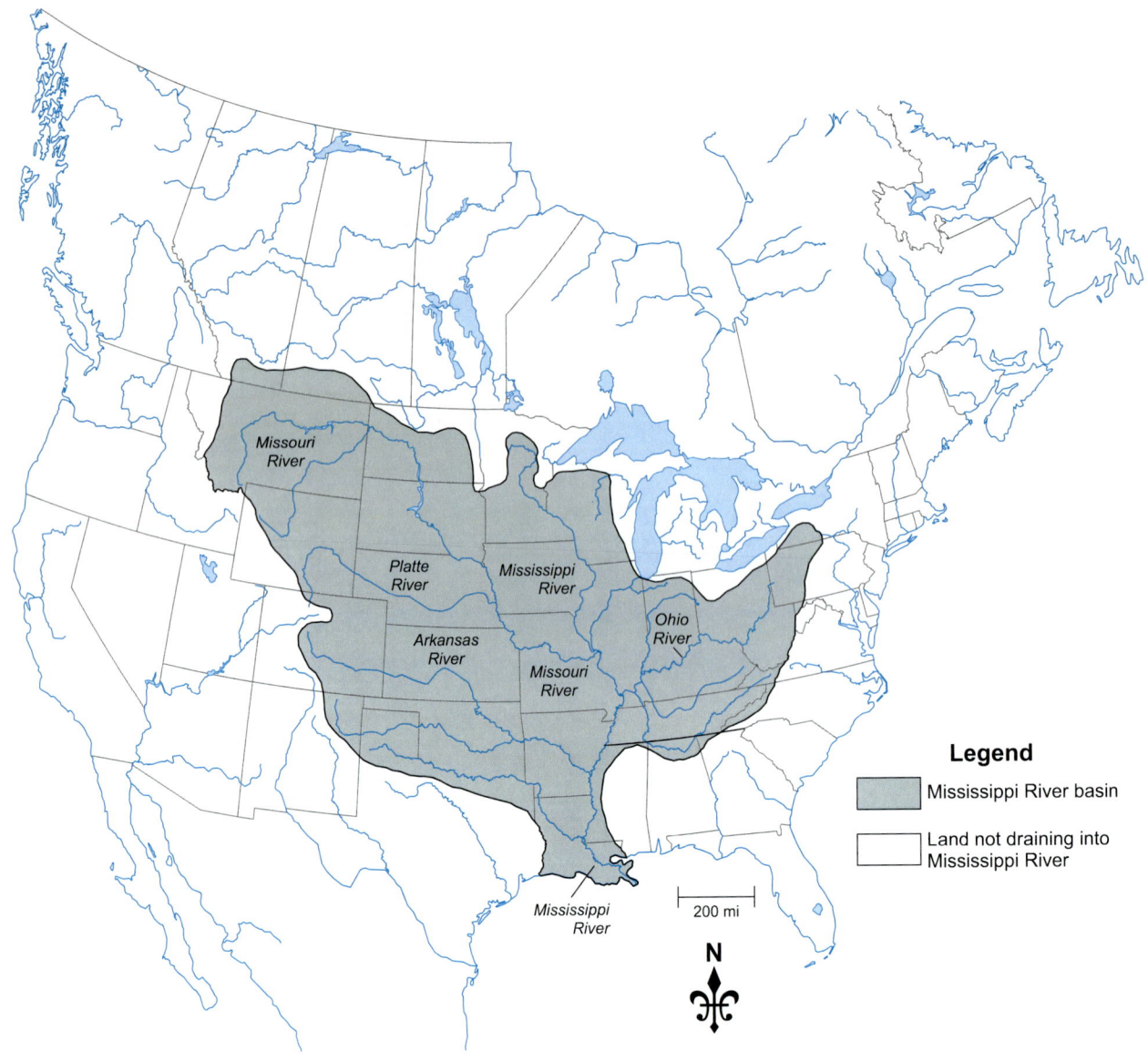

MAP 1.1 The location of the Mississippi and Ohio river basins, which occupy 41% of the continental United States.

The Great Flood of 2011 and drought of 2012 [3] well illustrate some of the vulnerabilities and unintended consequences that arise from designing and managing river systems without taking into account their changing nature and the need for more flexible-adaptive capacities [1, 4, 5]. Following the Great Flood of 1927, it became apparent that the extensive use of levees, channelization, and confinement of the rivers was inadequate to effectively contain these great rivers [6]. The subsequent addition of reservoirs upstream of the confluence of the Ohio and Mississippi rivers at Cairo, Illinois, and four downstream floodways below Cairo was a substantive shift by the US Army Corps of Engineers (USACE) to strategically incorporate a dispersion risk management strategy with confinement engineering [6, 7]. The underlying premise of dispersion is to replicate the natural floodplain functions of bottomlands, which historically served as outlets to rivers under flood conditions.

The 2011 induced levee breaching of the Birds Point–New Madrid Floodway reaffirmed the effectiveness of dispersion management and its capacity to protect the integrity of communities and land uses along the larger river system. However, many homeowners and landowners were unprepared for the consequences of opening the floodway. With the reemergence of social tensions and competing social values for the uses of river bottomlands, public policy makers, community leaders, environmental advocates, and government agencies are challenged to reassess the impacts of leveed structures that in recent history have protected urban and rural agricultural land uses. Although most river flooding has repetitive patterns that reoccur seasonally and over periods of years presenting known risks, floods are complicated in their range of intensity

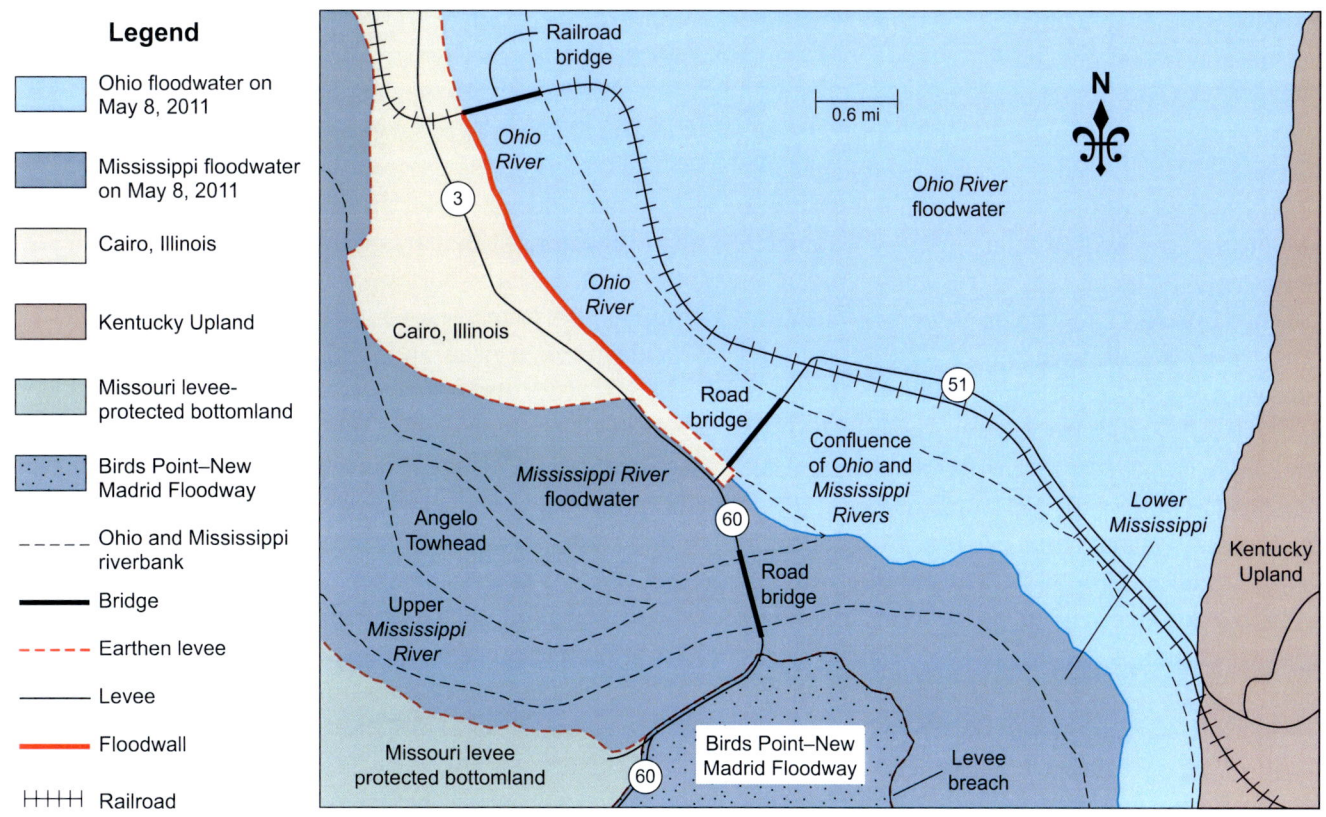

MAP 1.2 The confluence of the Mississippi and Ohio rivers during the flood of 2011. The sediment in the Mississippi has a much darker color reflecting the soil organic–rich sediment.

and duration and can result in unexpected consequences [8]. Levee breaching and other structural failures are often the result of unusually large runoff in a system cut off from its historical floodplain. Science is just beginning to understand the relationship between the river and its floodplain, the beneficial aspects of flooding, and the roles of wetlands and riparian corridors as well as the extensive social and economic damage floods wreak on the livelihoods of those along the river.

There is evidence that a new type of river management is needed, one that goes beyond the current confinement-dispersion strategy. Park et al. call for resilience management [1], an adaptive management approach to changing conditions that preserves the natural functions of the river ecosystem in ways that minimize catastrophic failure of engineered structures. This concept is just emerging, and there remain many practical details to work out. Some of these details involve better inventories and assessments of the soil resource immediately after levee breaches and subsequent flood events. Updated soil surveys and land scouring and deposition surveys can be used to create a better understanding of the ecosystem services the floodplain provides and guide restoration decisions when engaging and informing the public so as to come to politically acceptable agreement on management and land use decisions.

Early Attempts to Manage the Mississippi and Ohio River Landscapes

The first recorded attempt to manage the Mississippi and Ohio river landscapes occurred in 1717. New Orleans, a deep water port, was established by the French on the Mississippi River about 50 miles from the current Gulf of Mexico. The original settlement was 14 city blocks with drainage ditches around each block; these ditches were the first recorded attempt to manage the landscape. The first levee along the banks of the Mississippi River was allegedly erected in 1718, but this date has not been confirmed. Documented levees were built in 1722 by the French. The levees constructed in 1722 were four-foot-high earthen levees, which began a 300-year struggle to combat high water with embankments. The levees were privately maintained by area landowners, who used slaves, state prisoners, and poverty-stricken Irish immigrants to perform the deadly work. Situated on land with poorly drained soils and unfavorable topography, the New Orleans settlement was prone to periodic flooding by the Mississippi River. The city was a few feet above the sea on the deltaic floodplain of the Mississippi River, which was settling at a rate of between 2 and 10 feet per century.

Early US attempts to manage Mississippi and Ohio rivers can be tracked back to June 16, 1775, or shortly before the United States was established. The Continental Congress organized General George Washington's army with a chief engineer, Colonel Richard Gridley. The USACE, as it is known today, was established by President Thomas Jefferson in 1779. In 1803, the United States acquired New Orleans and 828,000 square miles mostly located in the Mississippi River valley from the French. The land transfer became known as the Louisiana Purchase (map 1.3) at the cost of $15 million. By 1811, steamboats started to arrive in New Orleans. When Lewis and Clark headed down the Ohio River in 1803, the water depth was very low. It was a dry year, and navigation was a challenge since locks and dams had not yet been built. The major navigation problem that delayed steamboat travel on the Ohio River was the Falls of the Ohio River near Louisville, Kentucky. Steamboats could only travel over the falls during times of flooding or high water. Consequently the steamboats dropped passengers and freight off at one end of the falls for overland transport to the opposite end of the falls, where they were picked up by another steamboat.

The General Survey Act of 1824 authorized the use of army engineers to survey roads and canal routes. In 1824 Congress also passed a river improvement act to promote navigation on the Ohio and Mississippi rivers and to remove sandbars on the Ohio and sawyers (fallen trees stuck on the bottom of a river) and snags on the Mississippi River. The act, often called the first rivers and harbors legislation, combined authorizations for both surveys and projects. In 1825, construction began on a canal to bypass the Falls of the Ohio, and by 1830 the privately financed Louisville and Portland Canal was completed. The canal was constructed by hand tools and animal-drawn scrappers and carts. When completed, the two-mile-long canal had three locking chambers with a total lift of 26 feet. Increasing steamboat trade on the Cumberland River by 1825 led Congress to survey the river and finance river improvements to transport eastern Kentucky coal, Tennessee produce, and lumber throughout the region.

In 1859, a levee breach near New Orleans flooded a hundred city blocks and displaced thousands of residents. In response, Congress passed the Swamp Act and sponsored the survey of the lower Mississippi River. The funds sparked a debate on how to best control the Mississippi River—more levees versus outlets and human-made outlets and spillways. In addition, the American Civil War between 1861 and 1865 damaged the levee system in New Orleans. After the war, the State Board of Levee Commissioners authorized the replacement of damaged sections of the levee system, but little work was completed by 1870.

In 1879, Congress created the Mississippi River Commission (MRC) to replace the State Board of Levee Commissioners. Still serving today, the MRC has a seven-member governing body. Three of the officers are from the USACE, including the chairman who is the final decision maker when it comes to opening the floodways. Another member is an admiral from the National Oceanic and Atmospheric Administration. The other three members are civilians, and at least two of the civilian members are civil engineers. Each member is appointed by the president of the United States. Senate confirmation of the selection is no longer necessary. The MRC is the lead federal agency responsible for addressing the improvement, maintenance, and control of the Mississippi River. The MRC and USACE sought to deepen the Mississippi River and make it more navigable and less likely to flood. In 1885, the USACE adopted a "levees-only" policy. For the next 40 years, the USACE

MAP 1.3 United States and Canadian lands were part of the 1803 Louisiana Purchase from the French.

extended the levee system, sealing many of the river's natural outlets, including the ones near New Madrid and Cape Girardeau, Missouri, along the way. By 1926, levees ran from Cairo, Illinois, to New Orleans.

Lock and dam construction on the Ohio, Cumberland, and Tennessee began in the late nineteenth century and continued into the twentieth century. In 1885, the first complete lock and dam project built by the USACE on the Ohio River was at Davis Island, a few miles south of Pittsburgh, Pennsylvania. The project proved its worth, and in 1910 Congress passed the Rivers and Harbors Act. The act authorized construction of a lock and dam system to provide a 9-foot channel for the entire length of the Ohio River. The "canalization" project was completed in 1929 and consisted of 51 movable dams with wooden wickets. The 600-by-110-foot lock chamber was used during low water to move boats up or down stream. During high water, the wickets were laid flat on the riverbed to allow vessels to use the main river channel and bypass the locks.

The Tennessee Valley Authority (TVA) was created in the 1930s as part of President Franklin D. Roosevelt's vision to address unemployment, rural poverty, and bring the country out of the Great Depression. The TVA, as authorized by Congress, is a unique public-private corporation with multiple missions, including hydroelectricity production, river navigation, flood control, malaria prevention, and land management (e.g., reforestation and erosion control). Built and managed by the TVA, the reservoirs and systems of locks and dams on the Tennessee and Cumberland rivers and their tributaries have continued to be social, cultural, and economic sources of prosperity for the region.

The Flood Control Act of 1936 made flood control management a federal policy and the USACE the major federal flood control agency. On December 1, 1941, the USACE mission was expanded to include civil works such as hydroelectric energy provision, recreation opportunity creation, natural disaster response, and environmental preservation and restoration. The USACE initiated the Ohio River Navigation Modernization Program in the 1950s. The new dams made of concrete and steel replaced the moveable wicket dams with permanent nonmovable structures. Each dam has two adjoining locks designed to accommodate 15 barges and a tow that can lock through in one maneuver. This has reduced locking-through time and the wait time for other vessels. In the 1940s, the TVA built the Kentucky Dam on the Tennessee River to better control the fast rise of the Ohio River during spring rains. The river has been dammed numerous times over the years, primarily by the TVA. The Barkley Dam, a 58,000-acre reservoir in Kentucky, was constructed by the USACE across the Cumberland River and completed in 1966. The lake is maintained at different levels throughout the year for flood control and navigation purposes.

The Mississippi and Ohio rivers have been managed since the 1800s by the USACE in partnerships with the MRC, TVA, and states with levee and drainage districts. Much of their efforts have been to reduce the effects of flooding on agricultural bottomlands and river cities and to create shipping channels that can function in droughts. Since the 1970s and into present time, the USACE river managers have invested substantively in infrastructure maintenance and replacement. The entire lock and dam system on the Ohio River will have been upgraded and replaced once the Olmsted Lock and Dam is completed in 2020 (see chapter 18). River siltation is an annual problem, and ongoing dredging is required to keep port city harbors open and assure navigation depths. A variable and changing climate continues to create natural and human catastrophes as evidenced by the 2011 record flood at the confluence of the Mississippi and Ohio rivers. This record flood, reaching 61.7 feet on the Cairo, Illinois, river gage (figure 1.2), was followed by a near-record drought in 2012 that reduced the Ohio River depth to 7.5 feet above the 9-foot-deep shipping channel, resulting in only 16.5 feet of water for deep drafting barges. Dredging to maintain the shipping channel on the upper Mississippi River near Thebes, Illinois, during the 2012 drought was extremely difficult because of the narrow, bedrock-lined navigation channel, a remnant of an ancient upland bridge [3]. In recent years the USACE has conducted extensive research on wetlands and river ecosystems to better understand the river-land relationship. They have restored, created, and enhanced tens of thousands of acres of wetlands yearly to increase floodplain storage capacities during high water and protect the biodiversity of the natural river ecosystem.

Managing Great River Landscapes for the Future

As we enter the twenty-first century, three major societal concerns have emerged: a changing climate; food insecurity; and homeland security associated with infrastructure, navigation, and water quality and supply. All three themes run throughout this book. Each chapter is a case study from which much can be learned to better plan for the future. These short documentaries focus on the Mississippi and Ohio rivers, how their confluence creates something far greater than the sum of their

FIGURE 1.1 The Birds Point levee breach created a crater lake that extended many feet through the levee.

FIGURE 1.2 The Cairo, Illinois, river gage on the Ohio River is used to determine river height and when it is necessary to open the Birds Point–New Madrid Floodway to relieve downstream river pressure.

flows, and the bottomlands that are sources of wealth and risk to those whose lives are intertwined with the rivers. They illustrate levee-protected agriculture and breach management when the river exceeds flood stage (figure 1.1); dredging in drought to assure a navigable channel; and locks, dams, aqueducts, and reservoirs engineered to tame the two great rivers and their tributaries for human uses. Collectively these chapters portray the multifunctional value of the rivers and human attempts to manage rivers and their bottomlands under intensified agricultural uses, changing settlement patterns, and shifting social values.

Each chapter presents historical geology and underlying soil and landscape features that frame the convergence of recent flood and drought events, the structures built to contain and manage the river system, and the resulting planned and unexpected consequences. The language of the river and its management represents a distinct culture with meanings that can inspire fear, confidence, and uncertainty: sand boils and sinkholes, river readings on the Cairo gage (figure 1.2), earthen levees, floodwalls, channel dredging, aqueducts (figure 1.3), swamp busting, diversions, levee districts, slurry trenches, relief wells, reservoirs, locks and dams, and floodways.

Maps, photographs, and diagrams are extensively used throughout the book and are central to understanding geography, time scales, and soil and water relationships. These visuals offer valuable illustrations and spatial orientations to the rivers and their surrounding landscapes and provide snapshots in time of historical and current geologic and geopolitical boundaries; levee boundaries; riparian corridors, swamps, and wetlands; and disappearing and emerging lands as the rivers change course.

The Human-Natural Systems of River Landscapes

Why recount the levee breaches of the recent past, the flooding impacts on agriculture and other land uses, and drought effects on navigation on the Mississippi and Ohio rivers? Despite attempts to control and manage the impacts of seasonal flooding and less predictable drought and extreme weather events, there is much unknown about coupled human-natural river systems [1]. Human history is the coevolution of learning how to govern ourselves, shape ecosystems, and learn from each other [9] to avert disaster and reduce hazards and vulnerability. Management of river landscapes under changing climates, population growth, global food insecurity, and threats to water scarcity and water quality will determine much of the future of civilization.

Although these case studies are intended to be accessible, engaging reading, there are a number of key themes for readers with an interest in learning a little river science and exploring the human-natural systems of river landscapes:

1. *Change is the only constant over the millennia.* Rivers and their landscapes are complex, dynamic, and ever changing.

FIGURE 1.3 One of two Sny River aqueducts transports floodwaters and sediment to settling basins.

2. ***There are many external drivers of change.*** Climate, population growth, settlement patterns, agriculture, industries, changing markets and economies, new technologies, and new scientific knowledge about water and soil and their interactions within river ecosystems exert pressure and present challenges and new opportunities.

3. ***Soil and water resources are essential assets but are highly vulnerable in modern-day river systems.*** Soil and water are the geologic legacies of the river landscape and represent the assets upon which past and current social, economic, and ecological well-being are built. How these resources are managed affects future opportunities and vulnerabilities.

4. ***Contested views make managing river landscapes difficult.*** People differ in their social values and what they consider the best functional uses of rivers and their floodplains. Managing river landscapes based on engineering and biogeophysical sciences alone will fail to reduce vulnerability and unforeseen risks. The diversity of social values, land use preferences, and human relationships with rivers and their floodplains must be better understood and made part of the management processes.

5. ***Resilience management can improve capacities to adapt and adjust to system disruptions and change.*** Effective management for future unknown risks and catastrophes will need new approaches beyond the confinement-dispersion strategies that current levee, floodway, and reservoir structures represent. While many river floodplains are likely to never be fully restored, the purposeful placement of wetlands and engineered structures can improve floodplain functionalities and rebalance competing human values and preferences for land uses with the natural behavior of the river ecosystem.

---

[1] Park, J., T.P. Seager, P.S.C. Rao, M. Convertino, and I. Linkov. 2012. Integrating risk and resilience approaches to catastrophe management in engineering systems. Risk Analysis 33(3):356-367, doi: 10.111/j.1539-6924.2012.01885.x.

[2] Melillo, J.M, T.C. Richmond, and G.W. Ohe. 2014. Highlights of Climate Change Impacts in the United States: The Third National Climate Assessment. US Global Change Research Program. Washington, DC: US Government Printing Office.

[3] Olson K.R., and L.W. Morton. 2014. Dredging of the fractured bedrock-lined Mississippi River channel at Thebes, Illinois. Journal of Soil and Water Conservation 69(2):31A-35A, doi:10.2489/jswc.69.2.31A.

[4] Olson, K.R., and L.W. Morton. 2013. Soil and crop damages as a result of levee breaches on Ohio and Mississippi Rivers. Journal of Earth Science and Engineering 3(3):139-158.

[5] Morton, L.W., and K.R. Olson. 2014. Addressing soil degradation and flood risk decision making in levee protected agricultural lands under increasingly variable climate conditions. Journal of Environmental Protection 5(12):1220-1234.

[6] Barry, J.M. 1997. Rising Tide: The Great Mississippi Flood of 1927 and How It Changed America. New York, NY: Simon and Schuster.

[7] Camillo, C.A. 2012. Divine Providence. The 2011 Flood in the Mississippi River and Tributaries Project. Vicksburg, MS: Mississippi River Commission.

[8] Wisner, B., P. Blaikie, T. Cannon, and I. Davis. 2004. At Risk: Natural Hazards, People's Vulnerability and Disasters, 2nd edition. London: Routledge.

[9] Dietz, T. 2013. Bringing value and deliberation to science communication. PNAS 110(Supplement 3):14081-14087.

Geologic and Climatic Impacts on Ancient Mississippi and Ohio River Systems

The Mississippi and Ohio rivers and their adjacent landscapes as we know them today have a long and dramatic geologic history stretching back billions of years. Shifts in the earth's crust, tectonic activity, and rising and falling sea levels altered landforms and river flows. Hydrologic cycles, sedimentation patterns, and warm, humid climates alternated with cooling climates in the ancestral Gulf Coastal Plain during the Cretaceous period (145 to 72 million years ago) and affected plant and animal life growth and extinctions, and expanded diversities [1]. Continental shifts and tectonic events in the Miocene and Pliocene epochs (figure 2.1) at the end of the Tertiary period initiated mountain building and land bridge development. More recent in the geologic time scale of the earth (2.6 million years ago to present) was the glaciation of the Northern Hemisphere and a period of rapid climate fluctuations with advances of massive ice sheets alternating with warmer interglacial periods [2]. As a result, fluctuations in meltwater flowing northward into Canada and the North Atlantic and Arctic oceans or south into the Gulf of Mexico repeatedly rerouted the Mississippi and Ohio river basin drainage areas and their channels as glaciers advanced during colder climatic periods and retreated in warmer intervals.

The earth's continuously changing climate and interior heat release from its core over the last 4.5 billion years have driven the formation and erosion of rocks, the lifting up and downwarping of land, and the rise and fall of sea levels and continents. The geologic processes of ice formation and heat have influenced hydrologic cycles, weathering, erosion, sedimentation, and lithification of rocks throughout the Mississippi and Ohio valleys. These processes set the context in which ancient rivers formed and the channels they cut through bedrock and plains created the landscape we see today. They help us understand river confluences: the Cumberland and the Tennessee as they drain into the Ohio, the continual relocation over time of where the Mississippi and Ohio rivers converge, the realignment of the upper Mississippi and Ohio rivers through Illinois, and the impacts of the ancient Teays River diverting Indiana meltwaters northward and then westward rather than southward into the Ohio River. They offer insight into the behavior of meandering rivers, the natural formation of the great swamplands of Missouri and Illinois, and the challenges of human efforts to drain these lands for agriculture and settlements. The

Era	Precambrian 4.5 billion to 254 million years ago	Paleozoic 541 million to 254 million years ago	Mesozoic 252 million to 72 million years ago			Cenozoic 66 million years ago to present						
Period			Triassic 252 million to 208 million years ago	Jurassic 201 million to 158 million years ago	Cretaceous 145 million to 72 million years ago	Tertiary 66 million to 3.6 million years ago					Mesozoic 252 million to 72 million years ago	
Epoch						Paleocene 66 million to 59 million years ago	Eocene 56 million to 38 million years ago	Oligocene 34 million to 28 million years ago	Miocene 23 million to 7 million years ago	Pliocene 5.3 million to 3.6 million years ago	Pleistocene 2.6 million to 126,000 years ago	Holocene 11,700 years ago to present
Events	Earth formation ~4.5 billion years ago	Mountain building on continents; incursions and retreats of shallow seas in interiors	Tectonic forces affect continent creation; rift valleys formed	Continued tectonic activity	Oceans and isolated continents; cool, moist climate	Early epoch cool and dry; transitioning to warm-humid tropical climate	Early epoch high precipitation; ice free world	Warm but cooling climate; increase in ice volume and lower temperatures	Shift in continents; increase in aridity through mountain building	Cool, dry climate; intensification of icehouse conditions	Ice Age begins; large polar ice caps develop; Ice Age glaciation and interglacial periods	Ice Age recedes and current interglacial begin

FIGURE 2.1 Geologic time scale of Earth. The International Commission of Stratigraphy, a scientific body in the International Union of Geological Sciences, defines global units of the International Chronostratigraphic Chart that are the basis for the units (periods, epochs, and age) of the International Geologic Time Scale.

geologic histories of these rivers and their landscapes are a valuable lens informing today's river management decisions about navigation, drainage and flood control, ecosystem protection, and agricultural land use.

The Mississippi Embayment and New Madrid Fault Zone

Fossils, coral reefs, and marine deposits of the shells of sea invertebrates reveal that the central Interior Lowlands (map 2.1) at the beginning of the Paleozoic era about 541 to 254 million years ago (figure 2.1) were once a shallow sea. As the central Interior Lowlands emerged from the sea to form a continent, water flowed off this land mass and through the Gulf Coastal Plain to the sea, depositing fluvial sediments on the continental shelf. With the appearance of land, swampy deltas grew wetland forests, ferns, and mosses; and amphibians and wetland plants thrived in the warm, humid climate. Over millions of years, these moist, stagnant swamps with huge volumes of decomposing vegetation left behind rich veins of coal deposits throughout Illinois, Ohio, Kentucky, and Tennessee.

The confluence of the Ohio and Mississippi rivers is the northernmost apex of the ancient Gulf Coast Sea and the Mississippi Embayment. This embayment is a north-south structural basin between the Appalachian Highlands in the east and the Ozark Plateaus in the west created by downwarping and downfaulting [1,3] as consequence of tectonic plate drifting of the Mississippi River valley in the late Cretaceous period (figure 2.1) [4]. A system of faults runs parallel to the axis of the Mississippi Embayment trough and underlies modern-day southeast Missouri and northeastern Arkansas. This area is an active fault zone associated with the New Madrid earthquakes in AD 1450 to 1470 and AD 1811 to 1812. New Madrid has been the center of seismic activity for thousands of years and was once at the edge of the ancient sea. Seismic activity rerouted the Mississippi and the Ohio rivers [5] as the land rebounded by as much as 13 feet in one thousand years following the glacial periods. Earthquakes in the New Madrid seismic zone are thought to have occurred as early as the Precambrian era at depths between 2.5 and 8 miles under the Mississippi Embayment [4].

The subsidence or downwarping of lands in the transition from the Cretaceous period to early Cenozoic (figure 2.1) allowed the inland sea to reinvade the Gulf Coastal Plain (map 2.1) and extend as far north as the southeast Missouri lowlands [1, 3]. The final retreat of this ancient sea from the northern Mississippi Embayment during the Eocene (figure 2.1) left fluvial deposits of marine clays, pumice, sands, and silts [1] approximately 165 feet thick [4]. The reemergence of a terrestrial landscape in the Mississippi Valley was one of river terraces descending topographically into flat lowlands interspersed with deltaic deposits and low-gradient drainage systems. Based on analyses of the distribution of river sediments in this region, it has been suggested that the Mississippi River main channel shifted eastward during the Quaternary to its current Holocene position [4]. The upland bluffs of Crowley's Ridge in Stoddard County, Missouri, and Benton Hills in Scott County, Missouri, reveal fluvial sands and gravels about 53 feet thick overlain with loess consisting of windblown silt as deep as 112 feet [4]. Fluctuations in upstream glacial meltwater discharge and downstream sea levels influenced the historic Mississippi River channel incisions and the many braid belts (channels) running on both sides of Crowley's Ridge in Missouri and Arkansas [6].

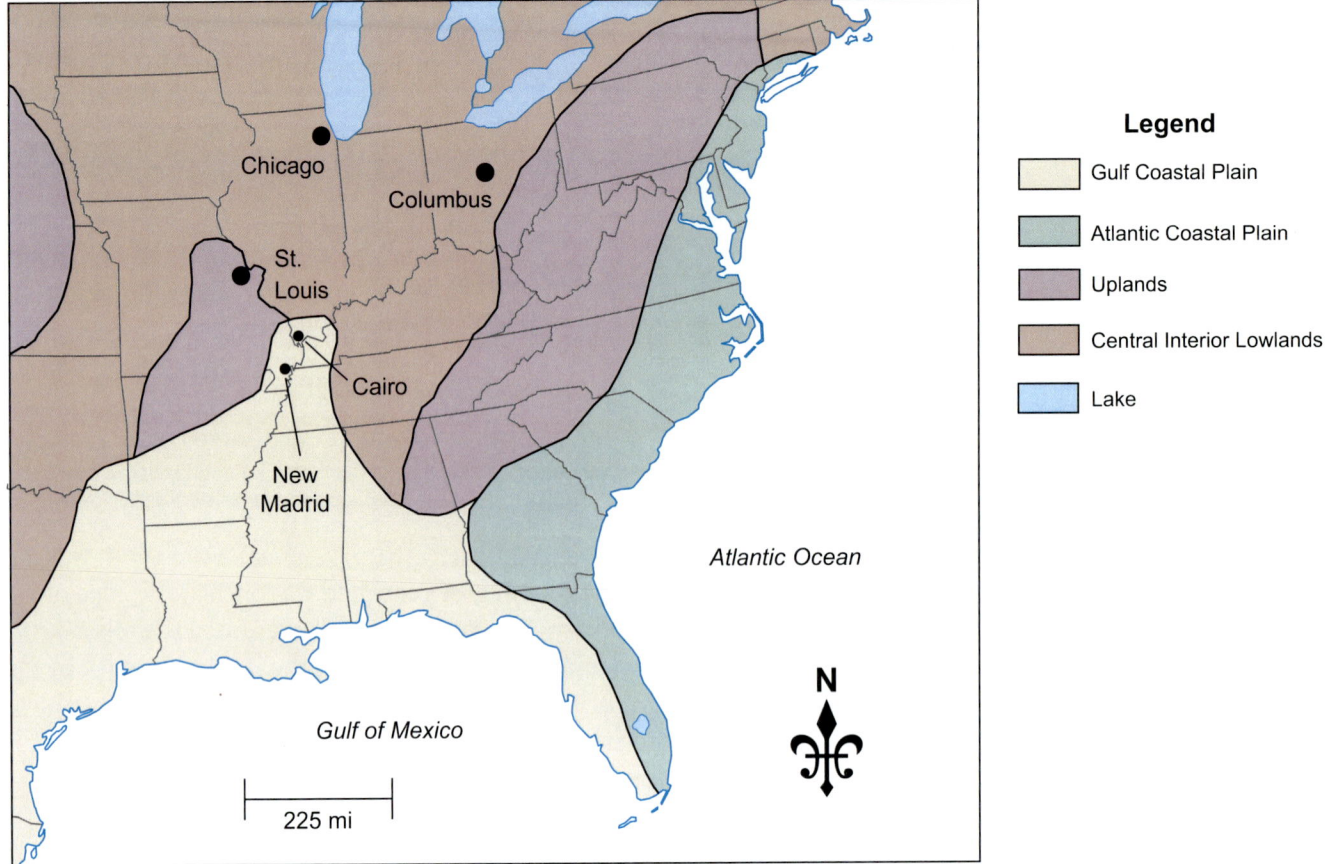

MAP 2.1 The Gulf Coastal Plain and the central Interior Lowlands of the United States.

Glaciation and Warming Intervals

The tropical climate of 66 to 59 million years ago shifted to a moderate, cooling climate, which led to the reglaciation of Antarctica and ice cap formation during the Eocene epoch into the Oligocene and marked the beginning of the Icehouse Earth climate (figure 2.1). This cooling trend continued over millions of years in which rapid evolution and diversification of animals and modern types of flowering plants occurred along with widespread growth of forests and development of modern mammal and bird families. With the intensification of the icehouse condition, the Ice Age began about 2.6 million years ago, and permanent ice sheets formed in the Northern Hemisphere (map 2.2). This marked the beginning of a period of rapid climate fluctuations, with the advance of ice sheets during deep cold alternating with warmer intervals and glacial retreat [2]. Some paleoclimatic models show a period of extreme aridification in western United States when an ice sheet receded (12,000 to 110,000 years ago) and displaced the jet stream. The eastern boundary of this dry climate ran from northern Illinois to eastern Texas and lasted about 11,000 years [2].

During the Pleistocene epoch, numerous glacial advances covered most of the upper Mississippi and Ohio river valleys. The four glacier stages, the Nebraskan (buried by other later glacier stages), Kansan, Illinoian, and Wisconsinan, are named for their southernmost advances (map 2.3). Meltwaters from these glaciers contributed to the realignment of the Mississippi and Ohio rivers. Before the Pleistocene glacial epoch, the ancient Mississippi River passed much farther to the east, as shown by the dashed lines on map 2.4 [7]. Today the lower Illinois River follows the ancient Mississippi River course. The ancient Mississippi River was blocked by the Wisconsinan glacier and its terminal moraine, which cut off the Illinois Valley and forced the river west into its present course through the ancient Iowa River valley to the confluence of the Illinois River [7, 8].

The Ancient Teays, Mississippi, Ohio, Cumberland, and Tennessee Rivers

The North American river systems developed during the late Pliocene and early Pleistocene (figure 2.1) with the ancient Teays River and its tributaries flowing northward from the western side of the Appalachian Mountains into Ohio and west through central Indiana and Illinois prior to joining the Mississippi River system (map 2.5) [2]. Glacial advances dammed the Teays-Mahomet in southern Ohio creating a vast lake for thousands of years before its overflow made new river channels [2]. These new rivers drained

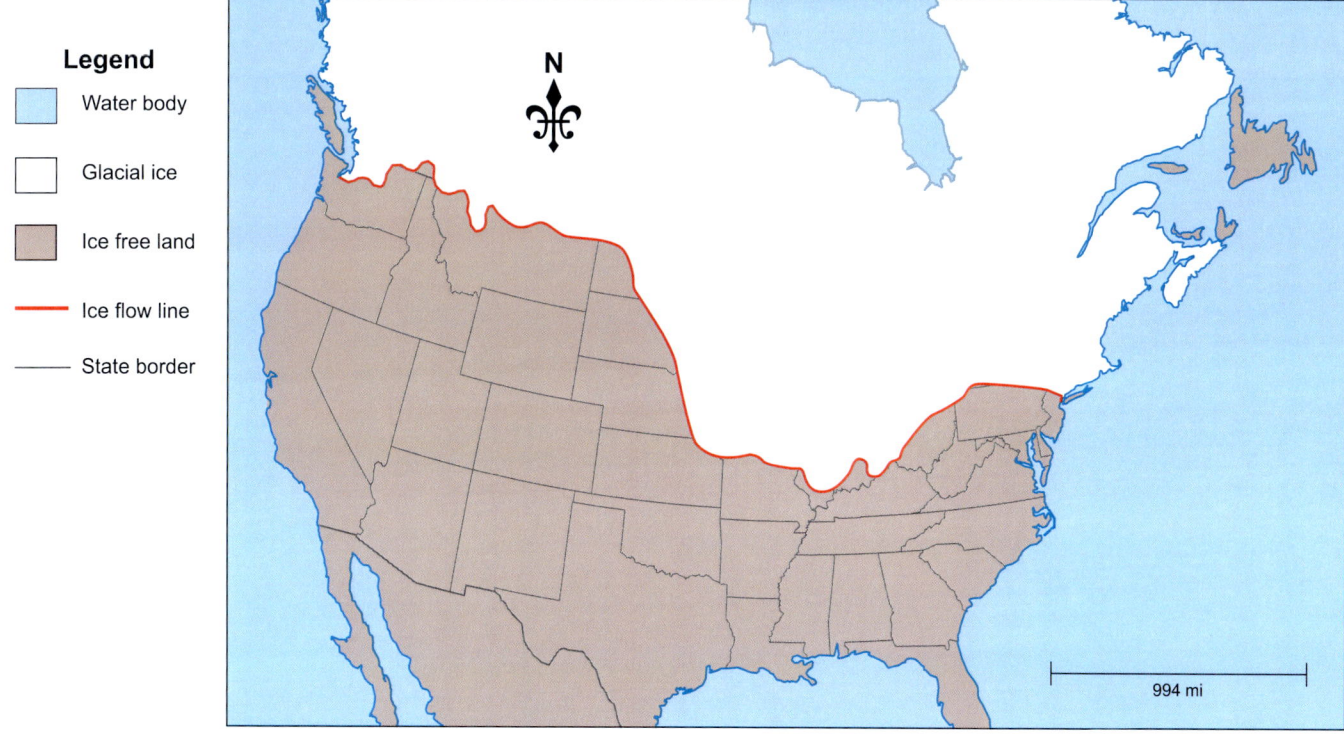

MAP 2.2 The southernmost glacial advance in the Northern Hemisphere during the Pleistocene epoch.

westward into the ancient Ohio system and formed the modern Ohio River.

Meltwaters from glaciers and seismic activity led to avulsions, the rapid abandonment of older channels and formation of new ones, throughout the ancient Ohio River basin. These events also affected the land bridge between the Cumberland River and Tennessee River, as erosion and sediment deposits altered the slope of river channels and made them unstable. During the Woodfordian period (30,000 to 800,000 years BP), the floodwaters from the Ohio River watershed drained into eastern Illinois and west through the ancient Ohio River valley (currently the Cache River valley) where it converged with the Mississippi River northwest of the current confluence (map 2.6) [9]. The middle Cache River valley is a wide valley as a result of the ancient Ohio River carrying a large amount of glacial meltwater through the valley, in addition to runoff from the Shawnee Uplands. The combined river flow through the current Ohio River valley in southern Illinois (map 2.5) left the ancient Ohio River valley without a major river, and the landscape only received water during extreme flooding events or from local sources. At the end of the glacial period, the meltwater eventually cut through the land bridge, and the ancient Cumberland and Ohio rivers joined the ancient Tennessee River. All these rivers now drain through the current Ohio River valley and into the main Ohio River channel.

The presence of the land bridge between the Ozark Highlands and the southern Appalachians prevented the Teays River from draining to the south and into the ancient Ohio River. The Teays River once flowed north and then west through southern Ohio, Indiana, and central Illinois to the ancient Mississippi River before turning south (map 2.5). Thus, the ancient Ohio River did not include the Wabash and White rivers but drained the Ohio, Green, and Cumberland rivers and flowed

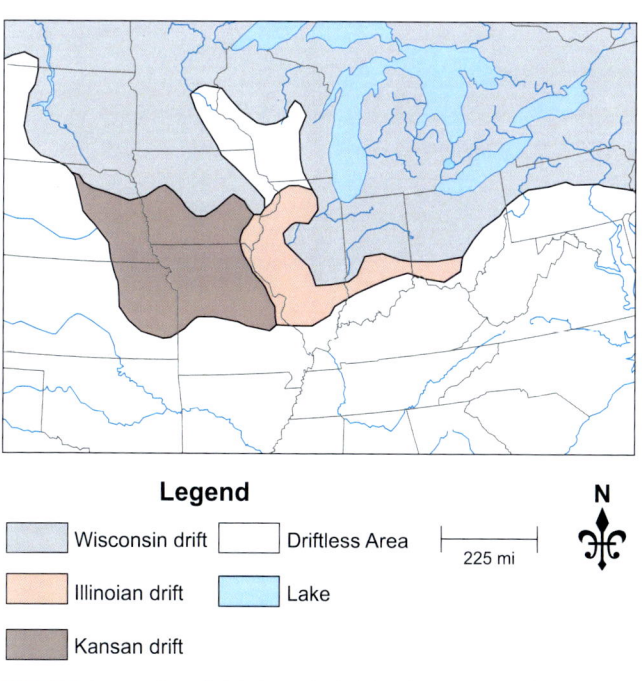

MAP 2.3 Areas of surficial deposits left by the Kansan, Illinoian, and Wisconsinan glaciers.

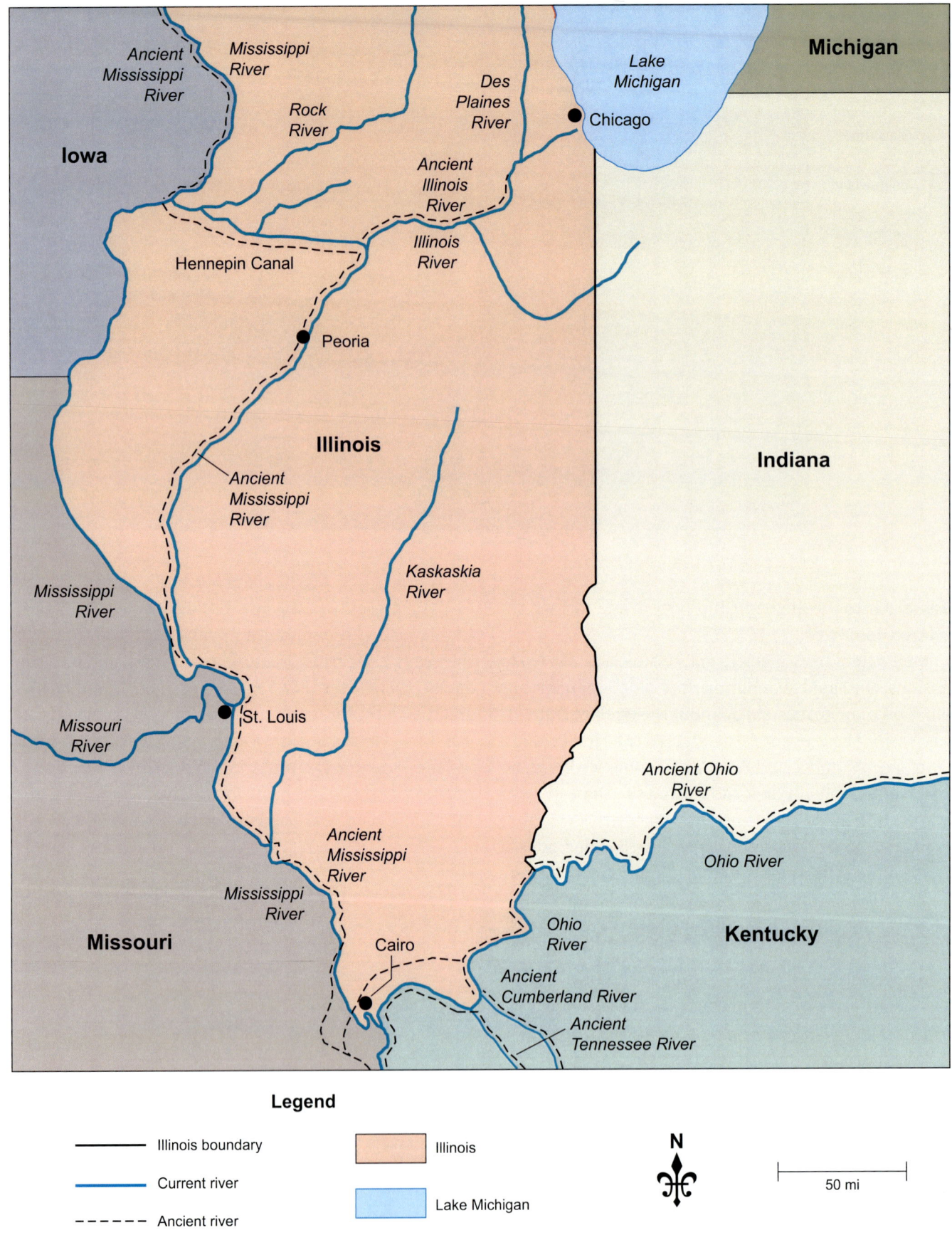

MAP 2.4 Ancient Mississippi, Ohio, and Tennessee rivers and their current river channels.

west through southern Illinois and north of the current Ohio River (map 2.5). The current Ohio River is in the ancient Tennessee River valley. The ancient Tennessee River joined the ancient Ohio River southwest of Cairo, Illinois, in the bottomlands of Missouri. The ancient Ohio and ancient Mississippi rivers maintained separate flow paths (map 2.4) through southern Illinois.

The upper Mississippi River was once part of the now-extinct glacial Lake Warren, which carved the valley of the Minnesota River and allowed the large glacial Lake Agassiz (map 2.7) to drain to the Gulf of Mexico. Before the last glacier, the ancient Wisconsin River drained the northern part of Wisconsin. About 18,000 years ago, the Green Bay Lobe of the glacial ice sheet pushed in from the east and butted up against the Baraboo Hills. The ancient Wisconsin River was closed, and water backed up, filling the basin to the north and west and creating glacial Lake Wisconsin. This glacial lake existed for a few thousand years with storms and ice scouring sand off the sandstone bluffs. About 14,000 years ago, the climate warmed, and the glacier retreated. The meltwaters raised the ancient lake level and opened a path around the Baraboo Hills. Eventually, the stream cut through a thin dam or plug in a few days near the Wisconsin Dells. In a catastrophic flood, most of the lake drained out to the south, and flowing floodwater cut new channels through the lake bottom sand and then cut canyons through the weakly cemented sandstone.

The upper Mississippi River valley is thought to have originated as an ice-marginal stream during the Nebraskan glaciation in the pre-Illinoian stage. The Driftless Area was not smoothed out or covered over by North American glacial processes but remained unglaciated at the height of the Ice Age. The Wisconsinan glaciation formed lobes that blocked the river where the Mississippi River now flows. It has been posited by scientists that there were instances of ice dams bursting because of the large volume of glacial meltwater that flowed into the Driftless Area and the absence of a lake bed. This may explain the history of glacial Lake Missoula.

Ancient Mississippi River Channel

The upper Mississippi River waterway runs from Minnesota to Cairo, Illinois (see map 1.1). Historically, the ancient Mississippi River entered Illinois south of Davenport, Iowa, and flowed east into the valley where the Hennepin Canal is located (map 2.4 and figure 2.2) [8]. Then, the ancient Mississippi joined with the Illinois River, flowing south from near the current city of Peoria, Illinois, toward St. Louis, Missouri. The end moraine from the Wisconsinan glacier blocked the flow of the ancient Mississippi River through the valley approximately 15,000 to 30,000 years ago. The upper Illinois River headwaters now start near Chicago, Illinois, and converge with the Mississippi River at Grafton, Illinois. Had the Wisconsinan glacier advance not plugged the ancient Mississippi River valley and caused the realignment to its current location, the land area south of the Hennepin Canal (map 2.4), west of current the Illinois River, and north of Alton, Illinois, would be west instead of east of the Mississippi River and part of the states of Iowa and Missouri (see chapter 3) [7, 8].

FIGURE 2.2 The Hennepin Canal was deactivated in 1951 but preserved at various locations along the ancient Mississippi River.

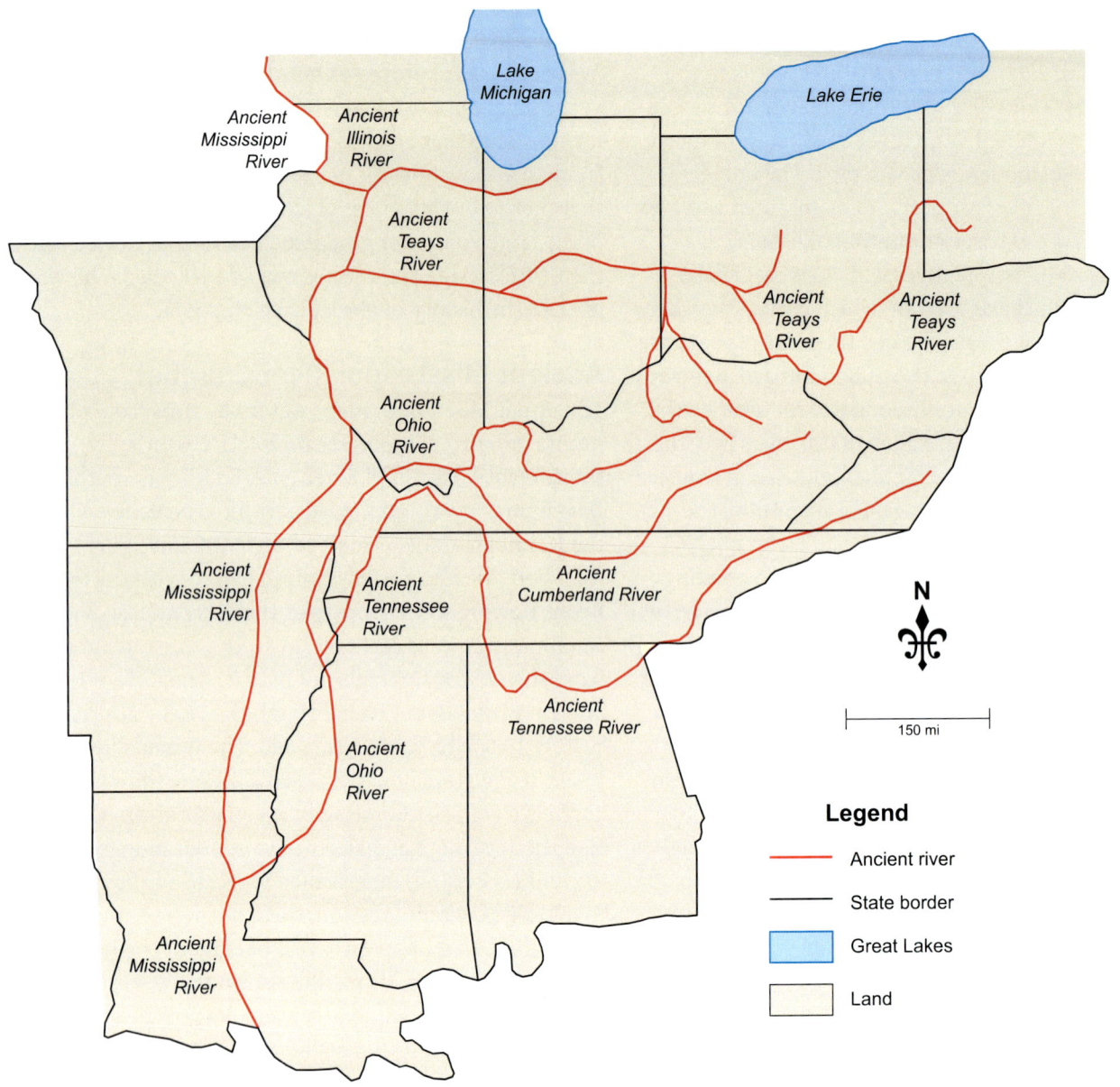

MAP 2.5 The flow paths of the ancient rivers of the eastern Mississippi and Ohio basin (the ancient Teays, Mississippi, Ohio, Cumberland, and Tennessee rivers) superimposed over modern state political boundaries.

Geologic History of the Mississippi River Valley South of Cape Girardeau

The rapid abandonment of old river channels and the creation of new channels are evident throughout the middle and northern lower Mississippi River valley. Fluctuations in glacier meltwater and sediment deposition during warming intervals affected river channel slopes and led to these avulsions. Channel slopes are reduced and become unstable when the river carries more sediment than the water can transport. With an increase in sediment deposits, the riverbed becomes higher than the floodplain, and it is easier for the river to breach its natural levees. Breaching, which occurs during flood events, causes the river to spill out of its old channel and create a new one as the water seeks a more stable slope and shorter route to downstream.

The Mississippi River was 26 to 69 feet below its present floodplain during the last interglacial warming interval [6]; however, with the fluctuations in glacial water discharge and sediment loads, the riverbed levels were raised up. A historic river channel of the Mississippi River just south of Cape Girardeau, Missouri, turned southwest into the current state of Missouri and traveled more than 30 miles to the west before turning south where it joined with the ancient Ohio River waters (map 2.6). During an early glacial Lake Agassiz drainage event, channel belts formed in southeast Missouri as floodwater deposited sediments in the ancient Mississippi River flow path at Cape Girardeau and incised the Thebes Gap where the current Mississippi flows [6]. The Mississippi River has historically entered the lower Mississippi Valley through three channel

MAP 2.6 The Illinois land bridge between the Southern Appalachians and the Ozark Highlands was created during late Tertiary period.

paths: (1) an opening cut between the Ozark Plateau and the northern end of Crowley's Ridge located southwest of Cape Girardeau, Missouri; (2) the ten-mile-wide break between the northern end of Crowley's Ridge and Benton Hills; and (3) the Thebes Gap through which the current river runs [6, 10].

The Mississippi River changed course at the end of the Ice Age. Prior to this, southeast Missouri and southern Illinois were engulfed in a shallow sea until the end of the Pennsylvanian period when the waters receded and regional elevation rose. After the last glacial advance, the melting ice waters flooded and altered the course of many channels and streams, including the ancient Mississippi and ancient Ohio rivers. With the change in the ancient Ohio and ancient Mississippi river paths, the location of their confluence also shifted. The confluence has been located in Morley, Missouri; Malden, Missouri; and now in Cairo, Illinois. Historically, this region has been a delta, confluence, and bottomlands dating back 30,000 to 800,000 years BP, with many of the Illinois lands shown on map 2.4 located on both sides of the upper Mississippi River as its channel changed positions over time.

Mississippi and Ohio River Landscapes Today

More than 40% of the continental United States is drained by the Mississippi and Ohio rivers and their tributaries. However, the largest basin, the Missouri River basin (see map 1.1), receives less rainfall and contributes less runoff flow to the lower Mississippi River than either the upper Mississippi River basin or the Ohio River basin. On average, the Ohio River basin receives the most rainfall and has steeper sloping lands, which contribute more runoff to the lower Mississippi River (60%) at the confluence than the combined upper Mississippi and Missouri river flows (40%).

The Mississippi and Ohio rivers converge at the northern edge of an ancient sea and lowlands of the Mississippi Embayment. A land bridge separates the central Interior Lowlands from the Gulf Coastal Plain. This upland land bridge, the Shawnee and Ozark hills, connects the Southern Appalachians and the Ozark Highlands (map 2.6). Plants and animals used this land bridge to migrate between the Southern Appalachians and the

MAP 2.7 During Ice Age warming intervals, the historic Lake Agassiz in the Northern Hemisphere drained north to Lake Hudson and to the Arctic Ocean and south toward the Gulf of Mexico.

Ozark Highlands. The ancient Ohio and Mississippi rivers shown on map 2.6 are overlain on the land bridge to help the reader visualize its location. The Ozark Highlands, including the St. Francois Mountains, were created by volcanic and intrusive activity approximately 1.5 billion years ago. By comparison, the Appalachian Mountains began forming about 460 million years ago [11].

The current Mississippi River flows southward from Minnesota as a meandering river of oxbows and cutoffs. The upper Mississippi River is upstream of Cairo, Illinois. The river flows approximately 1,250 miles from the headwaters at Lake Itasca, Minnesota, to Cairo, where its confluence with the Ohio River designates the start of the lower Mississippi River. On the upper Mississippi River, the floodplain is between 1 and 3 miles wide, but south of St. Louis, Missouri, the alluvial floodplain widens to approximately 20 miles. The upper Mississippi River has 29 locks and dams [8], and the Illinois River has 8 locks and dams, which allow year-round navigation. The dominant land use in the upper Mississippi River landscape today is agriculture. In addition, 285,000 acres of national wildlife refuges line the river. The largest refuge, the Upper Mississippi River National Wildlife and Fish Refuge, is part of the Mississippi Flyway and extends over 260 river miles.

Hydrologically, the Ohio River is the main tributary of the Mississippi River. It runs west from Pittsburgh, Pennsylvania, to the Mississippi River confluence at Cairo, Illinois, and drains lands west of the continental divide from the Appalachian Mountains. The Ohio River has 20 locks and dams to assure year-round shipping. The annual Ohio River flow at the confluence is greater than the annual upper Mississippi River flow, and its flood stage is measured by the Cairo gage along the Ohio River floodwall.

The ancient rivers, inland sea, and land bridge of the central Interior Lowlands affected the paths and confluences the current great rivers and their tributaries. As a result of a changing climate and seismic activity, melting glaciers realigned the Ohio and Mississippi rivers over time. With the settling of the Midwest, the locations of these great rivers were often used as state boundaries, and the size and shape of many states were affected by the geologic events which formed and reformed these rivers and changed their locations long before these states were established. The state with the most borders affected by the realignment of the ancient Ohio and Mississippi rivers was Illinois, but other states, including Kentucky, Missouri, and Iowa, gained or lost land as a result of the current locations of these two great rivers.

[1] Oboh-Ikuenobe, F.E., M.K. Spencer, C.E. Campbell, and R.D. Haselwander. 2012. A portrait of Late Maastrichtian and Paleocene palynoflora and paleo-environment in the northern Mississippi Embayment, southeastern Missouri. Palynology 36(supplement 1):63-79.

[2] Lemmon, E.M., A.R. Lemmon, and D.C. Cannatella. 2007. Geological and climatic forces driving speciation in the continentally distributed trilling chorus frogs (*pseudacris*). Evolution 61(9):2086-2103.

[3] Hosman, R.L. 1996. Regional Stratigraphy and Subsurface Geology of Cenozoic Deposits, Gulf Coastal Plain, South-Central United States. US Geological Survey Professional paper 1416-G. Washington, DC: US Government Printing Office.

[4] Van Arsdale, R.B. and R.K. TenBrink. 2000. Late cretaceous and Cenozoic geology of the New Madrid seismic zone. Bulletin of the Seismological Society of America 90(2):345-356.

[5] Olson, K.R., and LW. Morton. 2013. Soil and crop damages as a result of levee breaches on Ohio and Mississippi rivers. Journal of Earth Sciences and Engineering 3:139-158.

[6] Rittenour, T.M., M.D. Blum, and R.J. Globe. 2007. Fluvial evolution of the lower Mississippi River valley during the last 100 k.y. glacial cycle: Response to glaciation and sea-level change. Geological Society of America Bulletin 119(5/6):586-608.

[7] Olson, K.R., and F. Christensen. 2014. How waterways, glacial melt waters, and earthquakes re-aligned ancient rivers and changed Illinois borders. Journal of Earth Sciences and Engineering 4(7):389-399.

[8] Olson, K.R., and L.W. Morton. 2014. Runaway barges damage Marseilles lock and dam on the Illinois River and floods Marseilles, Illinois. Journal of Soil and Water Conservation 69(4):104A-109A, doi:10.2489/jswc.69.4.104A.

[9] Olson, K.R., and L.W. Morton. 2014. The 2011 Ohio River flooding of the Cache River Valley in southern Illinois. Journal of Soil Water Conservation 69(1):5A-10A, doi:10.2489/jswc.69.1.5A.

[10] Olson, K.R., and L.W. Morton. 2014. Dredging of the fracture bedrock lined Mississippi River channel at Thebes, Illinois. Journal of Soil and Water Conservation 69(2):31A-35A, doi:10.2489/jswc.69.2.31A.

[11] Cremeens, D.L., R.G. Darmody, and S.E. George. 2005. Upper slope landforms and age of bedrock exposures in the St. Francois Mountains, Missouri: A comparison to relict periglacial features in the Appalachian Plateaus of West Virginia. Geomorphology 70(1-2):71-84, doi: 10/1016/j.geomorph2005.04.001.

How Realigned Ancient Rivers Influenced the Border Locations of Eight Central States

3

The borders of Illinois were established when Illinois became a state in 1818. The western border was delineated using the Mississippi River, and the Ohio River was used as the southern border. The eastern border was formed by the Ohio and Wabash rivers as well as the line along longitude 87°31' connecting the Wabash River to Lake Michigan. As initially proposed, the northern border of Illinois would have been 51 miles to the south of the current latitude line 42°30'30". This 5,440,000-acre addition to Illinois ensured that the territory met the required minimum of 40,000 people to qualify as a state. The northern border was moved to allow the linkage of the Great Lakes shipping route to the Illinois and Mississippi river navigation channels. Illinois thus gained a valuable shoreline on Lake Michigan and a location for a shipping port hub, which became Chicago.

Illinois Territory

For more than 100 years prior to 1818, Illinois was known as Illinois Country or Illinois Territory (map 3.1). In the seventeenth century, the French built trading forts in Illinois Country. Louis Joliet and Father Pierre Marquette suggested a canal from the Illinois River to Lake Michigan to eliminate the portage at Mud Lake, but the canal was never built by the French. At the end of the French and Indian War in 1763, the area was ceded to the British and was then awarded to the new United States by the Treaty of Paris in 1783. When the state borders of Kentucky and Indiana were established, they formed Illinois's southern and eastern boundaries (the Ohio and Wabash rivers and the longitudinal line 87°31'30") by extending the eastern line north from the Wabash River to the southern tip of Lake Michigan. The proposed northern boundary in an 1817 plan considered by the US Congress (derived from the Northwest Ordinance) was a straight line from the southern tip of Lake Michigan in western Indiana to the Mississippi River near the Rock River confluence with the Mississippi River. Nathaniel Pope, Illinois Territory's congressional delegate, proposed modifying the northern border by moving it 51 miles to the north for economic reasons and to give Illinois access to Lake Michigan, the Great Lakes, and the St. Lawrence River. Another unstated reason for the northern border move was related to slavery. After the Missouri Compromise of 1820, Illinois became a Northern state and a key part

17

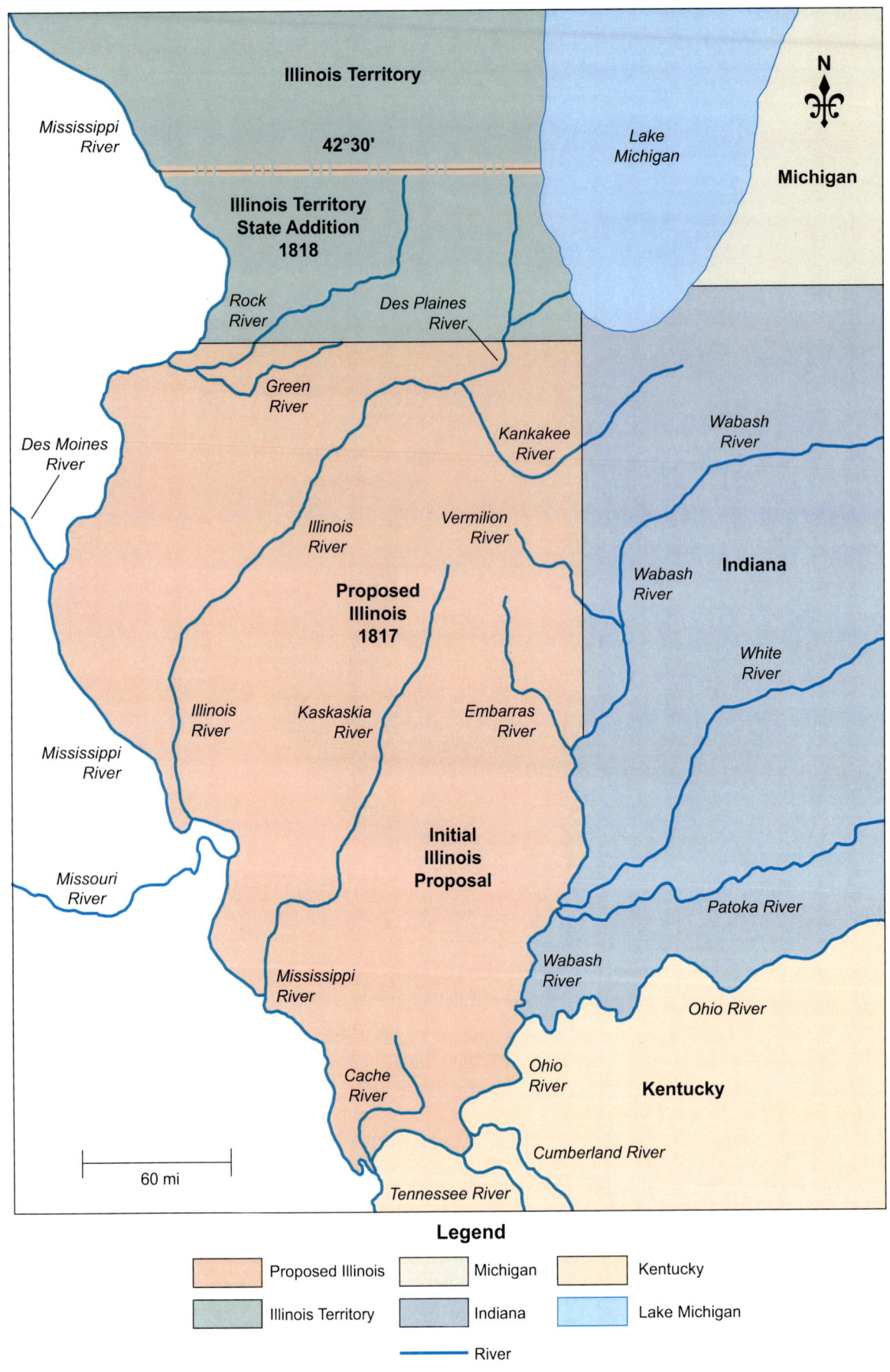

MAP 3.1 First proposed northern Illinois state boundaries within Illinois Territory did not touch Lake Michigan. When Illinois became a state in 1818, a portion of northern Illinois Territory was added to Illinois, and the balance north of 42°30' became the state of Wisconsin in 1849.

FIGURE 3.1 Starved Rock locks and dam help maintain the nine-foot navigation channel on the Illinois River waterway.

of the Union by 1860. While many in southern Illinois were sympathetic to the Confederate cause during the Civil War, most of the state of Illinois was not.

Connecting the Great Lakes and Mississippi River Waterways

Many inhabitants living in northern Illinois Territory, much of which later became the state of Wisconsin, objected to moving the boundary north, due to the loss of the Lake Michigan waterfront and the location for a shipping port. However, Nathaniel Pope was successful in persuading Congress to move the northern boundary of Illinois to its present-day location (map 3.2). The loss of land, water, and population delayed Wisconsin's development for 30 years. Wisconsin finally became a state in 1848. The port area on Lake Michigan became the future city of Chicago in 1837 and linked the Great Lakes and Mississippi River shipping routes with a portage between a small river that drained into Lake Michigan and the Illinois River waterway that flowed south to the Mississippi River (figure 3.1).

In 1848 the Illinois and Michigan Canal was completed, allowing the shipment of goods between the two waterway systems. With tensions rising and civil war a possibility, the canal provided the Union with a northern route to ship goods without having to use the Ohio River. After the railroads and the canal connected Lake Michigan to the rest of the state and regions beyond, Chicago grew with incredible speed. Chicago is now the largest city in Illinois, and the greater Chicago area includes three-quarters of the state's population. The ceding of 8,500 square miles of territory and the lakefront property on Lake Michigan by the US Congress to Illinois, as a result of Nathaniel Pope's efforts, altered the fortunes of Wisconsin and Illinois. The 5,440,000 acres added to Illinois as a result of the northern boundary shift included very productive soils. These soils are both Alfisols (timber), such as Hickory and Rozetta, and Mollisols (prairie), which include the Drummer and Sable soils. Today much of the rural landscape in the northernmost counties west of Chicago produce corn, soybeans, and wheat, and support a robust agricultural economy.

Prehistoric Location of the Ancient Mississippi River Channel

During the Pleistocene epoch, numerous glacial advances and retreats remade much of Missouri and Illinoian topography, drainage, and river hydrology, with the two most recent designated the Illinoian and Wisconsinan glaciations (see map 2.5). Meltwaters from these glaciers contributed to the realignment of the Mississippi River and affected the geopolitical location of the Illinois western state boundary [1]. Prior to the Pleistocene, the ancient Mississippi River passed much farther to the east than it does today, as shown by the blue dashed lines on map 3.2. During the Ice Age, this ancient Mississippi River was blocked by the Wisconsin glacier and its terminal moraine about 12,000 to 15,000 years ago. As result, the ancient Mississippi River realigned itself to its current position, and today's Illinois River flows in the ancient Mississippi River channel.

If the Mississippi River had not been realigned, the 7.5 million acres above St. Louis, Missouri (figure 3.2), would belong to the states of Missouri and Iowa. Before 1803, the land west of the current Mississippi River was controlled by the French and was part of the Louisiana Purchase in that year. After Iowa and Missouri became states, they had a border dispute that was settled by the US Supreme Court. The border between these two states was primarily the 40°35' latitude line, which if extended into the current area of Illinois between the Illinois and Mississippi rivers (map 3.2), would determine the acreage each state would have gained if the ancient Mississippi River had not changed course. A total of 3.5 million acres would have gone to Missouri and 4 million acres to Iowa. This area includes some of the most productive soils in Illinois. Most of these soils were formed under prairie in thick loess and include Muscatune, Osco, Tama, and Sable soils. Muscatune is the most productive soil in Illinois for corn and soybean production [2].

Subtraction of Illinois Land as a Result of Mississippi River Rerouting

The Ice Age also rerouted the Mississippi River south of the current city of Cape Girardeau, Missouri. After the last glacial advance, the melting ice deposited sediments and flooded the region, creating new channels and altering the courses of the ancient Mississippi and

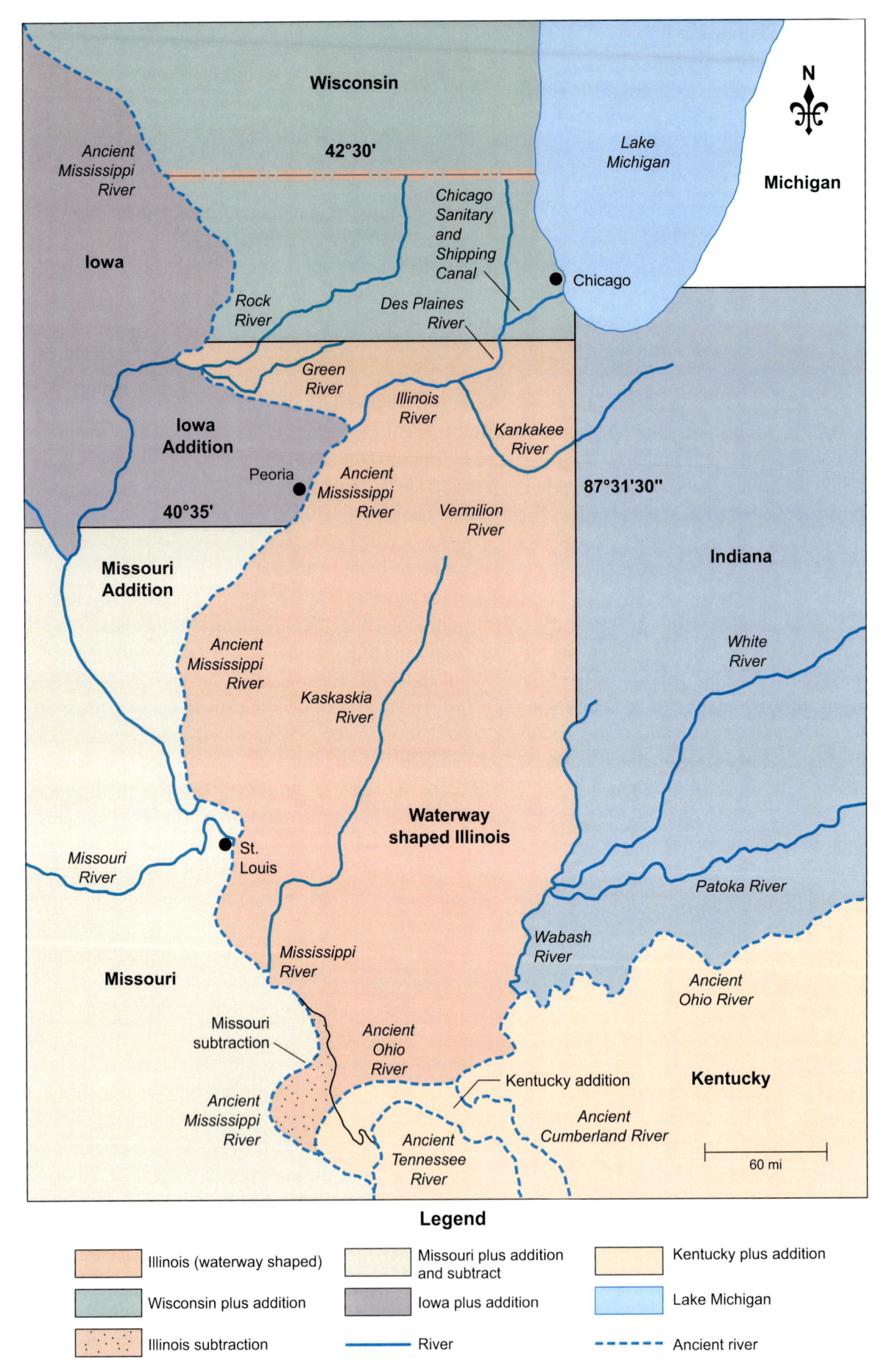

MAP 3.2 The ancient Mississippi River flowed east of the Quad Cities (on the Iowa and Illinois border) between the Rock and Green rivers to the Illinois River and south to St. Louis. The Mississippi River shift west of its ancient channel affected the borders of three future states, adding land to Illinois and subtracting land from Iowa and Missouri.

ancient Ohio rivers and their tributaries (map 3.2). Approximately 12 to 15 thousand years ago, the ancient Mississippi flowed west of the present Mississippi River channel (where Big Swamp is currently located) and joined the ancient Ohio as it flowed west through the Illinois alluvial bottomlands of the modern-day Cache River (map 3.3) [3, 4].

The six-mile upland stretch of the Mississippi River near Thebes, Illinois, is unique [3]. This section of the Mississippi River has a narrow valley with rock underlying the navigation channel. The Mississippi River cut through the Thebes Gap about 12,000 to 15,000 years ago after seismic activity along the Commerce geophysical lineament (a northeast-trending basement and gravity anomaly) [5, 6]. Sedimentation studies further find that repeated overflow from upstream glacial meltwaters changed the slope of the older river channel southwest of Cape Girardeau, Missouri, by filling it with sediment. They posit the Mississippi River incision into the narrow bedrock gorge of Thebes Gap was driven by the lower elevation of the ancient Ohio River in comparison to the fan-like, sediment-filled older channel [5]. The Mississippi River currently forms the state boundary between Missouri and Illinois. At Thebes, Illinois, the Mississippi River is now located 30 miles to the east [3] of where the ancient Mississippi River flowed. Before the twentieth century, the Mississippi River migrated rapidly by eroding the outside and depositing on the inside of the river bend.

The New Madrid, Missouri, area has been the center of seismic activity for thousands of years. Early Holocene, late Wisconsin liquefaction features in the western lowlands were induced as a result of earthquake upheaval along the Commerce geophysical lineament running from central Indiana to Arkansas [6]. The land has rebounded after the glacial periods by as much as 13 feet in 1,000 years. This seismic activity affected the Mississippi River and perhaps influenced the Ohio River rerouting as well [3]. Floodwaters of the ancient Mississippi River were routed around the bedrock-controlled uplands near Scott City, Missouri, and north of Commerce and Benton, Missouri [3], to an opening in the upland ridge 40 miles to the southwest. Then, the river turned back to the south and merged with the ancient Ohio River near Morley, Missouri. The two historic rivers also once joined at Malden, Missouri. The confluence of these two mighty rivers created a very rapidly changing channel. It appears the bedrock-controlled upland was worn away by both rivers after seismic activity, and the creation of the Commerce Fault contributed to the opening of the bedrock-controlled channel [3] after the last glacial advance. The location of the confluence continued to change over time and is now located south of Cairo, Illinois, at Fort Defiance State Park.

FIGURE 3.2 The arch of St. Louis, Missouri, is located on the banks of the Mississippi River.

As a result of the Commerce Fault, the distance the Mississippi had to travel was shortened from 50 miles to 6 miles. The ancient Mississippi River flowing southward from Minnesota was a meandering river of oxbows and cutoffs, continuously eroding banks, redepositing soil, and changing paths. Its historical meandering is particularly apparent in western Alexander County, Illinois, where topographical maps show oxbow swirls and curves (map 3.3) where the ancient Mississippi River once flowed.

The Ancient Ohio River Valley

The ancient Ohio River, a southwestern flowing river, was formed between 2.5 and 3 million years ago when glacial ice dammed portions of north flowing rivers. About 625,000 years ago, the ancient Ohio River, fed by the Green and Cumberland rivers of Kentucky, flowed through the current Cache River basin and was smaller than the current Ohio River [7]. At that time, the Wabash and White rivers (Indiana) had not yet formed, and the Tennessee River was not a tributary of the ancient Ohio River but formed the main channel before the current Ohio River took shape. Hydrologically, the Ohio River is the main eastern tributary of the Mississippi River. Today it runs along the borders of 6 states 981 miles west from Pittsburgh, Pennsylvania, to the Ohio and Mississippi

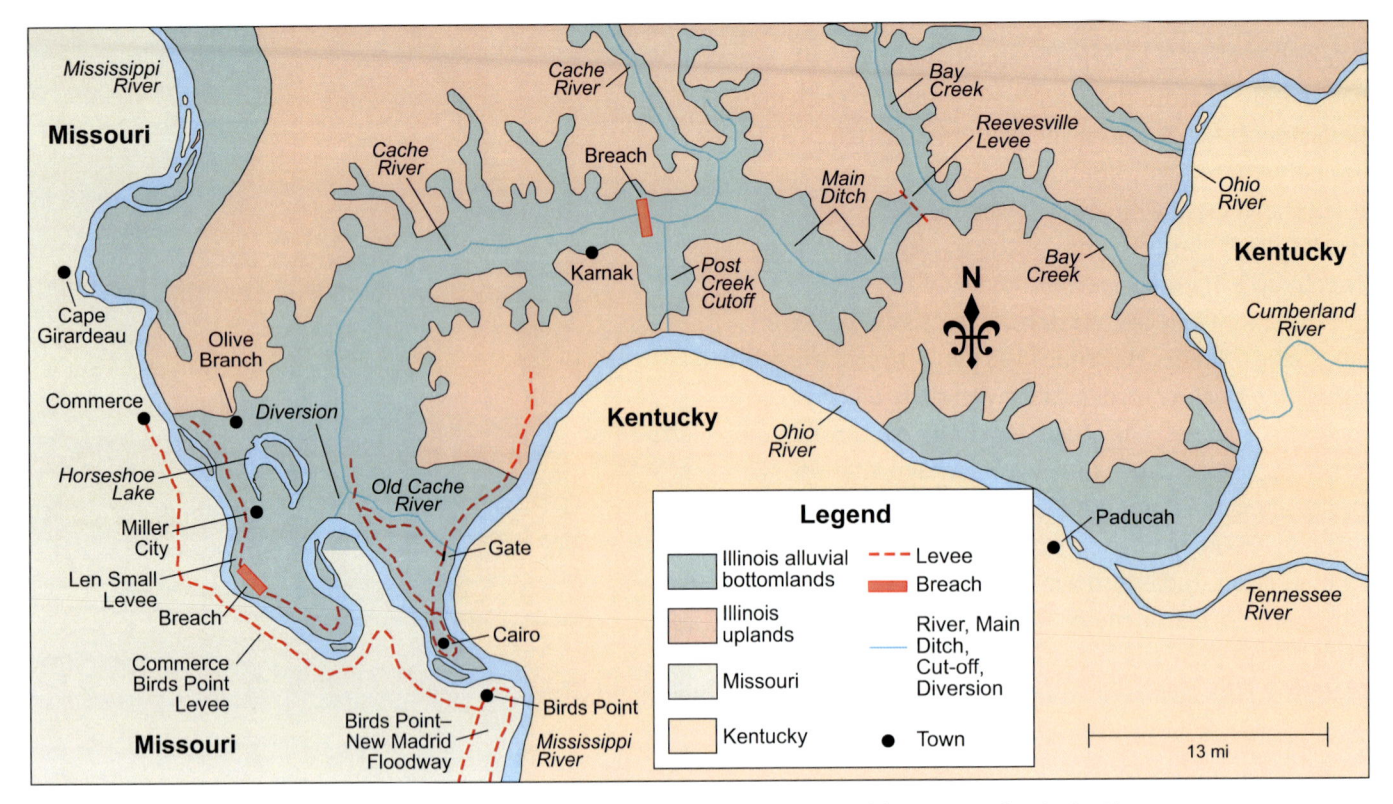

MAP 3.3 The path of the ancient Ohio River through southern Illinois and the location of the present-day Cache River.

river confluence at Cairo, Illinois, and drains lands west of the continental divide from the Appalachian Mountains encompassing all or part of 14 states.

New chronology and longitudinal profiles [5] provide insight into the response of this continental-scale river system to glacial, climatic, and base-level forcing during the last 100,000-year glacial cycle. The ancient Ohio River valley, 50 miles long and 1.5 to 3 miles wide, was formed by the meltwaters of northern glaciers as they advanced and retreated in numerous iterations over the last million years [5, 8]. With increasing sediment fill and changes in climate, the ancient Ohio River shifted away from the Cache River valley and into its present course. This event likely took place between 8,000 and 25,000 years before humans [7, 9]. As a result, the Cache River became a slow-moving stream with extensive isolated, low swampy areas (sloughs) and a water table that ebbed and flowed with seasonal precipitation.

The modern-day Cache River valley of southern Illinois (ancient Ohio River valley; map 3.3) has tupelo (*Nyssa sylvatica* L.)–bald cypress (*Taxodium distichum* L.) swamps, sloughs, and shallow lakes, remnants of the ancient river. The middle Cache River valley is 1.3 miles wide as a result of the previous river having been much larger since it carried waters from the ancient Ohio River valley and local waters from the upper Cache River valley to the Mississippi River. Extensive deposits of gravel and sand, some as deep as 160 feet, rest on the bedrock floor of the middle and eastern portions of the valley [8].

The New Madrid Fault runs under and near Karnak and Ullin, Illinois, and the Cache River valley elevation does not fit with the rest of the area. The Cache River valley is deeper at a lower elevation (between 320 feet and 340 feet) than would otherwise be expected in a slow-moving, swampy river system. It has been suggested [10] that a large section under the Cache River valley sank during a major earthquake at about AD 900. The presence of 1,100-year-old cypress trees in the Cache River valley swamps support this time estimate.

The Borders of the State of Illinois

The current geopolitical boundaries of the State of Illinois have been shaped by geologic realignments of its rivers over millions of years (map 3.4). If these historical waterway-related changes had not occurred, the State of Illinois would only have 22 million acres rather than its current 35 million acres. All but one of the changes in ancient river channel locations increased the area of Illinois by 30% and quadrupled the state's population, since Chicago and Rockford would be in Wisconsin; Cairo and Metropolis in Kentucky; Quincy in Missouri; and Rock Island, Moline, and Peoria in Iowa. Current state borders, such as the Mississippi and Ohio rivers, which were naturally realigned 12,000 to 15,000 years ago, dramatically increased the size of Illinois and decreased the size of Kentucky, Missouri, and Iowa. The relocation of these waterways affected the current border locations of these four central states.

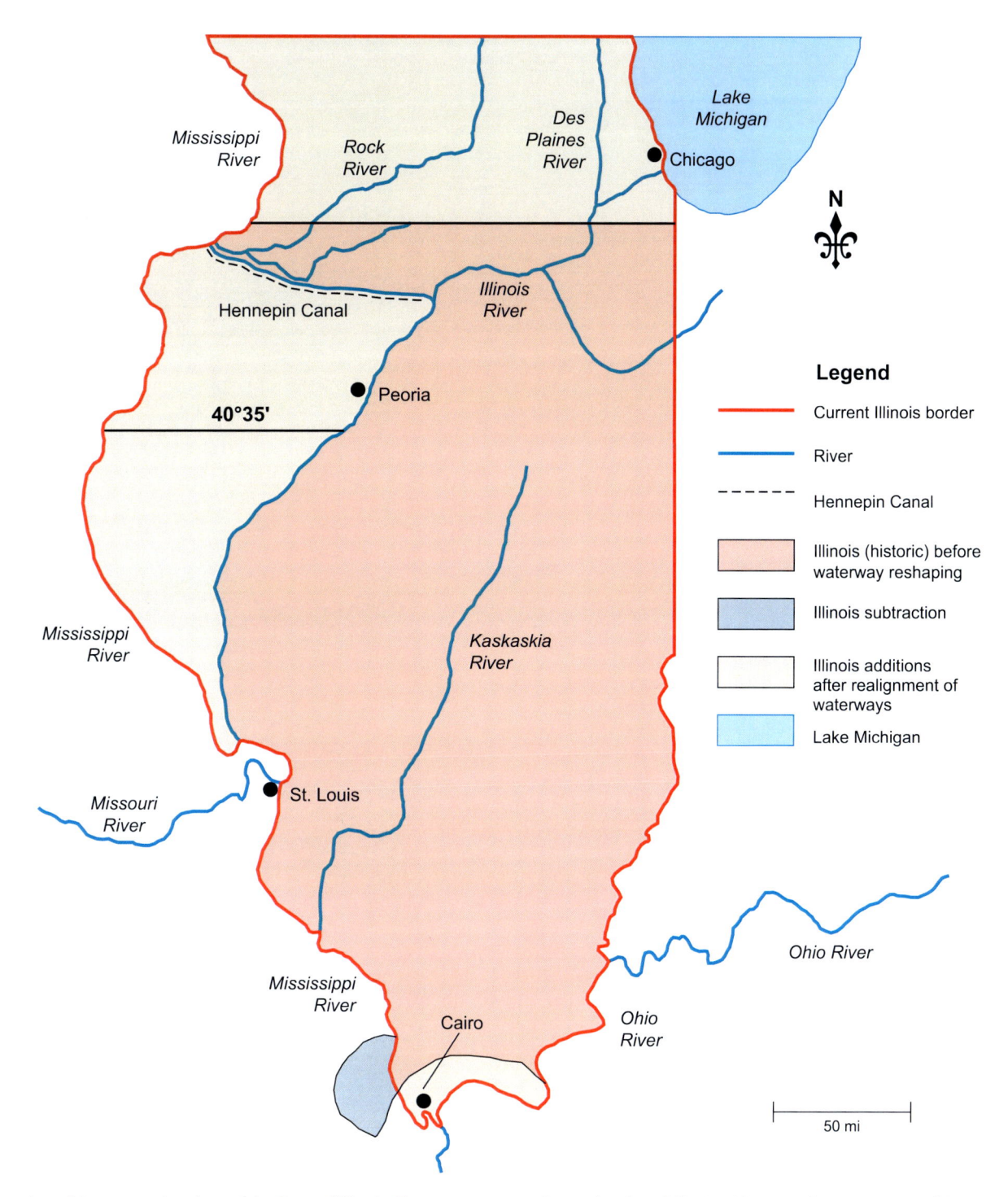

MAP 3.4 Location of the current borders of the State of Illinois. The orange area is the net border of Illinois without geologic rerouting of the ancient Mississippi and Ohio rivers and the US Congress decision to cede to Illinois the lake frontage on Lake Michigan and connecting waterways.

Ancient River Realignment Impact on Other State Borders

Map 3.5 shows the area along the lower Mississippi and Ohio rivers that would have been on the east side of the Mississippi River or the south side of the Ohio River if they had not been realigned naturally as a result of seismic activity and glacial meltwaters. Illinois would have lost 150,000 acres to Kentucky (map 3.5), and Missouri would have lost 150,000 acres to Illinois. Approximately 2 million acres of Missouri land in the Bootheel would be in Kentucky. Arkansas would have lost 1 million acres to Tennessee, and approximately 100,000 acres of Arkansas lands would be in Mississippi. Most of the 8 million acres affected are bottomland soils, and the realignment of the ancient Ohio and ancient Mississippi rivers influenced the location of the current borders of these south-central states located in the Mississippi and Ohio river valleys.

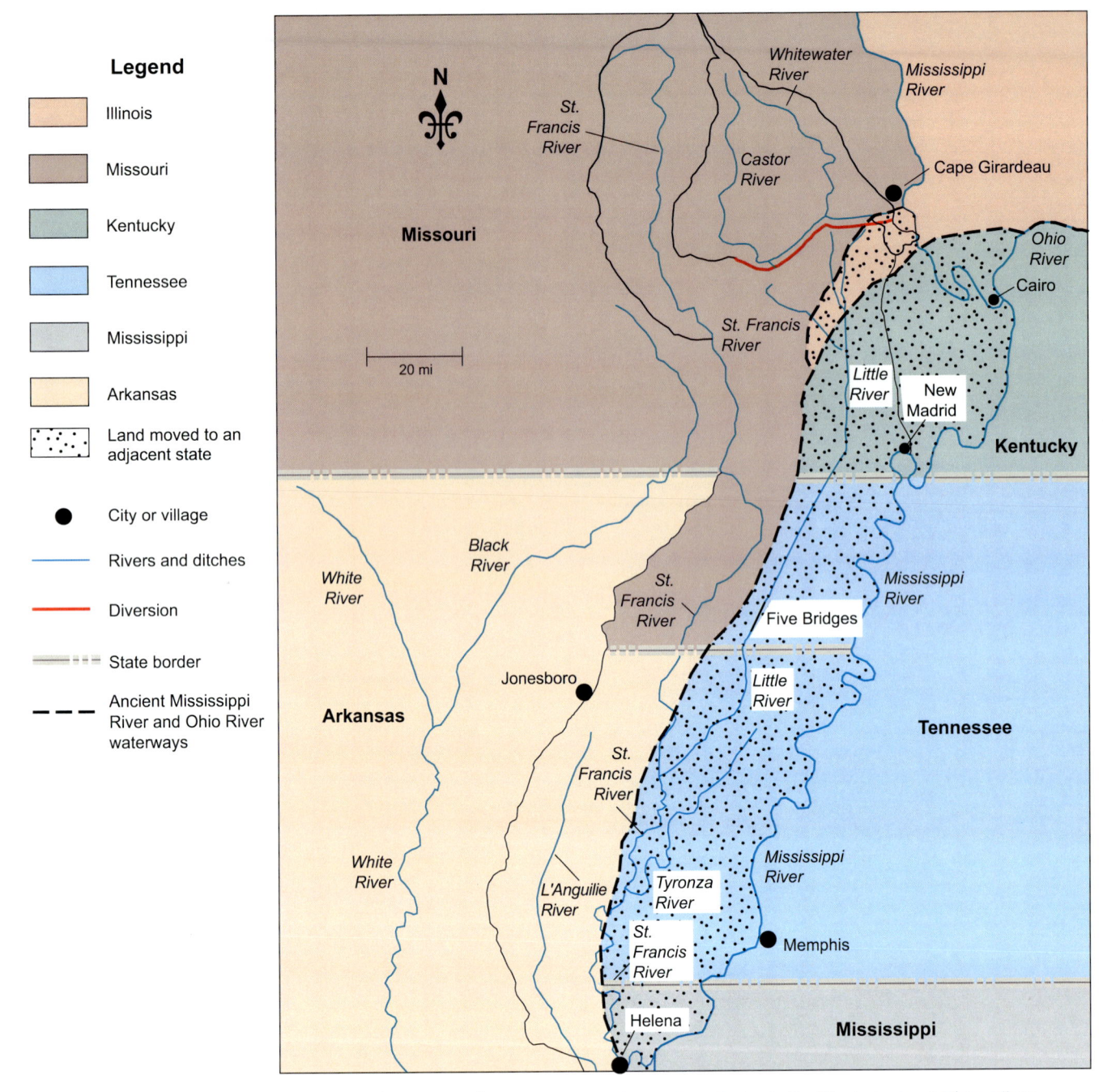

MAP 3.5 The effect of ancient Mississippi and Ohio river realignment on central state boundaries in the Mississippi and Ohio valleys.

[1] Olson, K.R., and F. Christensen. 2014. How waterways, glacial melt waters, and Earthquakes re-aligned ancient rivers and changed Illinois borders. Journal of Earth Sciences and Engineering 4:389-399.

[2] Olson, K.R., and J.M. Lang. 2000. Optimum crop productivity ratings for Illinois soils. Bulletin 811. Urbana, IL: University of Illinois, College of Agricultural, Consumer, and Environmental Sciences, Office of Research.

[3] Olson, K.R., and L.W. Morton. 2014. Dredging of the fracture bedrock lined Mississippi River channel at Thebes, Illinois. Journal of Soil and Water Conservation 69(2):31A-35A, doi:10.2489/jswc.69.2.31A.

[4] Olson, K.R., and L.W. Morton. 2014. The 2011 Ohio River flooding of the Cache River Valley in Southern Illinois. Journal of Soil and Water Conservation 69(1):5A-10A, doi:10.2489/jswc.69.1.5A.

[5] Rittenour, T.M., M.D. Blum, and R.J. Goble. 2007. Fluvial evolution of the lower Mississippi River valley during the late 100 k.y. glacial cycle: Response to glaciation and sea-level change. GSA Bulletin 119(5/6):586-608, doi:10.1130/B295934.1.

[6] Vaughn, J.D. 1994. Paleoseismological studies in the Western Lowlands of southeastern Missouri. Final Technical Report to US Geological Survey. Reston, VA: US Geological Survey.

[7] Cache River Wetlands Center. 2013. Cache River – State Natural Area. Cypress, IL: Illinois Department of Natural Resources. http://dnr.state.il.us/lands/landmgt/parks/r5/cachervr.htm.

[8] Alexander, C.S., and J.C. Prior. 1968. The origin and function of the Cache valley, southern Illinois. In The Quaternary of Illinois, ed. R.E. Bergstrom. University of Illinois College of Agriculture Special Publication 14. Urbana, IL: University of Illinois College of Agriculture.

[9] Esling, S.P., W.B. Hughes, and R.C. Graham. 1989. Analysis of sediment properties within the Cache Valley, southern Illinois and implications regarding the late Pleistocene-Holocene development of Ohio River. Geology 17:434-437.

[10] Gough, S.C. 2005. Historic and prehistoric hydrology of the Cache River, Illinois. Unpublished report to the Cache River Joint Venture Partnership (JVP). Murphysboro, IL: Little River Research and Design.

Multifunctional Bottomlands:
Sny Island Levee Drainage District

It was Red Rock, Saylorville, and Coralville lakes and the rivers that come out of Iowa that Mike Reed, superintendent of Sny Island Levee Drainage District in western Illinois, watched most closely as extreme weather events and flooding hit the upper Mississippi River basin in 2008, 2009, and 2010. The Sny Island Levee Drainage District, located along the eastern bank of the Mississippi River (map 4.1), controls runoff, holds water, and collects large volumes of sediment in basins using an elaborate maze of levees, basins, and diversion channels. The Sny River channel, which runs parallel to the Mississippi River, is the central control structure that channels upland waters prior to pumping into the Mississippi River at three separate locations.

The oldest drainage district in Illinois, officially established in 1880 shortly after the passage of the current Illinois Drainage Act in 1879, the Sny Island Levee Drainage District initially included approximately 110,000 acres of floodplain bottomlands with 4,000 acres of additional lands annexed later. The drainage district has operated for more than 130 years as a local government levee and drainage district and has been used by the Illinois Supreme Court as a model for the development of other drainage districts formed to enable public assessments for protecting agricultural land and valuable infrastructure from flooding. The US Army Corps of Engineers (USACE), which is responsible for the Mississippi River levees as we know them today, has been a key cooperator with the Sny Island Levee Drainage District. The circuit court sets the maximum tax assessment rate and gives the Sny Island Levee Drainage District the power to assess local floodplain farms and landowners who receive a benefit of being protected from Mississippi River and interior flooding by the levee, pump stations, and gravity outlets. Assessment funds are used to maintain the levees and drainage systems along the Mississippi River. Parcel assessment is based on elevation and spatial location within the district. The average 2011 assessment rate was $18.50 per acre. Three commissioners are elected in alternating years to the drainage district board as landowner representatives responsible for monitoring and managing the drainage district.

The 2008 flooding in the Mississippi River valley did not breach any of the levees of the Sny Island Levee Drainage District, and the 140,000 acres of protected bottomlands did not flood [1]. This was not the case in

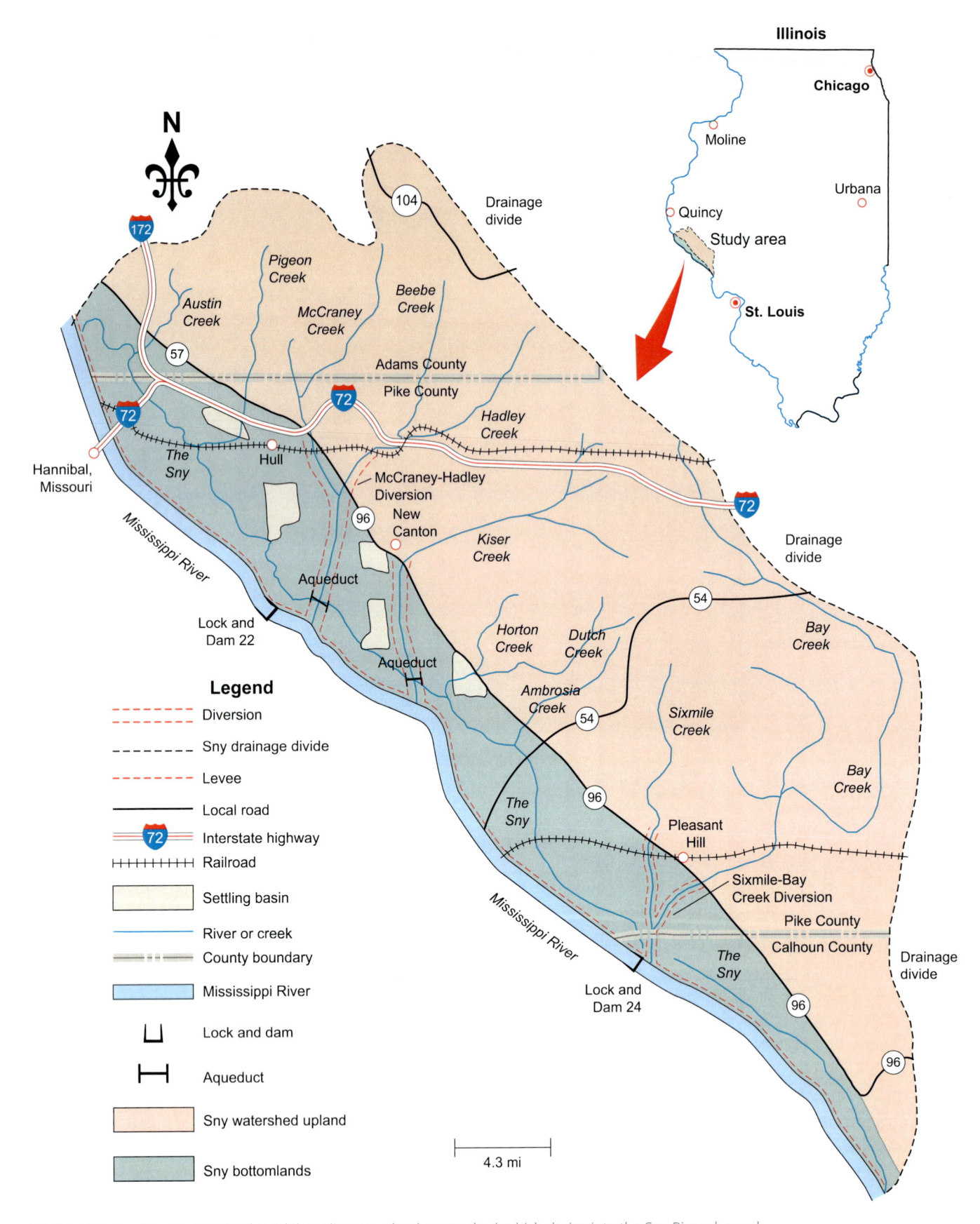

MAP 4.1 The Sny River bottomlands and the adjacent uplands watershed, which drains into the Sny River channel.

1993 Mississippi River basin flooding, when the northernmost 40,000 acres of the Sny Island Levee Drainage District–protected bottomlands flooded. However, the district levees and diversion levees in the southern section protected the remaining 100,000 acres of bottomlands from flooding.

The total watershed area of the Sny Island Levee Drainage District, including the acreage in the district itself, is 440,871 acres (map 4.1). Of that amount, 252,160 acres of watershed are diverted directly to the Mississippi River through Bay Creek (113,408 acres), Sixmile Creek (21,952 acres), McCraney Creek (32,128 acres), Hadley Creek (46,528 acres), and Kiser Creek (38,144 acres). The total watershed area diverted into the sedimentation basins is 67,804 acres. Of that amount, 18,432 acres are diverted into Pigeon Creek basin, 22,124 acres into Dutch Horton basin; 4,032 acres into Austin Creek basin, 5,824 acres into Fall Creek basin, 4,160 acres into Ambrosia Creek basin, 1,183 acres into Walnut Creek basin, 461 acres into Pothast basin, 1,703 acres into Grubb basin, 480 acres into Shewhart basin, 2,630 acres into Brewster/Brown basins, 3,400 acres into Atlas/Two Mile basin, 3,000 acres into Howell basin, and 375 acres into Johnson basin.

The Sny River Channel

More than 10,000 years ago, glacial floodwaters in the Mississippi River valley flowed in multiple channels. As the water receded, it left behind a river bottom terrace called Sny Island and the Sny River, a branching channel that intercepts the many creeks that run out of the uplands [2]. In 2007, these nutrient-rich bottomlands comprised more than a third of all Pike County, Illinois, acres in farmland [3]. The eastern border of the Sny Island Levee Drainage District is near state Route 96 and Route 57, which run parallel to the Mississippi River at the base of the limestone bluffs that separate the Mississippi bottomlands from the uplands (map 4.1). The official eastern boundary of the Sny Island Levee Drainage District is the high water mark of the 1851 flood, and the district includes several local stream floodplains to the east of Route 96 that were later annexed to better control upland runoff and flooding. The Pike County soils report number 11, published in 1915 by the University of Illinois Department of Agronomy [4], shows a Pike County levee extending all along the 54 miles of the Mississippi River in Pike County (map 4.1), and it is the western boundary of the Sny Island Levee Drainage District.

The Sny Island Levee Drainage District (map 4.1) currently protects 5,377 tracts with a total of 114,000 acres of floodplain bottomlands, which are mostly cultivated cropland (corn, soybeans, and some wheat). The bluffs are forested with little agricultural use, except for the valley bottoms adjacent to the creeks and streams. There are a number of small towns (including Kinderhook, New Canton, Rockport, Atlas, and Pleasant Hill) that are located on outwash from glaciers covered with alluvial fans created from past upland geologic erosion events. The Sny River channel was reengineered and reconnected and still parallels the current Mississippi River. It is often only several hundred feet to the east and was part of an old Mississippi River channel. The Sny River channel collects and stores water, sediment, and nutrients from local streams and watersheds before it is pumped or allowed to free flow into the Mississippi River in the upper three reaches of the district. The southernmost reach of the Sny River channel is now disconnected from the upper portion by the Sixmile–Bay Creek Diversion levee. The lower reach of the Sny River channel still drains by gravitational flow into the Mississippi River at river mile 269 in Calhoun County (map 4.1). The Sny River channel name was not used on the 1915 soil survey report since it was not continuous during dry periods. Instead, the maps showed a series of slough names such as Running Slough, Mud Slough, Salt Slough, and Burr Slough.

Settling Basins

In the 1920s and 1930s, the Sny Island Levee Drainage District became concerned about the sediment and nutrients that were entering and filling the Sny River channel and Mississippi River, as well as being deposited on Mississippi bottomlands as a result of local creek flooding events. Flooding of the Sny Island Levee Drainage District land continued to occur when local streams and creeks, such as Pigeon Creek, Hadley Creek, Kiser Creek, and Bay Creek, overflowed (map 4.1). This resulted in flooded cropland and sediment deposition on the agricultural bottomlands and in the Sny River channel. During some flooding events, the local runoff water was trapped behind (east side) the Mississippi River levees and resulted in crop loss even though the levees did not fail.

By the 1940s and 1950s, the Sny Island Levee Drainage District decided to create additional levees on local creeks and streams, as well as large settling basins, to protect agricultural land from local flooding and sediment deposition. Several hundred acres of these basins, including the 640-acre McCraney Creek basin located approximately three miles south of Hull, Illinois, and the 640-acre Kiser Creek basin located five miles west of New Canton, Illinois, were created (map 4.1). Since the land where the sediment basins were to be placed was on private property, a lease arrangement was negotiated with the landowner. Landowners were paid for the use of their land to store sediment and nutrients during the years the sediment basin was active and the land could not be farmed. These sediment basins captured eroded soils

FIGURE 4.1 A 640-acre settling basin (Kiser) filled up with sediments in the 1950s and is currently used for row crop production.

FIGURE 4.2 A 640-acre settling basin (McCraney) filled with sediments in the 1950s and is 7 to 10 feet above the bottomlands and 9 to 14 feet above the current drainage ditch.

carried by excess waters that were periodically diverted from adjacent streams, such as Hadley, McCraney, and Kiser creeks. As a result of the 1940s and 1950s cultivation practices (moldboard plowing) in the uplands of Adams, Pike, and Calhoun counties, these and other basins filled to 10-foot depth in less than 20 years.

Once filled, the basins dried out and the depth of sediment shrunk. The embankments were graded to the level of the dried sediment (figure 4.1) and returned to the private landowners to be farmed. Currently, all of these earlier basins are being cultivated for corn and soybeans. The sediment-filled basins are still at least 7 feet above the surrounding fields on the bottomlands (figure 4.2). These sediment basins were mapped in the 1999 Pike County soil survey [5] as soil number 815 (Udorthents, silty). The Department of Natural Resources and Environmental Sciences in the College of Agricultural, Consumer, and Environmental Sciences at the University of Illinois [6], assigned crop yield ratings and productivity indices (PIs) to these large, inactive sediment-filled basins. The original soils at these sites before the sediment basins were constructed consisted of Wakeland silt loam, Haymond silt loam, and Beaucoup silty clay loam, and had PIs between 128 and 132. They were adjusted down to PIs between 98 and 119 as a result of documented crop loss caused by local stream flooding. The PI adjustment varied by extent of crop loss within each county (Adams, Pike, or Calhoun). The optimum PI for these man-made soils on the top of the sediment basin was 108 (on a 147-point scale); these basins are now 7 to 10 feet higher than the original landscape, and the PIs are no longer in need of adjustment for local flooding.

By the late 1950s it became very apparent that the major east-west streams (Hadley Creek, McCraney Creek, Kiser Creek, Sixmile Creek, and Bay Creek) were transporting too much runoff water and sediment to the Sny River channel despite the construction of the individual 640 acre settling basins on the major streams. A new approach to controlling and holding water was undertaken in the 1960s by the USACE with the assistance of the Sny Island Levee Drainage District. In addition to the creation of 2,600 acres of new settling basins, such as the Horton-Dutch sediment basin (figure 4.3), three major diversions (levees on both sides of the major local creek or channel) were created (map 4.1). These diversions routed the water from the largest watersheds in the uplands through the Sny Island Levee Drainage District bottomland area and discharged directly into the Mississippi River. These diversions became known as the Hadley-McCraney Diversion located in northern Pike County, the Kiser Creek Diversion located in central Pike County, and the Sixmile–Bay Creek Diversion located in southern Pike and northern Calhoun counties, and include an upland watershed of approximately 252,160 acres. These diversions are approximately 6 to 12 miles long, with 10- to 15-foot-high levees on both sides of the channel, and can be between 200 and 300 feet wide with some cultivation, woodland, and wildlife in the diversion channels (figure 4.4). These diversions reduce the water, sediment, and nutrients going directly into the Sny River channel, while maintaining internal drainage control of the smaller watersheds in the uplands and any sediment and nutrients from other drainage ditches in the Mississippi bottomlands within the Sny Island Levee Drainage District. The 3,000 acres of land in the diversions include grassland,

FIGURE 4.3 The Horton-Dutch Creek settling basin is filling up with water, sediment, and nutrients. The basin levee is shown in the distance.

FIGURE 4.4 The wide Kiser Creek Diversion channel is timberland between the two levees just before it drains into the Mississippi River.

timberland, and cultivated land, or are covered by water and serve as a significant wildlife habitat.

Two aqueducts (map 4.1) under the Hadley-McCraney and Kiser Creek diversions carry the water from the Sny River channel, which crosses almost perpendicular to both diversions (figure 4.5). The Sixmile–Bay Creek Diversion blocks the Sny River channel flow from the upper 54 miles of the channel watershed, including the 114,000 acres of protected bottomlands in the Sny Island Levee Drainage District and the adjacent runoff from the bluffs and the smaller local creeks from the uplands (approximately 67,824 acres). The Sny River channel flow does not include the runoff water from the watersheds that drain into and through the three diversions. The Sixmile–Bay Creek Diversion levee directs the Sny River channel water to the lowest pump station (near where the Sixmile–Bay Creek Diversion levee joins the Mississippi River levee [figure 4.6] and south of Lock and Dam 24). The Sny River channel water can be stored in the channel to allow sediment to settle, and eventually, clear water is pumped or allowed to free flow into the Mississippi River, thereby reducing sediment and nutrient discharges. The last 5 miles of the Sny River channel flows to Pump Station 4 at river mile 269, where it can be pumped or allowed to free flow directly into the Mississippi River, draining a small portion of the western upland watershed in Calhoun County.

Currently 15 sediment basins, which cover 2,600 acres, have been built since 1970. Some sediment basins can be cultivated during dry periods, and others remain ponded on at least one end and provide wildlife habitat and numerous hunting opportunities for the entire region. To the delight of the Sny Island Levee Drainage District staff, the sediment basins, including Pigeon Creek basin (figure 4.7), are no longer filling up in less than 20 years. They have had a much longer active life (up to three times greater). There are two primary reasons for the settling basins filling much more slowly. One reason is the discharge of the three major diversions directly into the Mississippi River bypassing the Sny River channel drainage system, and the other reason is changes in farming and cultivation practices on both the bottomlands and uplands to better manage soil erosion. In the 1970s, the primary tillage method changed from a moldboard plowing system to a chisel plow system. By the 1990s, many farmers replaced the chisel plow system with no-till (figure 4.8). This switch in farming systems was in part a result of the tolerable soil loss (T) levels set by the 2000 Illinois erosion control program that became law in 1983, as well as the federal Food Security Act of 1985 (and later farm bills) that provided incentives and/or required farmers to reduce soil loss to T-tolerable levels of less than five tons per acre per year.

Observations

Strategically placed wetlands, settling basins, and levees are effective management practices for internal control of water and sediment, as well as nutrient filtering. Sny Island Levee Drainage District has been a pioneer in the development of ways to reduce local flooding and decrease the sediment and nutrient loads being discharged into the Mississippi River. Water is currently discharged at the rate of 0.25 inch per day. These measures allow the district to hold within the system as much as a month's worth of water, thereby reducing the peak flow of water discharged into the Mississippi River. The series of local levees, diversions, water storage basins, and settling basins slow the discharge of water to the Mississippi River, allow time for nutrients and sediment to settle out, provide multiple land uses for the settling basins during their active life, and convert 2,600 acres back into productive agricultural land

FIGURE 4.5 The Sny River aqueduct passes under the Kiser Creek Diversion channel.

(figure 4.1) after filling. Nutrient levels of agricultural lands vary considerably across the region and within tracts. Producers use grid sampling to deliver nitrogen and other nutrients where they are most needed and to avoid over application.

Managing the Sny Island Levee Drainage District requires constant attention to interior control of runoff water, soil erosion, and internal levee maintenance, as well as monitoring external conditions and Mississippi River pressures that could lead to breaches in the levees, which defend the farmland from the river. Extreme weather events experienced between 2008 and 2011 and expectations for future increased precipitation and more frequent four- to six-inch rain events present difficult challenges to the district. Three-quarters of their $2 million annual revenue is spent on fuel costs to run their three pumping stations continuously during periods of high water runoff.

The acreage and productive capacity of the agricultural land in the Sny Island Levee Drainage District have been maintained by the sediment-filled basins built in the 1940s and 1950s, which were reclaimed in the 1970s for crop production. It is extremely rare for such a large block of land, previously in agricultural use, to be converted to another land use, such as sediment-filled basins, for 20 or 30 years and then be returned to agricultural use and production. The replacement settling basins built after the 1970s are at sites with diverse land uses, including stream channels, timberland, ponds, agricultural lands, wetlands, and wildlife habitat. It is too soon to know what the future land use of the second set of sediment basins will be, but it is likely that some will be returned to agricultural use and production. The need for a third set of settling basins could decline as landowners in both the uplands and bottomlands increase conservation practices that control soil erosion and reduce the amount of sediment and nutrients being transported into the Sny River

FIGURE 4.6 The merged Mississippi River and Kiser Creek Diversion levee blocks the Sny River channel flow, and water has to be pumped over the levee to the Mississippi River.

FIGURE 4.7 The Pigeon Creek settling basin west of Hull, Illinois, is currently active during high runoff and flooding.

FIGURE 4.8 Many of the Sny River watershed uplands have no-till corn planted on very steep slopes, which reduces soil erosion into drainage ditches.

channel and eventually pumped or allowed to free flow into the Mississippi River.

The combination of purposefully created wetlands and settling basins alongside agricultural lands protected by levees provides diverse habitats for wetland species, fishing, and recreational duck and deer hunting. The high cost per acre of land assessment pushes producers on both bottomlands and uplands to select high-value crops and to farm right to the edge of their internal drainage ditches. Fast-moving, high water in these ditches increases bank erosion where soils are not held by vegetation, clogging the drainage system and increasing the need for more frequent ditching and Sny River channel dredging. Incentives to encourage 5- to 10-foot vegetative strips alongside these steep ditches are conservation measures that would hold soil in place and reduce the movement of soil from field to water. Much of the upland and bottomland farmland is already in no-till (figure 4.8), and producers should be encouraged to continue this practice and expand it to row crops currently not utilizing this management practice. In addition, the conversion of very steep slopes from row crops to perennial cover would help reduce further soil loss. This, however, does not address an underlying concern—the need for landowners to produce sufficient revenues to cover the drainage district per acre assessment to protect the region. When fuel oil prices increase, high agricultural productivity coupled with high commodity prices are needed to assure economic stability for the district. One possible way to mitigate this treadmill is to build on the diverse habitat created by this system of wetlands, sediment basins, and levees, and purposefully develop an economic tourism plan to increase the recreational use of this region.

---

[1] Olson, K.R. 2009. Impacts of 2008 flooding on agricultural lands in Illinois, Missouri, and Indiana. Journal of Soil and Water Conservation 64(6):167A-171A, doi:10.2489/jswc.64.6.167A.

[2] Gard, W.T. 2002. The Sny Story. The Sny Island Levee Drainage District and Sny Basin. North Richland Hills, Texas: Smithfield Press.

[3] USDA National Agricultural Statistics Service. 2009. The Census of Agriculture. Washington, DC: USDA National Agricultural Statistics Service. http://www.agcensus.usda.gov.

[4] Hopkins, C.G., J.G. Mosier, E.V. Alstine, and F.W. Garrett. 1915. Pike County Soils. Soil Report no. 11. Urbana, IL: University of Illinois, Agricultural Experiment Station.

[5] Struben, G.R., and M.E. Lilly. 1999. Soil Survey of Pike County, Illinois. Washington, DC: USDA Natural Resources Conservation Service.

[6] Olson, K.R., and J.M. Lang. 2000. Optimum Crop Productivity Ratings for Illinois Soils. Bulletin 811. Urbana, IL: University of Illinois, College of Agricultural, Consumer and Environmental Sciences, Office of Research.

Conversion of Missouri's Big Swamp to Fertile Agricultural Land

More than a century ago, American swamps and river lowlands were considered wasteland of no value and a hindrance to land development. The United States Swamp Land Acts of 1849, 1850, and 1860 granted states the right to reclaim 64.9 million acres of swamps through the construction of levees and open channels (ditches) to control flooding; to encourage settlement, land cultivation, and commerce; and to eliminate widespread mosquito breeding [1]. Southeast Missouri, once one of the world's largest tracts of forested bottomlands, was a vast wilderness of bald cypress (*Taxodium distichum* L.), tupelo (gum; *Nyssa* L.), hardwoods, and water (figure 5.1), barely accessible to settlers migrating west. Starting in the early 1890s, these historic river floodplains and their tributaries were drained and transformed into fertile agricultural lands [2, 3].

Today this vast network of ditches (map 5.1), channels, and levees in southeast Missouri bottomlands makes possible an intensive system of agriculture that produces almost a third of Missouri's agricultural economic output. This drainage feat has also changed the hydrology, nutrient cycling, biodiversity, and structure of the entire ecosystem [3]. Unified and managed by the Little River Drainage District (LRDD; map 5.2) with support from the US Army Corps of Engineers (USACE) [4], this region drains 620,000 acres of Little River basin bottomlands (figure 5.2). It is the drainage outlet for 1.2 million acres of bottomlands and uplands to the Arkansas border and combines with 1 million acres of the St. Francis River basin as it flows into the Mississippi River near Helena, Arkansas.

Swamps, Sloughs, and Fertile Mississippi River Valley Bottomlands

Historically, the path of the Mississippi River just south of Cape Girardeau turned southwest into the current state of Missouri and traveled more than 30 miles to the west before turning south toward Morley, Missouri (see map 6.2) [5], where it joined with the ancient Ohio River waters that drained through the Cache River valley (ancient Ohio River valley) [6]. In the New Madrid area, thousands of years of seismic activity [7] have affected the Mississippi River channel.

The shifting of the ancient Mississippi River to a new channel left behind an expansive network of perennial streams, swamplands, sloughs, bayous, and fertile river-

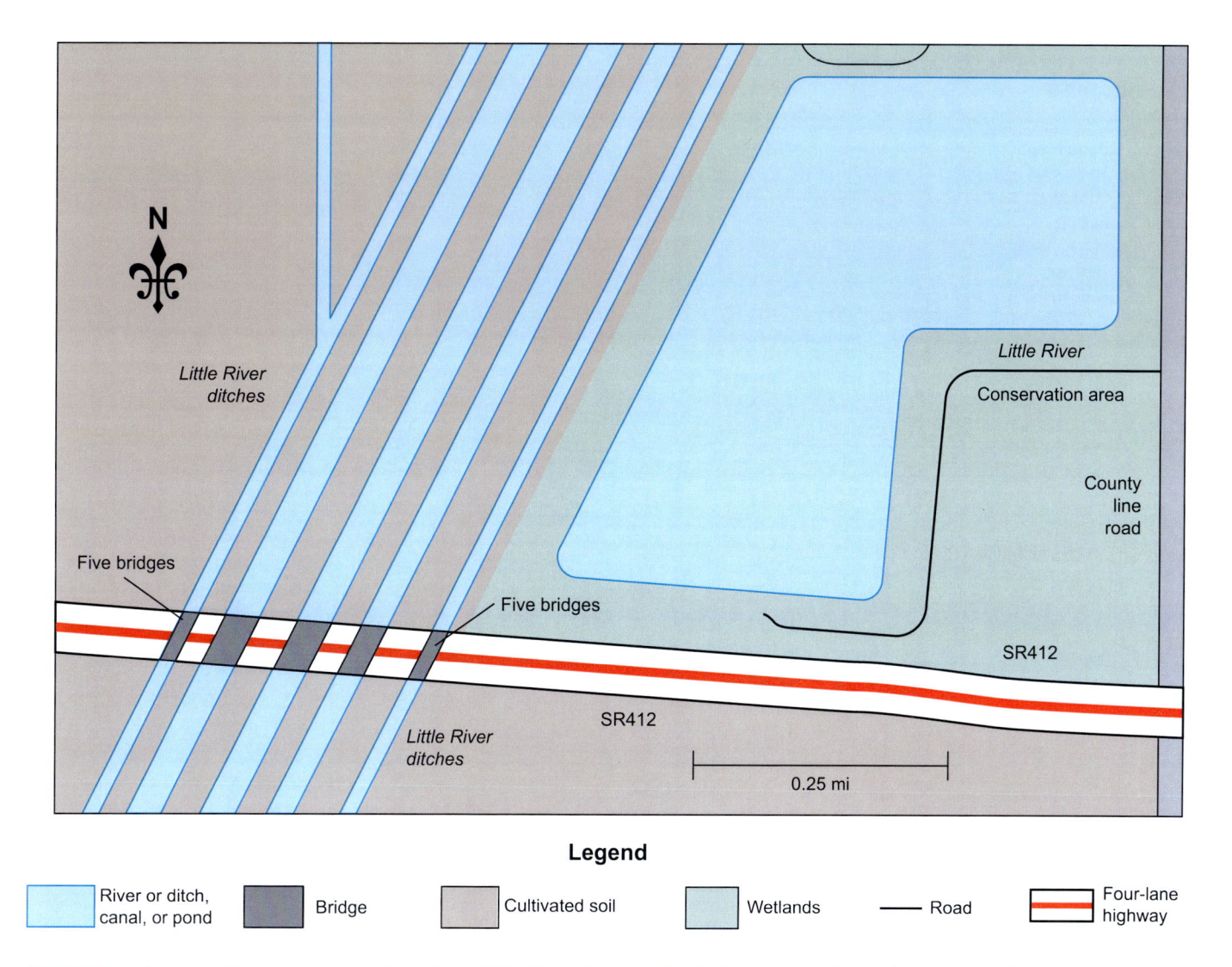

MAP 5.1 Five drainage ditches run parallel in southern Little River Drainage District just north and east of the Arkansas border.

ine bottomlands inundated during much of the year. This Mississippi alluvial valley, dissected by Crowley's Ridge, produced Big Swamp and bottomlands extending from just below the southern boundary of Cape Girardeau, Missouri, to the Arkansas state line (map 5.2), and as far west as the escarpment of the Ozark Plateau. Because of the swampy conditions, less than 15% of the land was suitable for cultivated agriculture. Plans to turn the two million acres of swampland (including the Little River, St. Johns Bayou, and St. Francis basins) into farmland date back to the 1840s, but the task was too big for individual farmers to undertake. Not even the federal government, which owned the swamplands at that time, had ever undertaken a project of that magnitude [8].

When the European settlers arrived in southeastern Missouri in 1820s, they settled on Crowley's Ridge (figure 5.3), which stands 250 to 550 feet above the Mississippi River bottomlands and extends 150 miles from southeast Missouri to Helena, Arkansas (maps 5.2 and 5.3) [9, 10]. Many settlers built homes on top of the

FIGURE 5.1 The Mingo National Wildlife Refuge, a 21,676 acre bottomland preserve, is a small restored remnant of the original 1.2 million-acre Missouri Swamp that was transformed into fertile agricultural lands by extensive drainage systems in the early 1900s.

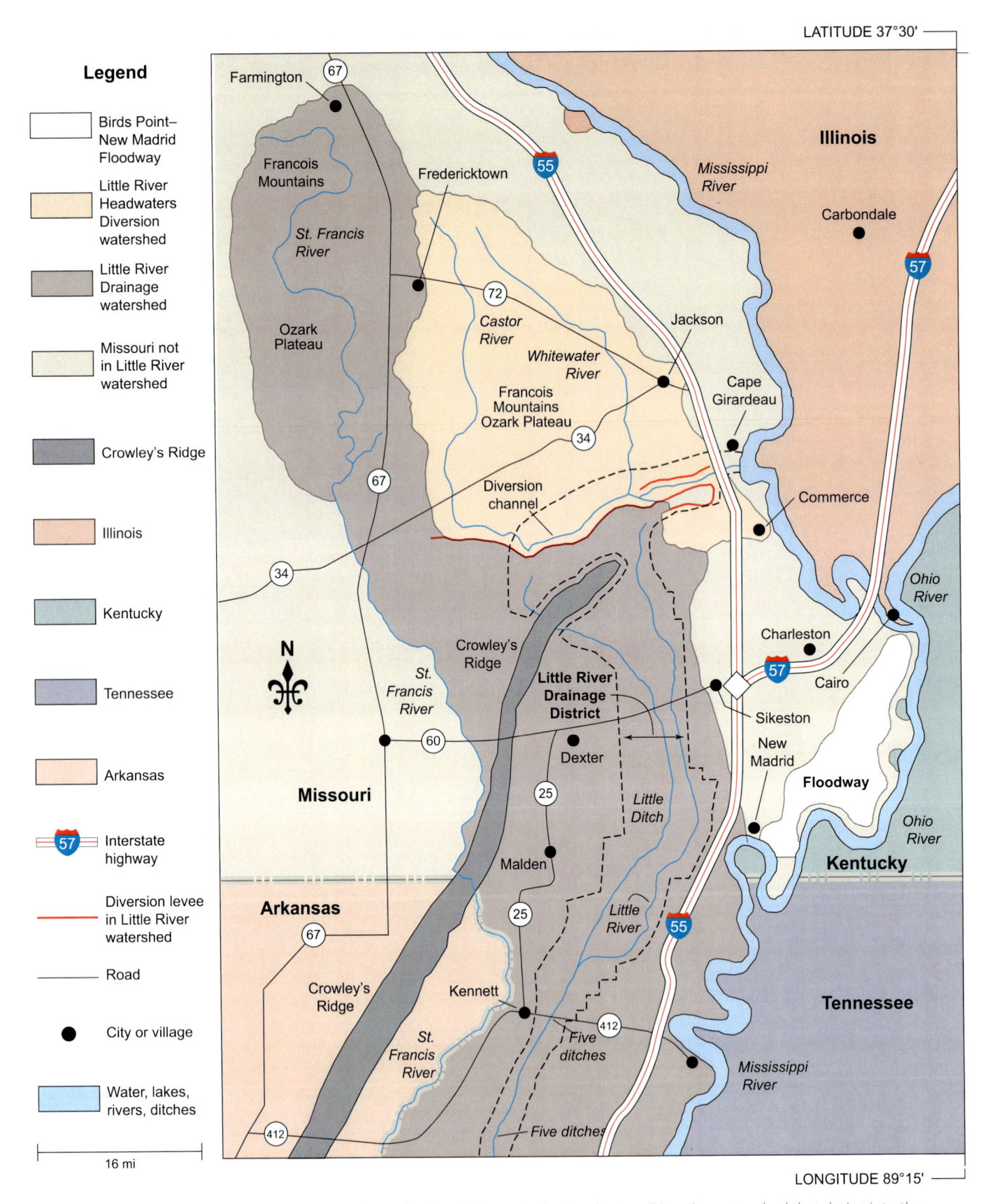

MAP 5.2 This Little River Drainage District map shows the Ozark Plateau in the Headwaters Diversion watershed that drains into the diversion channel and the Mississippi River south of Cape Girardeau. The upland and bottomland areas west and south of the Headwaters Diversion watershed drain into the St. Francis River.

semiforested ridge and out of the way of the Mississippi River floods. They used the trees to build homes and make furniture, and to provide fuel for heating and cooking. The gently sloping ridge had spring-fed creeks and soils suitable for cultivation (Memphis silt loam), with the richest soils at the base of the ridge (figure 5.3). The

34

FIGURE 5.2 Agricultural lands created by draining Big Swamp.

settlers who farmed the uplands of the ridge found that the loess soils eroded very easily and large gullies formed on some of the slopes of Crowley's Ridge [11].

National Swamp Land Acts

The National Swamp Land Acts of 1849, 1850, and 1860 transferred public domain swampland considered of no value to states to help them control river flooding and incentivize land development [1]. Although the continuously flooding Mississippi River bottomlands were the original target, over a 100-year period 15 states were authorized to carry out wetland reclamation programs under this legislation. Swamplands were categorized according to capacity of the land to be made fit for profitable agriculture [1]. The categories included (1) permanently wet and not fit for cultivation, even in favorable years, unless cleared and levee protected; (2) wet pasture for livestock, with forage often of inferior quality; (3) subject to periodic overflow by streams, but at times able to produce crops; and (4) too wet for profitable crops during above-normal rainfall, but usable during seasons of light or medium rainfall.

National inventories of wetlands documented over 3.4 million acres of swamp and overflow lands in Missouri as authorized for reclamation. The Swamp Land Act of 1850 gave the swampland southwest of Cape Girardeau to the State of Missouri. The transfer of land stipulated that the land be reclaimed for the benefit of the nation. By 1890 the state had conveyed most the land to counties. The counties sold the land to private companies and owners. Initially, the bottomlands were not habitable as ordinary runoff and floodwaters regularly spilled across much of southeast Missouri. This region, known as the Bootheel, is a natural basin that caught upland waters and pooled them in sloughs and lakes and created the Missouri swamplands. After the Civil War (1861 to 1865), the opening of the American West by railroads offered greater settlement opportunities in western lands that were more suitable for cultivation than the southeast Missouri swampy bottomlands. However by the 1890 census, it was clear that the United States no longer had

FIGURE 5.3 At the eastern base of Crowley's Ridge are fertile soils that are irrigated during dry periods.

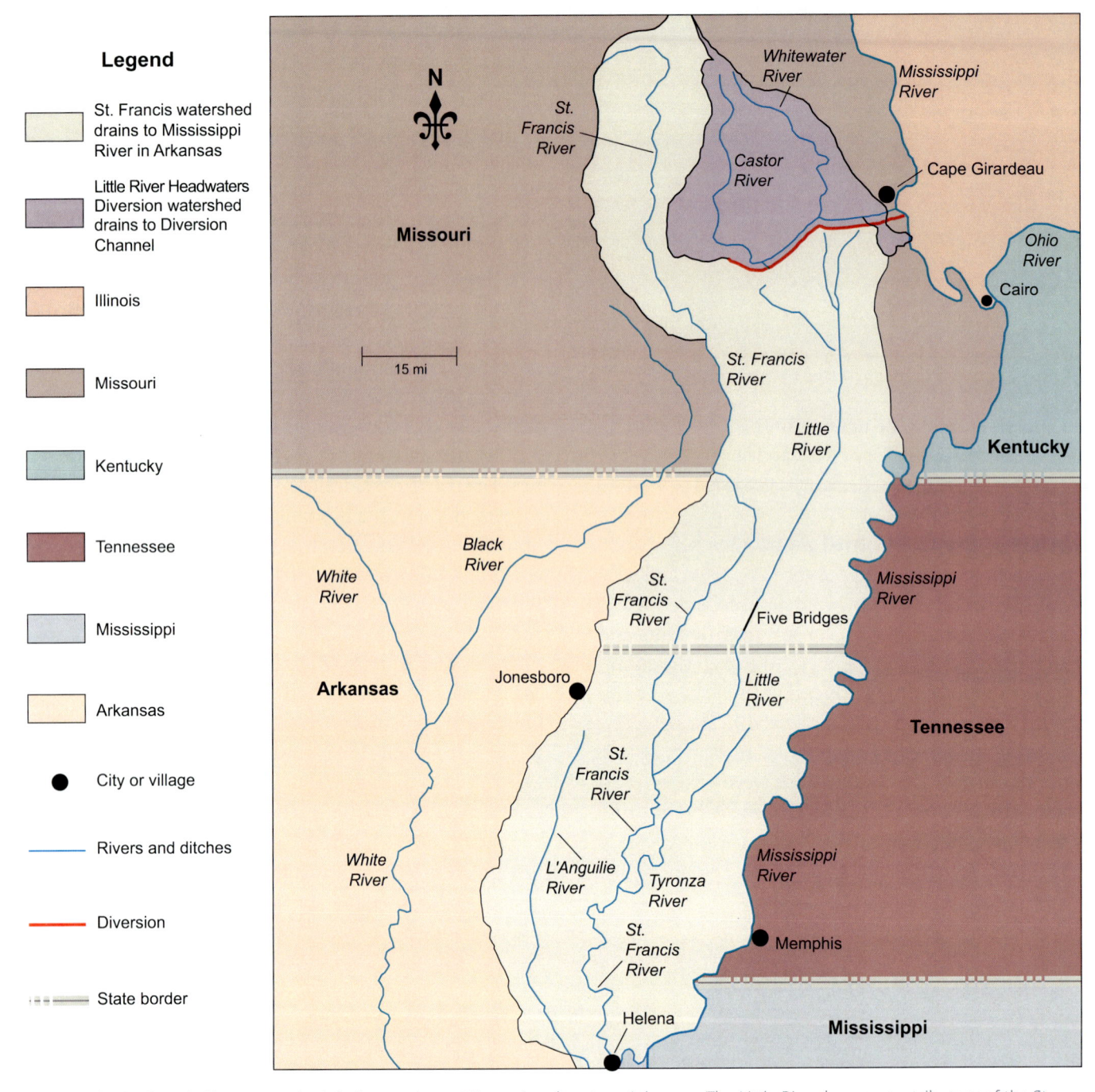

MAP 5.3 The St. Francis River watershed drains southeast Missouri and eastern Arkansas. The Little River becomes a tributary of the St. Francis River in Arkansas before draining into the Mississippi River at Helena, Arkansas.

a frontier, and American wetlands became the next area of opportunity for settlement and cultivation.

In 1875 two Kochtitzky brothers landed in New Madrid, Missouri, to conduct surveys of the region [12]. Their father Carl Kochtitzky, a Laclede County clerk in south-central Missouri, learned of the 1850s legislation that granted the swamplands to the State of Missouri and then to the counties. The brothers, in talks with the local county judge, discovered that the old Pole Road started before the Civil War was no longer in use but had charged $0.75 per wagon and $0.25 per bale of cotton. The Pole Road bed consisted of cut logs laid down side by side to distribute ground pressure for traversing wagons, mules, horses, and people to keep from sinking into the mud. An alternative line for transporting goods and services was the Crowley Ridge road stretching 200 miles from Cape Girardeau to Helena, Arkansas.

Carl Kochtitzky and his sons saw an opportunity to move cotton to market more quickly via railroad and formed with four other businessmen the Little River Valley and Arkansas Railroad partnership to install 27 miles of rail from New Madrid to Malden on the Dunkin-New Madrid county line. The winter of 1877 to 1878 was so cold that oxen dragged timbers across ice to the bridge sites. The first year's operation freight volume from cotton only paid the operating costs. To increase the volume of freight, the partners looked at drainage reclamation to create cotton cropland. A new partnership was contracted in 1885 to dredge and straighten the Little River from the railroad to the county line. The contract was for $0.14 a cubic yard with payment in land valued at $1.25 per acre. This brought new land into agriculture and boosted cotton production. Farmers in Pemiscot, Dunkin, New Madrid, and Stoddard counties in 1890 reported to the US Department of Agriculture 41,491 cotton acres, almost double the 22,981 acres reported in 1880 [13]. Another rail venture, started in 1894 and completed in 1899, ran from Cairo, Illinois, to Poplar Bluff, Missouri. The freight for this line was oak staves for wine caskets. During this time 150,000 acres of land became available for reclamation when the title was transferred from New Madrid County in 1899 to the private rail company. The National Swamp Land Act, in concert with railroad construction, connected timber and agricultural interests to new markets and fueled the draining of wetlands for settlement and agricultural expansion.

Stumps in the Swampland

The swamps of southeastern Missouri were full of valuable bottomland hardwood trees. These forests were mostly oak, hickory, and cypress trees; some of the oak trees had a circumference as large as 27 feet and some cypress trees had an 11-foot circumference [14]. Timber companies built drainage ditches, roads, and rail systems to harvest and market this valuable commodity. The 1905 construction of the Thebes railroad bridge between Illinois and Missouri [15] enabled railroads to build into Big Swamp (figure 5.1) and carry the timber to eastern markets. Lumber plants and sawmills like Himmelberger and Harrison in Morehouse, Missouri, provided jobs and made settlement possible. The three largest towns in the Bootheel were New Madrid, Caruthersville, and Charleston, which grew quickly into hubs of trade on the Mississippi River. However, their growth was short lived, and their populations peaked in 1890.

By 1910, land purchased by timber and railroad companies was deemed wasteland again, with only tree stumps and water remaining. Lumber companies were left paying taxes on thousands of acres of cleared swampland and did not know what to do with it [12]. With broad expanses of the region under 3 or 4 feet of water year round and some portions under as much as 15 feet near the riverbed, local levee and drainage boards attempted sporadic and uncoordinated efforts to build levees and drainage ditches [16]. Lacking engineering skills, financing, and water management experience, these boards made little progress in controlling flooding or reclaiming the swamplands for farming. Although the alluvial soils beneath the water were some of the most fertile soils in Missouri and well suited for farming [11], these bottomlands seemed impossible to drain.

The Little River Drainage District

It is commonly thought that the LRDD was the first attempt to develop the swamplands of southeast Missouri. However, historical records indicate that by 1905 one-half of the basin had been partially reclaimed and assessed by earlier organized drainage districts. Some of these attempts were successful and others failed since runoff water from the Ozark Plateau and from Mississippi River flooding events continued to flood into Big Swamp. These prior drainage projects became the building blocks for unifying the financing, governing, and engineering efforts under the LRDD. Due to the difficulties of draining these swamplands, it took almost 50 years to develop the necessary laws, business models, and engineering techniques needed to create a successful drainage district model. Engineering innovations, willing investors, and men and women with perseverance settling the region were important factors in building the roads, rails, diversion channels, and bridges that became the basic infrastructure for draining the region.

Enactment of the Drainage District Law in 1899 was the final legislation enabling comprehensive tax assessments that made it possible to organize drainage districts on a landscape level [12, 17]. This laid the financial infrastructure for a group of visionaries with timber interests and substantial resources to undertake a unified effort to convert the swamp into lands suitable for cultivation. In 1905 a meeting of large landowners and local leaders was held in Cape Girardeau, Missouri, to create the LRDD. Publication the same year of the Little River drainage map by Otto Kochtitzky helped promote the project and offered a region-wide vision that inspired investors to purchase bonds and landowners to agree to pay taxes for the land reclamation effort. The LRDD became the mechanism that motivated individual landowners to come together as one to implement and construct a "Plan of Drainage" to meet the federal mandate. Two years later, the state circuit court approved the LRDD as a not-for-profit organization that encompassed 620,000 acres in seven Missouri counties (Bollinger, Cape Girardeau, Dunklin, New Madrid, Pemiscot, Scott, and Stoddard). The district was given the ability to set benefit tax assessment levels up to a maximum of 10%.

Once formally organized, the five member LRDD board of supervisors and their chief engineer Otto Kochtitzky developed an engineering plan to drain the entire region. They had a vision for constructing an east-west channel across the top of the drainage system (see map 6.1) [9] and a series of smaller parallel ditches running north to south to the Arkansas border that would drain the great Missouri wetlands [9]. According to Larry Dowdy, chief engineer of LRDD, Islam Randolph, an eminent civil engineer from Chicago who had worked on creation of the Chicago Sanitary and Ship Canal and the Panama Canal, was employed as a consulting engineer and approved the drainage plan [18].

The new drainage and levee system opened the land for settlement, agriculture, and industrial land uses, and became one of the largest human transformations of a landscape in world history.

FIGURE 5.4 The diversion channel flows east from the Castor River to the Mississippi River and outlets south of Cape Girardeau.

Constructing the Headwaters Diversion System

On November 27, 1912, the LRDD awarded the $1.25 million project to D.C. Stephens Company of Buffalo, New York [11]. The contract called for creation of a 30-mile drainage channel with an outlet at the Mississippi River just south of Cape Girardeau, Missouri. The channel was to be 100 feet wide and 20 feet deep (figure 5.4). The project required the clearing of 4,000 acres of timber and creation of 40 miles of levees (figure 5.5) on the south side of the Headwaters Diversion channel. The moving of 8.8 million cubic yards of soil made it the largest single contract for earth movement in the world, even larger than the Panama Canal. Headwaters from the Castor River were to be diverted into the new

FIGURE 5.5 The diversion embankment protects agricultural bottomlands from the Missouri Ozarks upland stream runoff and Mississippi River backwater at flood stage.

diversion channel starting near the town of Greenbrier in Bollinger County (see map 6.1).

According to *Engineering Record*, published in 1914 [19], the maximum service the diversion channel was designed to handle was based on the maximum storm discharge simultaneously with flooding in the Mississippi River equal in height to that which occurred in 1844, or 12 feet out-of-banks. Engineers based their maximum calculations on precipitation patterns and capacity to handle extreme storm events in the upland region. Prior to 1914, weather stations in the area had recorded extreme rain events 8.3 to 9.5 inches in 48 hours and 10.5 to 13 inches in 72 hours. The highest one-day record rainfall was in July of 1905 with 6.45 inches on July 20, followed by 0.91 inches on July 21.

The original diversion channel consisted of three detention basins (West, Middle, and East) located at points where major headwater streams (Castor River, Crooked Creek, and Whitewater River) intersected the channel. These basins with the main channel running through them were designed to store excess water—first receiving the flood pulse, then breaking and reducing the crest by holding the water for a short time and slowing the release into the next basin. The "...crest was assumed to enter the West and Middle basins at the same time, resulting in backflow from the Middle to the West basin with no outflow from the latter until 31 hours after the storm began" [19]. The Crooked Creek (between the two basins) hydrograph was placed 8 hours back to allow for the crest through the West Basin. The headwaters of the Castor River (see map 6.1) ran directly into the West Basin, which extended east about 12 miles and encompassed an area of 15 square miles with capacity to hold 5 billion cubic feet of water at an elevation of 367 feet. The basin was formed by an earthen levee on the south side running 12 to 20 feet high, with a 12-foot crown and one-to-two side slopes, and the Ozark foothills on the north. Flood crests passing through the West Basin were to be delayed about 8 hours with peak discharge from where the Castor River entered at its outlet.

Crooked Creek and Whitewater River flowed into the Middle Basin and made up the main diversion channel (see map 6.2). This basin was bounded on the south by elevated ridges and on the north by the uplands. It had a maximum capacity of more than 1 billion cubic feet and covered an area of 4.5 square miles. As the diversion channel flowed east into the Mississippi River floodplain, the East Basin was leveed on both sides with 16- to 20-foot embankments. The north embankment east of the East Basin was designed to be 2 feet lower than the south embankment so as to serve as a spillway if the maximum flood capacity was exceeded. While use of this spillway would flood 2,000 acres, it would protect the levee on the south side of the diversion channel from failing.

Not all landowners supported the efforts of the LRDD to drain the swampland and divert upland runoff to the Mississippi River. The Headwaters Diversion levees and channel system was built on land owned by Louis Houck, a prominent civil leader and landowner. In 1910 he was concerned that the land he owned south of Cape Girardeau and his railroad interests would be damaged. He did not want to have to pay a benefit tax assessment to the LRDD. Twice Houck fought the LRDD all the way to the US Supreme Court to keep his land from being involved in the LRDD. Both times he lost [18]. In 1913, the LRDD financed construction by issuing $11.1 million in bonds. The project was paid for by collecting a benefit tax assessment from bottomland landowners of $4 per acre.

North-South Drainage into Arkansas

The LRDD drained the Missouri Big Swamp south of the Headwaters Diversion between 1909 and 1928 by constructing 957 miles of ditches (figure 5.2) and 304 miles of levees [20]. With the west-east diversion in place, the LRDD could take advantage of the one-foot-per-mile drop from Cape Girardeau to the Arkansas border and construct a north-south gravity system of levees and ditches. The largest, 100-mile-long Ditch 1, collected the runoff from all other ditches in the district and carried it south into Arkansas where it entered though the Little River, ran into the lower St. Francis River, and later emptied into the Mississippi River, a distance of 250 miles (map 5.3). By 1928, the district had completed construction of the levees and ditches and had drained 520,000 acres of bottomland and transported runoff water from 620,000 acres of uplands. Two detention reservoirs and one sedimentation basin were created to catch sediment and runoff from other creeks. The area not covered by water increased from 10%—mostly the upland hills remaining in the former Mississippi Valley bottomlands—to 96% (map 5.2).

The LRDD draining of Big Swamp by straightening the Little River and creating a latticework of laterals and north-south flowing ditches was not well received by landowners in Arkansas. The drainage districts of Arkansas were unprepared to manage the full brunt of the LRDD plan to discharge their runoff into Big Lake, Arkansas, just south of the Missouri border. In 1917, the Mingo Drainage District and Inter-River Improvement District were created in Missouri and provided the final provocation for the inhabitants of eastern Arkansas who were to receive the drainage waters. According to engineers' calculations, the planned improvements would increase the flow of the St. Francis River into Arkansas during flood events 5 to 10 times its previous volume. If the plans were completed, Arkansas newspaper articles promised nothing short of "ruination for the people of the St. Francis River valley" [16]. Further, it was claimed that the improvements would drive the people of Arkansas out of their homes and away from their farms and factories. Nothing posed a more serious threat to the people of Arkansas than these two Missouri drainage districts [16]. Senator Thaddeus H. Caraway joined the fray, and in a letter to the "prominent men in Eastern Arkansas," he warned that the Missouri-intended diversion of the overflow from the Ozarks would turn the entire water column loose on Arkansas. Senator Caraway counseled Arkansans to seek an injunction against the Missouri improvement districts until the US government canalized or dredged the St. Francis River (map 5.2), to control the increased volume of water [16]. In hindsight, these estimates did not seem to account for the reduced flow to Little River and St. Francis River from the Francois Mountains. No credit was given to the Headwaters Diversion channel located south of Cape Girardeau, Missouri [8], which by 1916 diverted water that historically drained into the Little River and not directly into the Mississippi River south of Cape Girardeau, Missouri (maps 5.2 and 5.3).

With local and state protests growing, Congress finally responded to the increased need for a consistent flood control policy across state lines. The Ransdell-Humphreys Flood Control Act of 1917 reaffirmed the federal government's commitment to flood prevention and control but only appropriated $45 million, which was insufficient to address the problem. However, local and state interests were encouraged by a provision of the act that reduced their contribution to half the

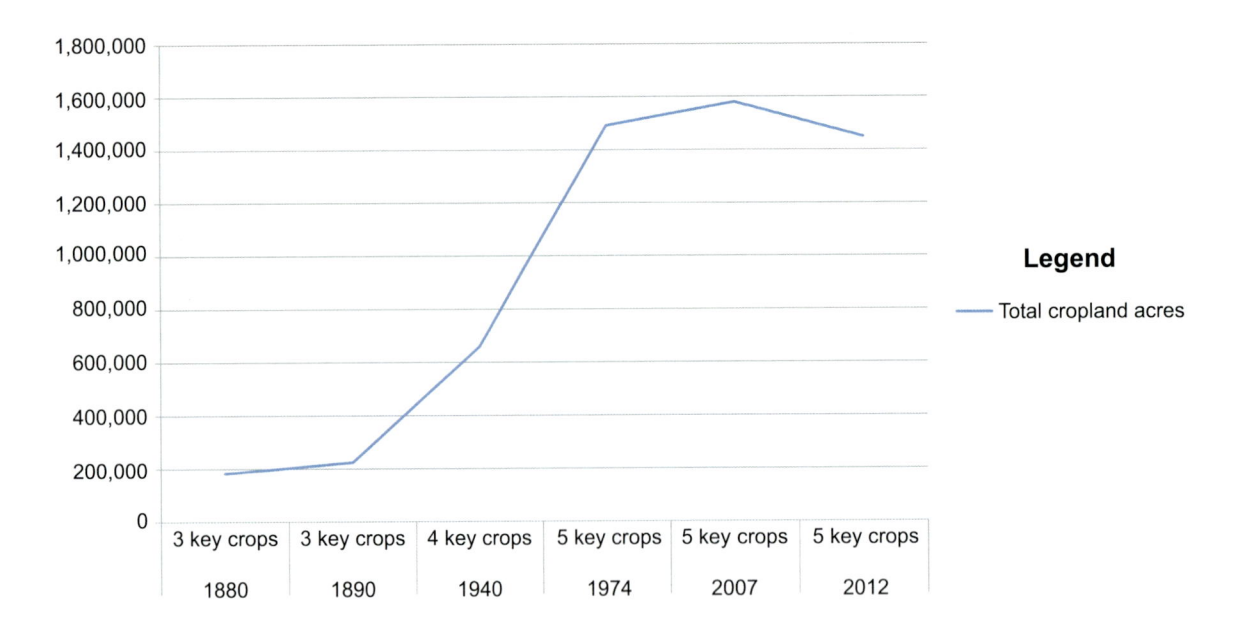

FIGURE 5.6 Total cropland acres from 1880 to 2012 for five key crops in seven (Bollinger, Cape Girardeau, Dunkin, New Madrid, Pemiscot, Scott, and Stoddard) counties in lower southeast Missouri [13].

cost of flood prevention works in their area, from the previous two-thirds local match required. In the same year, the Arkansas General Assembly authorized the establishment of six drainage districts by direct legislation. These districts established a series of ambitious, interrelated projects designed to reclaim their lands by drainage ditches and levees. As part of the improvements, the Arkansas and Missouri drainage districts straightened and deepened a shallow, sluggish creek to handle the increased flow. The improved channel was renamed Ditch 4. Drainage districts' plans in both Arkansas and Missouri were revised between 1917 and 1926 as construction continued. Following the Great Flood of 1927, plans had to be revised again, but the draining and land development continued. Flood control legislation in 1928 authorized the USACE to work in the St. Francis River basin to assist the LRDD in improving and maintaining the Headwaters Diversion channel and levee as well as the lower north-south channels that collected district water for discharge into Big Lake Wildlife Refuge in northern Arkansas [16]. Today, through a collection system in Arkansas not operated by the LRDD, the water flows into the Mississippi River at Helena, Arkansas (map 5.3).

Swampland to Agriculture

Much of the former Mississippi River valley southwest of Cape Girardeau, Missouri, was settled by 1930, and the newly drained alluvial bottomland soils were converted to farming (figure 5.2). Over time, many acres of publicly owned swampland were transferred to private owners who invested in drainage and land development, consolidating the LRDD into fewer and larger parcels. In 2007 about 3,000 landowners had holdings ranging from half an acre to 50,000 acres [11].

The increase in acreage available for agricultural production since drainage efforts in the lower LRDD were undertaken has been dramatic (see figure 6.3). Cotton, corn, and wheat were the primary crops of southeast Missouri in 1880, with a total of 23,304 acres, 122,788 acres, and 35,523 acres, respectively, reported for seven counties (Pemiscot, Dunkin, New Madrid, Stoddard, Scott, Cape Girardeau, and Bollinger) [13]. The 1890 US Census of Agriculture shows an increase in total acres in southeastern Missouri of these three crops from 181,615 acres in 1880 to 397,480 acres, a more than 200% increase. By 1940, cotton acres increased substantially, and soybean acres became a fourth key crop (figure 5.6). Much of the increase in crop acres and diversification of primary crops can be attributed to a combination of the Swamp Land Acts, increased settlement, and local drainage districts' efforts. Following the completion of the upper and lower LRDD drainage system and post-Depression years, almost 778,000 total acres were reported in 1940 for corn (340,643 acres), cotton (250,164 acres), wheat (102,300 acres), and soybean (84,849 acres), a 196% increase from 1890 (figure 5.7). By 1974 soybean acres topped 1.15 million acres in the seven counties, and wheat was reported on 351,469 acres, cotton on 283,015 acres, and corn on 94,361 acres. The addition of soybean to the cropping mix enabled farmers to double crop with wheat and take full advan-

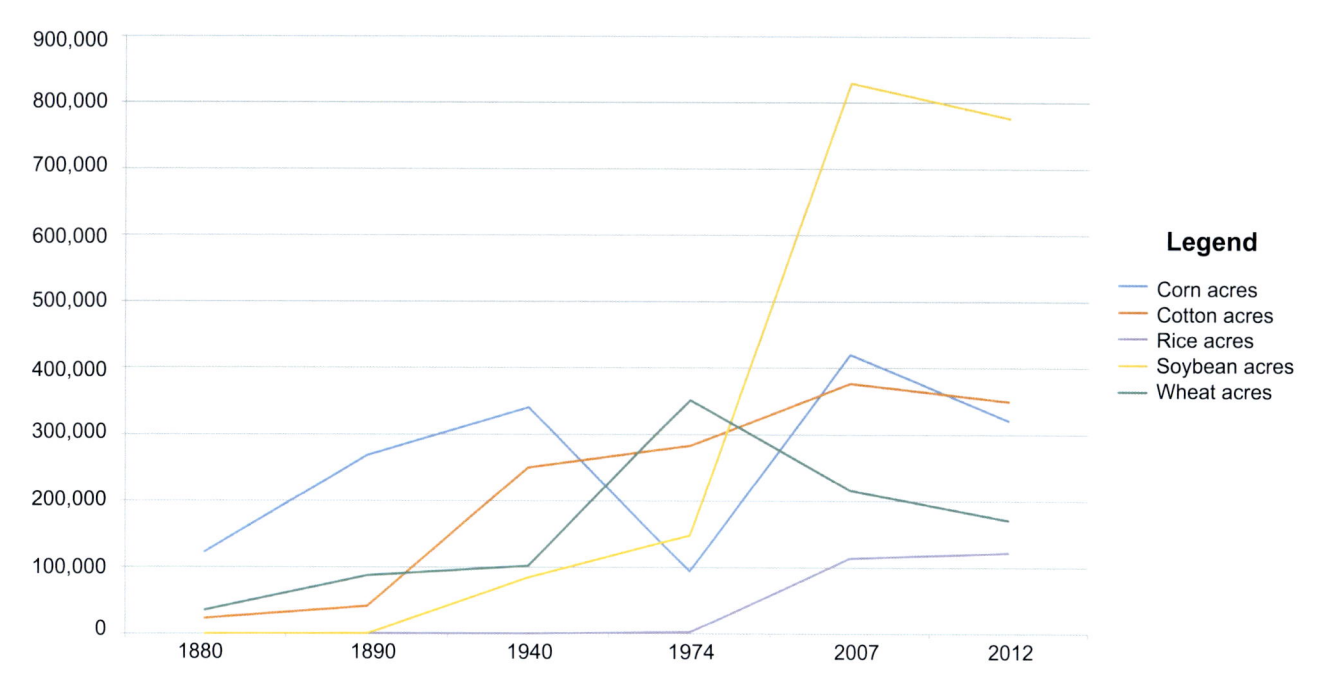

FIGURE 5.7 The completion of the upper and lower Little River Drainage District affected cropland acres of five key crops from 1880 to 2012 in seven counties of southeast Missouri (Bollinger, Cape Girardeau, Dunklin, New Madrid, Pemiscot, Scott, and Stoddard) [13].

tage of the continental climate and soils that were no longer under water during the long growing season. Rice acres, which were only recorded in Stoddard County in 1974, expanded throughout the region with a total of 114,034 acres and 121,836 acres reported in the lower five counties (Pemiscot, Dunkin, New Madrid, Stoddard, and Scott) in 2007 and 2012, respectively. Since 1974, these five commodity crops in seven southeast Missouri counties (figure 5.7) have totaled 1.7 million acres (2012) to 1.9 million acres (2007). Today, in addition to these commodities, the LRDD-drained farmlands produce sorghum, vegetables, peaches, watermelons, and other fruits. Missouri's top wheat, soybean, and sorghum producing counties are all found in this region and report some of the highest corn, wheat, and sorghum yields in the state.

---

[1] US Geological Survey. 2014. Wetland Resources. http://www.nwrc.usgs.gov/topics/wetlands/wetlandResources.htm.

[2] Pracht, C., and J. Banks. 2002. The Little River Drainage District collection. Collection Building 21(1):10-12.

[3] Pierce, S.C., R. Kroger, and R. Pezeshki. 2012. Managing artificially drained low-gradient agricultural headwaters for enhanced ecosystem functions. Biology 1:794-856, doi:10.3390/biology1030794.

[4] Camillo, C.A. 2012. Divine Providence: The 2011 Flood in Mississippi River and Tributaries Project. Vicksburg, MS: Mississippi River Commission.

[5] Morton, L.W., and K.R. Olson. 2013. Birds Point–New Madrid Floodway: Redesign, reconstruction and restoration. Journal of Soil and Water Conservation 68(2):35A-40A, doi:10.2489/jswc.68.2.35A.

[6] Olson, K.R., and L.W. Morton. 2014. The 2011 Ohio River flooding of the Cache River Valley in southern Illinois. Journal of Soil Water Conservation 69(1):5A-10A, doi:10.2489/jswc.69.1.5A.

[7] Vaughn, J.D. 1994. Paleoseismological studies in the Western Lowlands of southeastern Missouri. Final Technical Report to US Geological Survey, 1-27.

[8] Olson, K.R., L.W. Morton, and D. Speidel. 2016. Little River Drainage District conversion of the Big Swamp to fertile agricultural land. Journal of Soil and Water Conservation 71(2)37A-43A, doi:10.2489/jswc.71.2.37A.

[9] Joiner, J. 2008. Little River drainage district work was completed in 1928. Rural Missouri, Associate of Missouri Electric Cooperatives.

[10] Stroud, H.B., and G.T. Hanson. 1981. Arkansas geography: The physical landscape and geography. The Encyclopedia of Arkansas History and Culture. Historical-cultural Setting. Little Rock: Rose Publishing.

[11] Blackwell, S. 2007. A landscape transformation by the Little River Drainage District. Southeast Missourian. Nov. 4, 2007.

[12] Kochtitzky, O. 1931. Memoir. In The Story of a Busy Life (published in 1957). Cape Girardeau, MO: Ramfre Press.

[13] National Agricultural Statistics Service. 2014. USDA Census of Agriculture.

[14] Anderson, J., and E. Samargo. 2007. Bottomland Hardwoods. Morgantown, WV: West Virginia University, Division of Forestry and Natural Resources. http://forestandrange.org/new_wet-lands/index.htm.

[15] Olson, K.R., and L W. Morton. 2014. Dredging of the fracture bedrock lined Mississippi River channel at Thebes, Illinois. Journal of Soil Water Conservation 69(2):31A-35A, doi:10.2489/jswc.69.2.31A.

[16] Harrison, R.W. 1961. A study of state and local efforts toward land development in the alluvial valley of the lower Mississippi River. In Alluvial Empire. Little Rock, AR: USDA Economic Research Service.

[17] Olson, K.R., M. Reed, and L.W. Morton. 2011. Multifunctional Mississippi River leveed bottomlands and settling basins: Sny Island Levee Drainage District. Journal of Soil and Water Conservation 66(4):104A-110A. doi:10.2489/jswc.66.4.104A.

[18] Little River Drainage District. 1909. Final Report of Board of Engineers of Little River Drainage District.

[19] Engineering Record. 1914. The Little River drainage improvements. Engineering Record 70:8:204-206.

[20] Levis, K. 2012. Work on the Little River Drainage District. Semo News Service. Oct. 5, 2012. http://www.dddnews.com/story/1901027.html.

Upland Diversions and Bottomland Drainage Systems:
Intended and Unintended Consequences

The Little River Drainage District (LRDD) was one of the first drainage districts to be formed in the United States [1], and currently, in partnership with the US Army Corps of Engineers (USACE) and Mississippi River and Tributaries Commission, the LRDD manages 960 miles of ditches and 304 miles of levees protecting over 1.2 million acres of land [2, 3]. Construction of levees, diversions, and floodways; and land use conversion from wetlands to agriculture for the last 200 years, have substantively altered the hydrologic cycle of the region. The reddish yellow clay soils of the forested Missouri Ozark Uplands underlain by impervious hardpan grade south into the alluvial lowland soils of the ancient Mississippi River floodplain. These timbered bottomlands of sloughs, bayous, and the slow-moving Little River became one of the largest tracts of wetlands in the United States drained to create agricultural lands and rural settlements. Much of southeast Missouri caught runoff water from these adjacent uplands and held them in the floodplain basin. Before the LRDD diversion channel and a series of large levees were constructed, floodwaters regularly spilled into the basin and created a swamp unsuitable for human settlement.

The Castor and Whitewater rivers and Crooked Creek flowed off the elevated plateaus onto the second bottomlands, with waters pooling in the extensive lowlands and depressions of historic Big Swamp and into the tributaries of the Little River (maps 6.1 and 6.2) [2]. The Little River and St. Francis River naturally drained the entire Bootheel region of southeast Missouri into the Arkansas-White-Red River basin (see chapter 2). One-third of this heavily forested alluvial plain was permanently under water, and 70% was under water from two to six months during the year, allowing only 15% of land to be cultivated. The upper portion of the Castor River basin, diverted from its natural channel in the 1910s, now drains into the westernmost basin of the diversion channel, which carries water east directly into the Mississippi River. The disconnected southern portion of the Castor River has become part of the extensive network of channels and ditches of the lower LRDD.

The Headwaters Diversion, completed in 1915, redirected all of the Ozarks Plateau headwaters runoff east into the Mississippi River just south of Cape Girardeau, Missouri. These upland runoff waters had previously

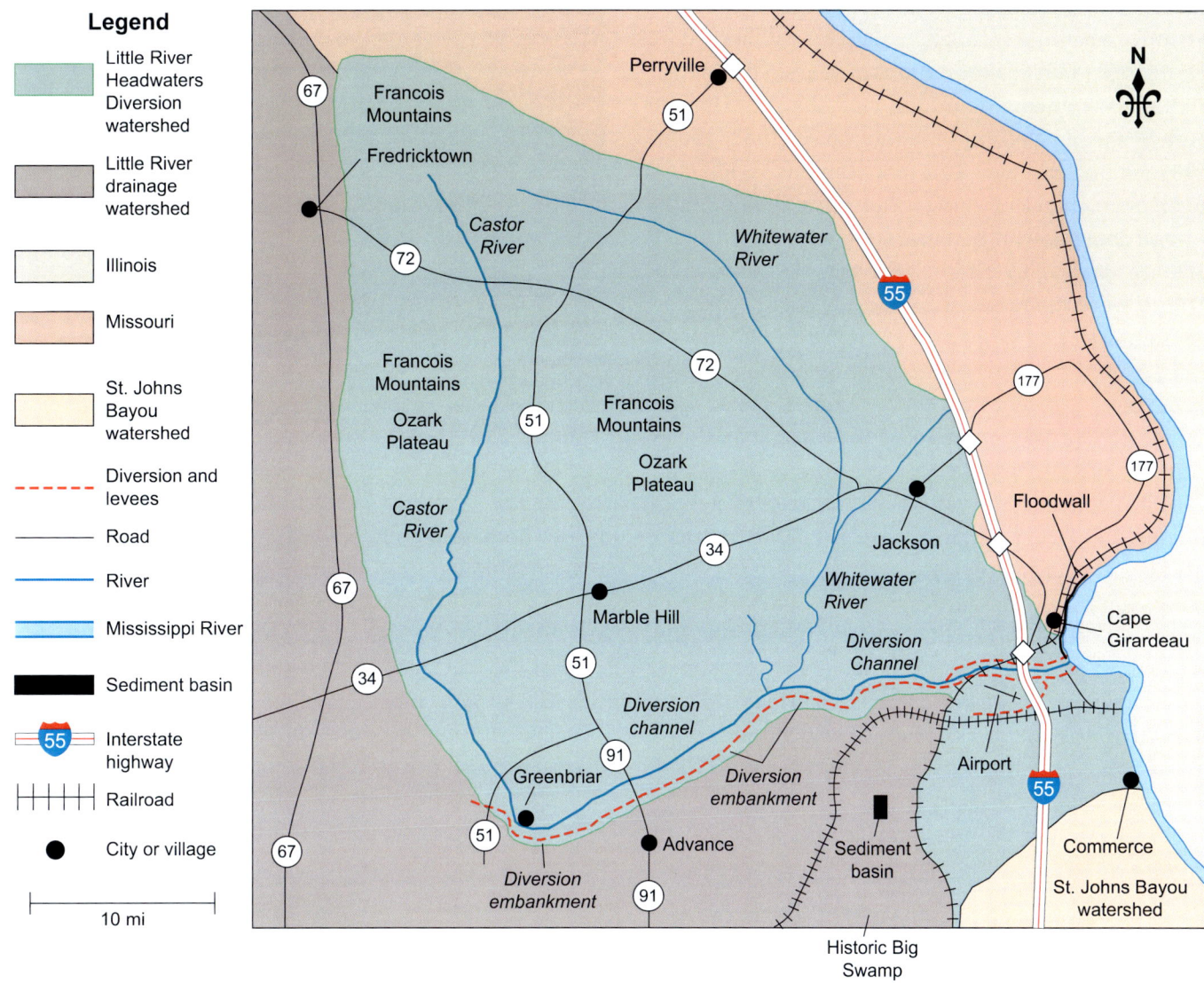

MAP 6.1 The Headwaters Diversion watershed drains Missouri Ozark upland streams southward where they are diverted directly into the Mississippi River south of Cape Girardeau, Missouri, via the diversion channel.

drained south via the Little and St. Francis rivers. The Headwaters Diversion system, consisting of three large basins, 50.3 miles of channels, and 44.7 miles of levees, diverts and temporarily stores ordinary and floodwaters running off 720,000 acres of uplands (Ozark Plateau including Francois Mountains) [4, 5, 6]. Today, the Headwaters Diversion helps drain and protect 1.2 million acres of agricultural lands (map 6.1) in southeast Missouri from internal seasonal flooding and Mississippi River backflow at flood stage (see chapter 5). The Little River levee and LRDD Headwaters Diversion channel built in the 1910s successfully permitted the drainage of the 1.2 million-acre Big Swamp in the Bootheel of Missouri. However, it also had the unintended consequence of increasing the flow and peak of Mississippi River water south of Cape Girardeau through the Thebes Gap and south to Helena, Arkansas, a distance of approximately 360 river miles. When the Ozark Uplands and Francois Mountains experience above-average rainfall for extended periods of time, the additional runoff transported by the diversion channel (2 million acre-feet per year) increases the chances of Mississippi River levee breaches south of Commerce, Missouri, and adds to the peak river height at the confluence of the Ohio and Mississippi rivers [7].

Missouri Ozark Plateau and the Headwaters Diversion Watershed

The LRDD Headwaters Diversion watershed is triangular in shape with the apex about 10 miles northeast of Fredericktown, Missouri (map 6.1). Located in the Missouri Ozark Plateau, the watershed was part of the upland land bridge—the Shawnee and Ozark hills—which connected the Southern Appalachians and the Ozark Highlands (Ouachita and Francois mountains; see map 2.2). This land bridge separated the central

Interior Lowlands to the north and the southern Mississippi Embayment trough, which was submerged by the Coastal Plain Sea for billions of years. The Ozark Highlands, Francois Mountains, and Shawnee Uplands were formed by volcanic and intrusive activity about 1.485 billion years ago. This mountainous region is much older than the younger Appalachian Mountains formed about 460 million years ago [8]. Granite and rhyolite, two highly visible igneous rocks found in the Ozarks, were the result of a series of volcanic eruptions and ash flows that melted and cooled preexisting rocks from the earth's crust. Other igneous rocks made from volcanic magna contain crystallized minerals rich in silicon, aluminum, sodium, potassium, calcium, iron, and magnesium [9].

The geologic history of the land bridge left a wealth of minerals in the Shawnee Uplands in Illinois (figure 6.1; see map 2.2) and the Francois Mountains of the Ozarks (map 6.1). The Ozarks were the center of the Lead Belt in the 1700s and 1800s, a mining region yielding lead, iron, barite, zinc, silver, manganese, cobalt, and nickel ores. The historic Mine La Motte near Fredericktown, Missouri, was the site of lead mining activity by the French as early as 1702. The area today accounts for over 90% of the primary lead production in the United States. Granite has been commercially quarried since 1869 in the vicinity of Elephant Rocks State Park. The red architectural granite quarried in the area has been used in buildings in St. Louis, Missouri, and other cities and is currently marketed as "Missouri Red" monument stone [10]. Nearby Taum Sauk Mountain is the highest peak (1,772 feet) in the range and the highest point in the state of Missouri.

The continental climate of this region produces an average annual precipitation about 45 to 48 inches with warm moist air from the Gulf of Mexico flowing northeast and dropping moisture as it interacts with cold air

FIGURE 6.2 The headwaters of the Castor River flowing through the Ozark Plateau have the same aqua green color as Shawnee Upland lakes at mining sites.

FIGURE 6.1 Illinois Shawnee Upland mining operation in the land bridge between the Southern Appalachians and the Ozark Highlands.

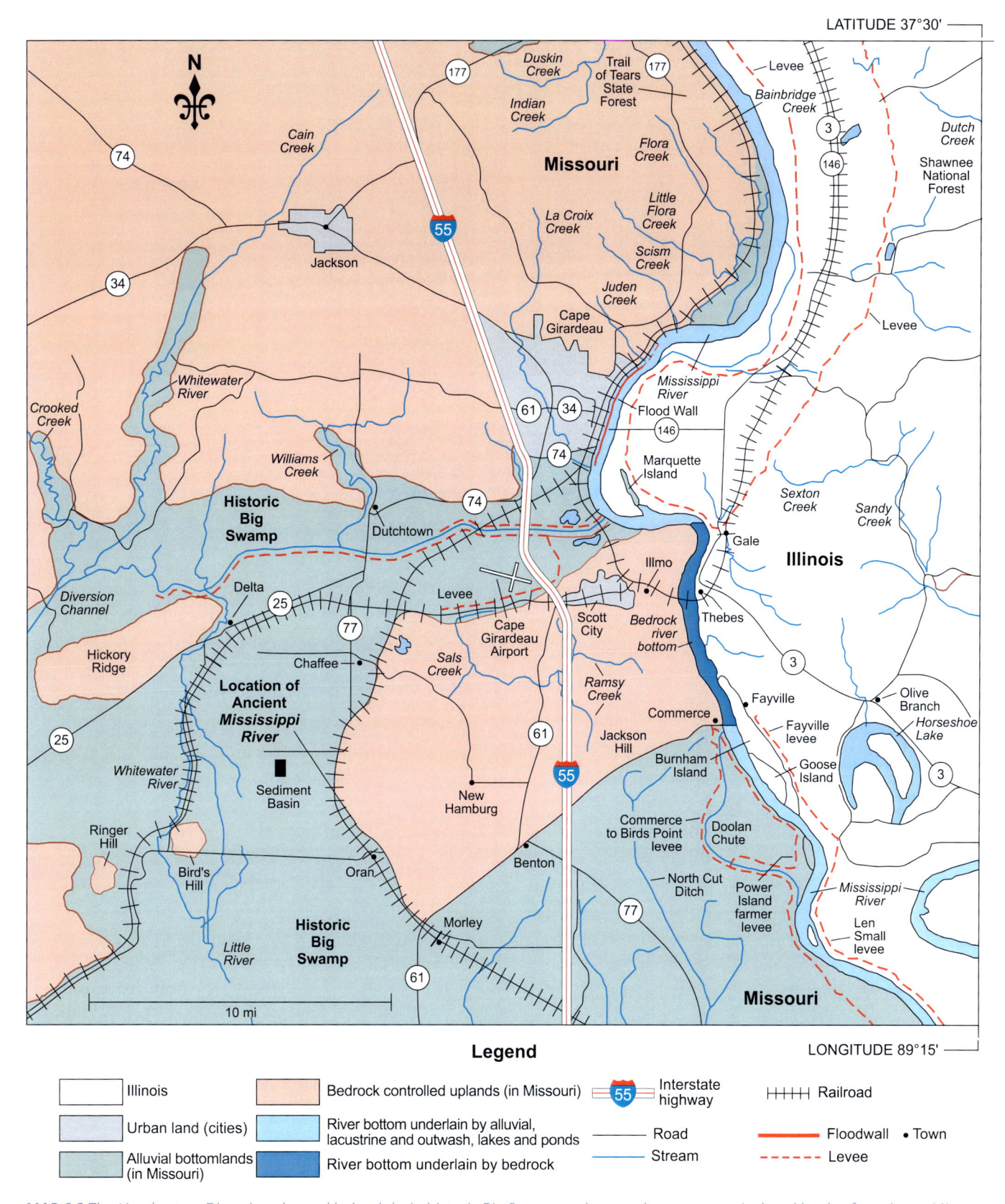

MAP 6.2 The Headwaters Diversion channel helped drain historic Big Swamp and currently protects agricultural lands of southeast Missouri from flooding.

from Canada. The Ozark Plateau is a rugged landscape of narrow valleys 100 to 3,000 feet in width with steep three-to-one side slopes and rock strewn whitewater rivers that are only navigable in or near flood stage conditions [5]. Today, this is a favorite recreational area for hiking, kayaking, and whitewater paddling. Over time, the Castor (figure 6.2) and Whitewater rivers in this region (map 6.1) have carved through the soft

FIGURE 6.3 Reductions in woodland acres and increases in land in farm acres from 1880 to 2007 for seven counties in lower southeast Missouri (Bollinger, Cape Girardeau, Dunkin, Pemiscot, Stoddard, New Madrid, and Scott) [23].

limestone and dolomite surface rocks, leaving exposed pink granite and rhyolite igneous rocks and creating narrow "shut-in" streams with unexpected ledge drops into pools, rocky shoals, and riffles that experienced paddlers use to test their skills [11].

Economic Growth and Development

Many attempts were made to drain sections of Big Swamp in Missouri from 1850 to 1905. These efforts often failed because the entire volume of the Ozark Plateau runoff waters continued to drain into Big Swamp. Further, during major flooding events the Mississippi River often breached local farmer levees, re-entered the ancient Mississippi River valley south of Cape Girardeau, and added floodwaters to Big Swamp. The successful drainage of Big Swamp could not have happened without the creation of the system of impounding basins, channels, and levees that carries the waters of the eastern Missouri Ozark Plateau hill streams eastward to the Mississippi River south of Cape Girardeau (map 6.1). The redirected Ozark Plateau runoff from the Little River to Mississippi River made possible the drainage of Big Swamp.

Three important developments paved the way for the LRDD project success: (1) Missouri transferred ownership of swamplands to counties, and counties sold the land to entrepreneurs; (2) entrepreneurs built railroads in and out of swamps; and (3) large-scale lumbering operations cleared the thick forested bottomlands. Once the timberland was cleared, the timber companies could sell land for a profit to those with agricultural interests. The decrease in timber acres and the extensive draining of lower southeast Missouri enabled more of the land to be converted into farmland (figure 6.3).

Other events contributed to the new drainage district's success and coincided with LRDD efforts to construct levees and drainage channels in southeast Missouri. First, railroads were built into Big Swamp from Illinois to carry the timber out as a result of the 1905 construction of the Thebes railroad bridge (map 6.2) [12]. The Thebes reinforced concrete railroad bridge allowed railroad lines to be constructed deep into Big Swamp and enabled the huge logs extracted to be transported by railroad cars to eastern markets. The second event was the building of a lock and dam on the Mississippi River at Keokuk, Iowa, in 1913. The dam, fought by the timber interests, transformed the industry in the upper Midwest because it stopped log drives from coming downriver. Timber companies, needing transportation to move timber from their hardwood forests, became sources of new revenue and enabled the railroad system to expand. Third, after construction of the Panama Canal ended in 1913, large numbers of skilled workers sought new jobs, and they found them in the earth-moving business of southeast Missouri [13]. Thousands of workers participated in the difficult and dangerous work of clearing, mostly by hand, the swamp of trees and stumps. The workers and their families came from many states, including the adjacent states of Kentucky and Illinois. Houseboats were homes for many of the men and

FIGURE 6.4 The murals on the Cape Girardeau floodwall document the historic development of the city and its relationship with the Mississippi River.

women who dug drainage channels and built the levees; others lived in tents in temporary settlements.

The flood of 1927 [14] and the Great Depression from 1929 to 1935 made it impossible for the LRDD to make payments on construction loans. The Flood Control Act of 1928 authorized the USACE to improve the Headwaters Diversion channel and mainline levee to protect against future Mississippi River flooding and levee breaching. This new partnership provided a critical infusion of resources that helped the district maintain the newly constructed systems of levees and channels protecting human settlements and agricultural lands. However, the district carried an outstanding debt of almost $7.6 million on the $11 million originally financed in 1913. Crop prices dropped so low that many crops went unharvested, and the timber business struggled to survive during the Depression. Many landowners defaulted on their mortgages and their $4 an acre benefit tax assessments while land prices dropped to $15 an acre. In 1937, after both the Depression and the flood of 1937, the LRDD received a $2.4 million loan from the Reconstruction Finance Corporation (RFC), a New Deal agency that provided assistance to businesses. Forty-three years after beginning construction, in April of 1952, LRDD settled the bond debt and paid off the construction phase of the drainage system [6].

The Headwaters Diversion channel and levee system was built to expand agricultural lands and protect cropland from seasonal and extreme flood events. By 1937 much of the former Mississippi River valley southwest of Cape Girardeau, Missouri, was settled, and the alluvial bottomland soils became productive lands growing corn, grain sorghum, wheat, cotton, and rice. The diversion channel and levee system was intended to be self-maintaining, relying primarily on gravity, with no mechanical devices in the channel to control flow. In dry years the Castor and Whitewater rivers, Crooked Creek, and the diversion channel continued to flow and add to the Mississippi River depth in the Thebes Gap, which reduced the need for dredging in the bedrock-lined, nine-foot-deep shipping channel [12]. However, there is evidence that the additional volume of water the diversion channel carried to the Mississippi River during major flooding events raised the peak flow at the Cape Girardeau, Cairo, and New Madrid gages and often as far south as Helena, Arkansas. After the construction of the LRDD diversion in 1915, flooding of Cape Girardeau became a common occurrence (e.g., 1922, 1927, 1943, 1944, 1947, and 1951). A floodwall and levee system was constructed between 1956 and 1964 to protect Cape Girardeau. The current levee system and floodwall (figure 6.4) was built by the USACE [15] at a cost of $4 million and requires additional funds each year to maintain. It is 7,210 feet long including a 4,000-foot floodwall that can handle a Mississippi River 54-foot peak flow on the Cape Girardeau gage.

Intended and Unintended Consequences

Today the west-east LRDD Headwaters Diversion system drains 720,000 acres of the Francois Mountains and Ozark Plateau directly to the Mississippi River at Cape Girardeau [16]. South of the Headwaters Diversion the extensive Little River north-south drainage system collects runoff from 500,000 upland acres and 620,000 acres of bottomlands (Little River basin) and has 957 miles of ditches draining south from the Headwaters Diversion channel to the Arkansas border (see map 5.2). The LRDD has about 3,000 landowners and parcels varying from 0.5 to 50,000 acres in size. These landowners in 2012 paid over $1.2 million in benefit tax assessments for levee mowing, channel clearing and excavation, brush spraying, ditch mowing, and overall district operations [6]. Mississippi River floods have challenged the LRDD to be vigilant in maintaining these levees and channels. The Mississippi River rose to a record height in 1973, threatening the Headwaters Diversion levee. The levee held, but the district discovered numerous weak spots and has in recent years invested considerable resources to assure long-term protection of the region [17]. The levees constructed by the USACE were built to an elevation to safely pass a 500-year flood event. In 2010 the USACE made their 5-year inspection of the LRDD levee system and certified that it met all requirements of the National Flood Insurance Levee Evaluation [6] and passed capacity to withstand a 100-year flood event. This accreditation means that landowners behind LRDD levees are not subject to purchasing mandatory flood insurance.

The Headwaters Diversion levee held during another great flood, that of 2011, which required the USACE to activate the New Madrid Floodway to protect Cairo and downstream communities against extensive uncontrolled breaching damage along the frontline levee of southeast Missouri. Record rainfall in March and April of 2011 occurred throughout the entire LRDD watershed, and the Headwaters Diversion upland watershed received more than 30 inches of rainfall, causing two district detention basins to overtop and their levees to crevasse [3]. Post-2011 flood inspection of the Headwaters Diversion channel and levees on its south side uncovered 11 sites of erosion and bank caving in the West basin levee and damage to the channel levee where slow-velocity sand boils occurred [17]. The USACE has addressed these weaknesses and is working on a redesign of the levee to prevent future sand boils.

The formation of the LRDD enabled Big Swamp to be drained, cleared of trees, and cultivated. The region now has some of the most productive land in Missouri. The LRDD provided social, economic, and physical infrastructure for southeast Missouri, offering economic benefits to timber and farming interests; making dry land for homes, towns, and industries; and incentivizing an extensive transportation system. Without the Headwaters Diversion channel and levee, thousands of acres of agricultural bottomlands would not be protected from ordinary or floodwater runoff from the Castor and Whitewater rivers. The Cape Girardeau Regional Airport area site (map 6.2) was often covered by 20 feet of water prior to the construction of the Headwaters Diversion. Before 1907, Bloomfield Road, the only road going south from Cape Girardeau, followed the ridges. The railroad, bridges, and road infrastructure was central to the economic and social development of southeast Missouri.

However, the consequences of the redirected floodwaters on Cape Girardeau, Cairo, and the bottomlands of Illinois, Missouri, and Kentucky have not been well addressed. The intent of the diversion in the northeastern corner of the LRDD was to isolate the upper basin and prevent overloading of the lower drainage system constructed to drain the low-gradient, slow-moving waters in the historic river floodplain to the south. The additional water from the diversion contributed to the need for the construction of the Len Small–Fayville farmer levee in Illinois and the Commerce farmer levee in Missouri (map 6.2), both designed to protect agricultural lands from flooding. When the Ozark Uplands and Francois Mountains experience above-average rainfall in short periods of time, the additional runoff transported by the diversion channel increases the chances of Kentucky, Illinois, and Missouri farmer levee breaches south of Cape Girardeau.

The increase in Mississippi River peak flow placed additional river pressure on levees and led to increased flooding, especially during the floods of 1927, 1937, and 2011. The Kentucky, Illinois, and Missouri farmers' and landowners' response to the additional volume and height of the Mississippi River from the diversion channel valley and the prevention of the Mississippi River floodwaters from flowing into the ancient Mississippi River valley and Big Swamp was to build floodwalls and levees. The Great Flood of 1927 resulted in Cairo building a floodwall and strengthening levees and the creation of the Birds Point–New Madrid Floodway. The Len Small (built in 1943)–Fayville (built in 1969) farmer levee in Illinois [15, 18], the Hickman levee in Kentucky, and the Commerce farmer levee in Missouri were also constructed and strengthened after the completion of the Headwaters Diversion. Over time, the Kentucky landowners were able to get the USACE to strengthen the mainline Hickman levee, which did not fail in either 1993 or 2011. However, the Len Small–Fayville levee failed in 1993, and both the Commerce farmer and Len Small–Fayville levees failed during the 2011 flood. In late December of 2015 and early January of 2016, a rare winter storm over the Ozarks dropped more than 10 inches of rain over three days and set a record Mississippi River crest at Thebes of 47.74 feet, almost 2 feet over previous record flood stage (figure 6.5). Down-

FIGURE 6.5 Mississippi flooding at Thebes, Illinois, on January 5, 2016. The floodwaters covered the playground and the riverbanks and flooded homes and business that were not on stilts. A tugboat is pushing barges upriver in the background.

FIGURE 6.6 After the Len Small levee breached on January 2, 2016, Mississippi River floodwaters flowed into Miller City, Horseshoe Lake, and Olive Branch. Many of the Olive Branch home and business owners both north and south of Route 3 built sandbag barriers to protect their structures from the floodwaters.

stream, the river pressure caused the Len Small levee to fail again and flooded farmland, buildings (figure 6.6), and homes within miles of the breach.

Mitigated Impacts on Arkansas

The Mississippi River realigned and passed through the Thebes Gap as a result of seismic activity and meltwaters from glacial events about 12,000 to 15,000 years ago. Between creation of the Thebes Gap and 1915, all Francois Mountain and Ozark Plateau runoff water (Castor River, Crooked Creek, and Whitewater River) and any surface overland water flowed into Big Swamp and south into Arkansas via the Little River. During major spring floods, the Mississippi River would reclaim its ancient channel to the west, and additional Mississippi River floodwater would flow into Big Swamp. Thus, during this period Arkansas received both upland runoff as well as Mississippi River floodwater overflow from Big Swamp.

After the LRDD Headwaters Diversion channel was constructed (1913 to 1916), upper Mississippi River floodwaters were blocked by earthen levees and could no longer return to the ancient path. This reduced some of the water flow that Arkansas had historically received. The Headwaters Diversion also diverted about 720,000 upland acres (Ozark Plateau and the Francois Mountains) of runoff into the Mississippi River south of Cape Girardeau away from Little River basin. As a result, only 520,000 acres of Little River bottomlands south of the Headwaters Diversion drained into Arkansas during periods of high rainfall and local flooding. Additional runoff from 0.62 million acres of adjacent agricultural and forest land in the Little River watershed from Big Swamp drained south into Arkansas.

Some of the Headwaters runoff, if the Headwaters Diversion channel had not been constructed, would have stayed in Big Swamp and perhaps evaporated or infiltrated, so not all of it would have flowed across the border into Arkansas as overland and subsurface flow. Surface and subsurface velocity and volume through the alluvial materials to the south into Arkansas would have increased as a result of the 1910s Little River network of ditches. The construction of the Headwaters Diversion channel is estimated to have reduced annual runoff into the Little River basin about 2,332,000 acre-feet (assuming 20% soil retention in the mountains from 4 feet of annual rainfall). The 520,000 acres of Little River bottomlands drained through ditches (assuming a 60% to 80% soil retention in bottomlands from 4 feet of annual rainfall) most likely carried annually about 416,000 to 832,000 acre-feet of water south toward Arkansas. The 620,000 acres of adjacent agricultural land and forest land from the Big Swamp area are estimated to have contributed another 600,000 to 1,200,000 acre-feet per year of flow to Little River drainage ditches into Arkansas. Thus, the construction of the Headwaters Diversion would have reduced the amount of water flowing to the Arkansas border via the Little River basin by as much as 320,000 to 1,300,000 acre-feet per year.

The volume of water in the upper St. Francis River (watershed of more than 1 million acres; see maps 5.2

and 5.3) and to the north and west of the Little River basin was not affected by the construction of the Little River Diversion, but the lower St. Francis River was affected. When the Little River drainage ditches were being constructed, Arkansas landowners and politicians claimed that the flow from the Little River and ditches would increase 5- to 10-fold and result in the ruination of Arkansas bottomlands. This estimate does not seem to account for the Headwaters Diversion and the upland water from the Francois and Ozark Plateau being diverted directly into Mississippi River and no longer draining into the Little River. The ditches did improve the drainage of the Little River basin by draining Big Swamp and resulted in an increase in water flowing into Arkansas and the lower St. Francis River. However, a 5- to 10-fold increase appears to have been an exaggeration that did not take into consideration the flow reduction effects of the LRDD Headwaters Diversion of Ozark Plateau runoff.

Little River Drainage District Legacies and Future Challenges

Without the unified investment of the LRDD, the main interstate north-south highway, Interstate 55 (map 6.2), would have been much more costly to build and the development of industries needing transportation delayed since the Mississippi River was 15 to 45 miles to the east. The history of the land use change in southeast Missouri reveals that its benefits have not been equally distributed [19]. The convergence of the Depression and the draining of the swamp concentrated ownership of the land. Sharecropping became a way of life for the region's underclass. Once the timber was cut down, the workers were stuck in the Bootheel without jobs and with skills that were no longer needed. Agriculture turned into a landlord and sharecropper arrangement [19]. People could not pay their property and drainage district taxes. Bootlegging became one of the few ways to make a living. To this day some of the most impoverished counties in America are in southeast Missouri. In Pemiscot, Mississippi, and New Madrid counties, more than 20% of the population lives in poverty. Populations in New Madrid and Pemiscot counties are smaller today than in 1920s and 1930s. Regional populations have also declined since the surge when land was first drained and cleared.

The US Geological Survey [20] estimates that almost all of the 65 million acres of wetlands given to the states under the swamp acts are now in private ownership. In 1900, the swamplands of the Mississippi River alluvial plains were considered wasteland of little to no value, obstacles to settlement and the development of commerce. Soils' capacity to grow row crops has been the historical metric used to determine whether land is valuable or marginal [21]. The agricultural productivity of this region is testimony to the effectiveness of the LRDD elaborate system of ditches, retention basins, and levee-protected lands. Only in recent years have scientists understood the multifunctional uses and ecosystem services that saturated soils and wetlands provide and what is lost when they are converted to other land uses. Not only does agricultural drainage alter hydrologic patterns, but it also alters the biodiversity of habitat and reduces water quality when soil erodes and farm nutrients run into the ditches [22].

The LRDD intricate system of drainage ditches and diversion channels moves more than 31.5 million gallons of water annually off the land and into the Mississippi River. The drainage district's investments in the Headwaters Diversion channel (see chapter 5), levees, and ditches, along with the construction of the Thebes railroad bridge (see chapter 22), created an economic and technological engine that resulted in the successful conversion of Big Swamp to productive agricultural lands during the past 111 years. Had the conversion been delayed for 100 years, the area would probably have become a federally protected wetland preserve [20]. Advances in agriculture and water management have helped engineers better understand how to manage artificially drained, low-gradient agricultural lands. However, the tension between wetland services and agricultural land uses will continue to increase and be a future challenge that the LRDD will need to address in order to find some level of balance among competing economic, social, and biophysical conditions.

The LRDD oversight of so many thousands of acres well positions it to evaluate and address the landscape-wide vulnerability of drained and levee-protected lands from interior and river flooding as precipitation and extreme and variable weather stress their infrastructure and affect downstream communities. Without continual maintenance and repair of ditches and levees, these Mississippi River bottomlands have a predisposition to flood farmland seasonally and revert back to the wetlands of the past. Further, the straightening of streams and channelization to move water off fields are sources of soil erosion and bank scouring that need constant attention to control sedimentation throughout the system.

One of the big challenges that Mississippi River drainage districts and the USACE face is how to best manage the water velocity and emergent vegetation in their drainage systems to control seasonal flooding, keep the channel beds stable, and reduce off-field and landscape-wide nutrient losses while creating habitat diversity under seasonal drought and uncertain climate patterns [6].

The 2007 LRDD report [6] celebrating their 100-year anniversary well summarizes the changes that have occurred and the challenges that continue into the future:

> *This region was once a bog, a swamp, a lowland, a morass, a hunting and fishing paradise, a no man's land, an endless stretch of virgin hardwood timber, a stillness that stole the sleep from the tired, a hideout, a barrier, a challenge to do the impossible. Today a traveler driving across the areas could never know of these things as he or she passes field after field laid out on a perfect grid, interrupted only occasionally by a narrow bridge every mile or so. Below them, however, the swamp does not sleep and labors to regain itself, held only at bay by the never-ending work of men and women of the Little River Drainage District.*

---

[1] Olson, K.R., M. Reed, and L.W. Morton. 2011. Multifunctional Mississippi River leveed bottomlands and settling basins: Sny Island Levee Drainage District. Journal of Soil and Water Conservation 66(4):104A-110A, doi:10.2489/jswc.66.4.104A.

[2] Camillo, C.A. 2012. Divine Providence: The 2011 Flood in Mississippi River and Tributaries Project. Vicksburg, MS: The Mississippi River Commission.

[3] Pracht, C., and J. Banks. 2002. The Little River Drainage District collection. Collection Building 21(1):10-12.

[4] Burns, B.F. 1919. Board mattress for preventing scour in drainage channel. Engineering and Contracting, eds. H.P. Gillette, Sir H. Davy, and J. Davy. Vol LII(September 10):299.

[5] Engineering Record. 1914. The Little River drainage improvements. Engineering Record 70:8:204-206.

[6] Little River Drainage District. 2007. Little River District of southeast Missouri 100 year report 1907-2007. Cape Girardeau, MO: Little River Drainage District.

[7] Olson, K.R., and L.W Morton. 2012. The effects of 2011 Ohio and Mississippi River valley flooding on Cairo, Illinois, area. Journal of Soil and Water Conservation 67(2):42A-46A, doi:10.2489/jswc.67.2.42A.

[8] Cremeens, D.L., R.G. Darmody, and S.E. George. 2005. Upper slop landforms and age of bedrock exposures in the Francois Mountains, Missouri: A comparison to relict periglacial features in the Appalachian Plateaus of West Virginia. Geomorphology 70(1-2):71-84, doi:10/1016/j.geomorph2005.04.001.

[9] LeMaitre, R.W., A. Streckeisen, B. Zanettin, M.J. LeBas, B. Bonin, and P. Bateman. 2002. Igneous Rocks: A Classification and Glossary of Terms. Cambridge: Cambridge University Press.

[10] Elephant Rocks State Park, Missouri. 2014. Visitor center exhibit. Belleview, Missouri: Elephant Rocks State Park.

[11] McCord, M.W. 2014. Castor River, Missouri. Southwest Paddler. http://southwestpaddler.com/docs/castor2.html.

[12] Olson, K.R., and L.W. Morton. 2014. Dredging of the fracture bedrock lined Mississippi River channel at Thebes, Illinois. Journal of Soil and Water Conservation 69(2):31A-35A, doi:10.2489/jswc.69.2.31A.

[13] Blackwell, S. 2007. A landscape transformation by the Little River Drainage District. Southeast Missourian. Nov. 4, 2007.

[14] Barry, J. M. 1997. Rising Tide: The Great Mississippi Flood of 1927 and How it Changed America. New York: Simon and Schuster.

[15] Olson, K.R. and L. Morton. 2013. Impact of 2011 Len Small levee breach on private and public Illinois lands. Journal of Soil and Water Conservation 68(4):89A-95A, doi:10.2489/jswc.68.4.89A.

[16] Olson, K.R., L.W. Morton, and D. Speidel. 2016. Little River Drainage District conversion of the Big Swamp to fertile agricultural land. Journal of Soil and Water Conservation 71(2)37A-43A, doi:10.2489/jswc.71.2.37A.

[17] Pierce, S.C., R. Kroger, and R. Pezeshki. 2012. Managing artificially drained low-gradient agricultural headwaters for enhanced ecosystem functions. Biology 1:794-856, doi:10:3390/biology1030794.

[18] Olson, K.R., and L.W. Morton. 2013. Soil and crop damages as a result of levee breaches on Ohio and Mississippi Rivers. Journal of Earth Science and engineering 3(3):139-158.

[19] Gellman, E.S., and J. Roll. 2011. The Gospel of the Working Class. Urbana-Champaign, IL: University of Illinois Press.

[20] US Geological Survey. 2014. Wetland Resources. http://www.nwrc.usgs.gov/topics/wetlands/wetlandResources.htm.

[21] Hatfield, J.L., and L.W. Morton. 2013. Chapter 2: Marginality principle. *In* Advances in Soil Science, eds. R. Lal and B.A. Stewart. New York: CRC Press.

[22] Blann, K.L., J.L. Anderson, G.R. Sands, and B. Vondracek. Effects of agricultural drainage on aquatic ecosystems: A review. Critical Review Environmental Science Technology 39:909-1001.

[23] National Agricultural Statistics Service. 2014. USDA Census of Agriculture.

St. Johns Levee and Drainage District Attempts to Mitigate Internal Flooding

When the Mississippi River reaches 290 feet above sea level near New Madrid, Missouri, bottomlands adjacent to the river, farmland, roads, ditches, and wetlands begin to flood. Concurrently, at the lower end of the New Madrid Floodway, the rising Mississippi backs up into Main Ditch (figure 7.1), the 1,500-foot gap in the frontline levee designed to drain the floodway and St. Johns Levee and Drainage District to the river (map 7.1). When this occurs, the Main Ditch gates on the setback levee (figure 7.2) are closed to protect the St. Johns Bayou basin from Mississippi River backflow. However, with the Main Ditch gates closed, precipitation within the basin has no outlet as it drains to Main Ditch and, as a result, backs up tributary streams and floods a substantive portion of agricultural lands and the town of East Prairie, Missouri. For example, after the 7.5 inches of rain during the first three days of May of 2011 and the closing of the gates, the internal drainage system backed up floodwaters throughout the St. Johns Bayou basin all the way to East Prairie.

The construction of the Commerce to Birds Point to New Madrid levee system (map 7.1) artificially separated lands within St. Johns Bayou basin and the New Madrid Floodway from their natural drainage paths to the Mississippi River. While much of the farmland behind these levee systems is protected from flood-swollen waters of the Ohio and Mississippi rivers as they converge at Cairo, Illinois, and flow south, thousands of acres of bottomlands are flooded through the gap in the lower New Madrid Floodway frontline levee (figure 7.1). In addition, the closure of the setback levee gates at the gap (figure 7.2) results in the internal flooding of many more thousands of acres in St. Johns Bayou basin.

Whenever heavy snowmelt and prolonged rains occur in the upper Mississippi and Ohio river valleys, the people of this region watch the river gage numbers with concern for high water river pressure threats to their levee systems and internal flooding of low-lying areas from drainageways. Figure 7.3 illustrates how different Mississippi River high water events affected the acreages that were flooded in St. Johns Bayou basin and New Madrid Floodway in 1997, 2008, and 2011. The United States Army Corps of Engineers (USACE) has responsibilities for many aspects of the Mississippi River including flood control, river navigation, and floodplain ecosystems. In recent years USACE flood mitigation planning in the Bootheel of Missouri has attempted

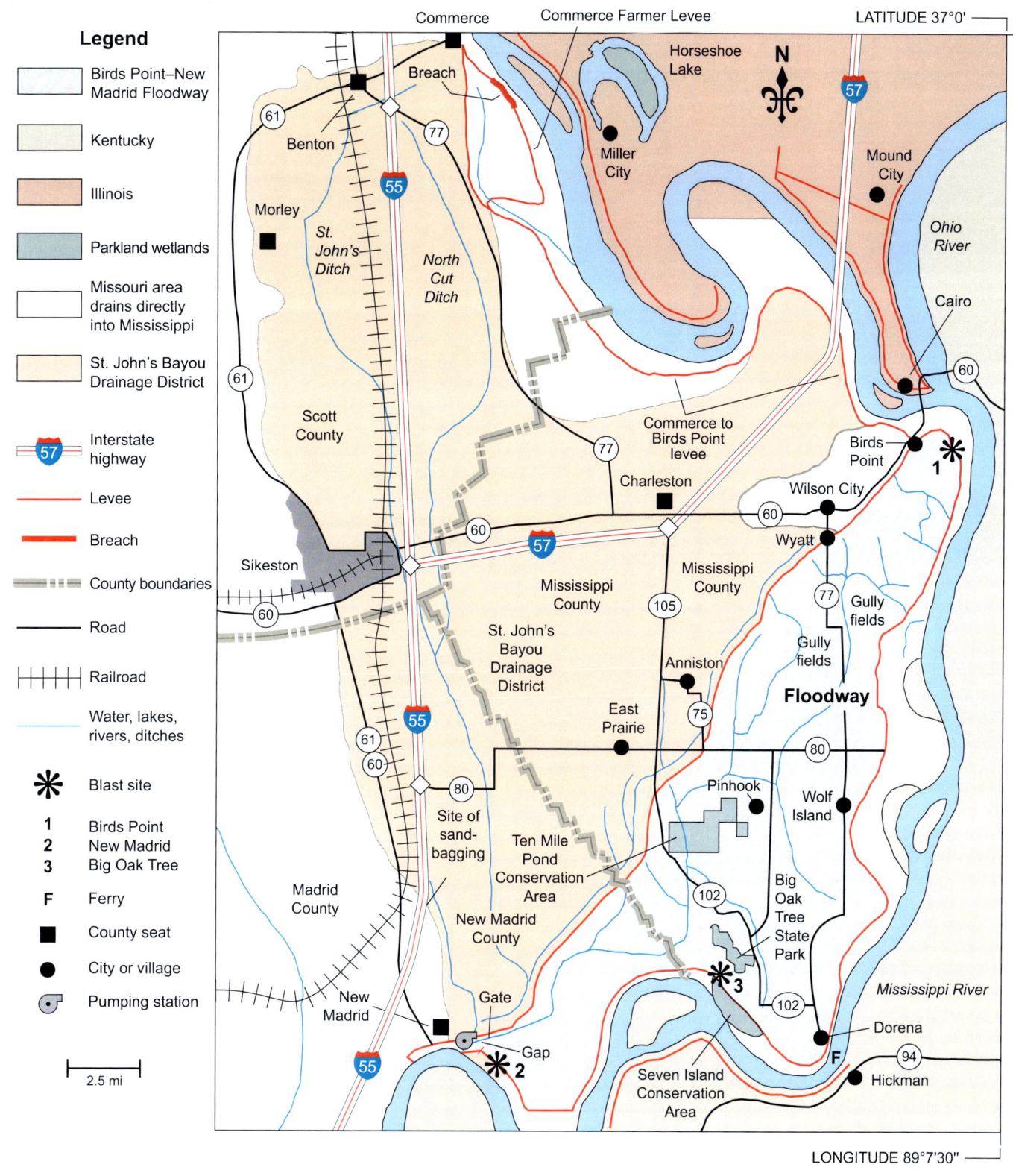

MAP 7.1 St. Johns Bayou Drainage District and the New Madrid Floodway in Missouri and their systems of levees protect agricultural lands from Mississippi River flooding but have altered the natural internal drainage of both of these basins.

to reduce the number of days that communities are isolated by floodwaters, limit crop and non-crop losses, and reduce damage to critical infrastructure by adopting flexible strategies that reconnect the hydrology of the floodplain to the river.

Geologic History

St. Johns Bayou basin lies to the east of the Little River Drainage District basin, west of the New Madrid Floodway, and south of the Little River Diversion and the upland land bridge. Its sloughs, bayous, and network of low-gradient streams are part of the Mississippi Embay-

FIGURE 7.1 A riverboat on the Mississippi River can be seen passing the outlet of Main Ditch that drains through the 1,500-foot frontline levee gap at New Madrid, Missouri, to the Mississippi River.

ment trough that runs north-south through southeast Missouri. The upper portion of St. Johns Bayou near Morley, Missouri (map 7.1), was once a confluence of the ancient Mississippi River and the ancient Ohio River as it drained west through the Cache River valley (ancient Ohio River valley) [1]. After the last glacier advance, the melting ice created new streams and channels, deposited sediments, and altered the courses of the Mississippi and Ohio rivers [2, 3, 4]. One of several historic confluences, the Morley site is 30 miles north of the current confluence of the Mississippi and Ohio rivers located south of Cairo, Illinois (map 7.1).

During the current Ice Age warming interval, the Mississippi River became a meandering river characterized by oxbows and winding flow paths created as the river current eroded banks on the outside of a river bend and deposited sediments on the inside of the bend. Map 7.1 shows the present-day meandering course of the Mississippi River as well as historic remnants of older, abandoned river channels in the alluvial floodplain. As the river migrated east to its current channel, the alluvial bottomlands located south of Commerce and Benton, Missouri, became a swamp called Tywappity Bottom.

The Bootheel of Missouri

Both topography and climate make the Bootheel the wettest region in the state of Missouri with annual rainfall averaging 55 inches. As an alluvial floodplain of the Mississippi River, the region experiences frequent headwater and backwater flooding. Headwater flooding is caused by runoff from surrounding uplands during heavy rain events (see map 6.2), and backwater flooding occurs during seasonally high water when snowmelt and rain from upstream tributaries raise Mississippi River levels and create backup into riverine bottomlands. When Spanish explorers in the 1700s passed through the area, these southeast Missouri lowlands supported 2.5 million acres of bottomland hardwood and swamp forest with an occasional slightly higher dry zone of bottomland planted in cultivated crops. The soils are rich and fertile, but settlers had difficulty harvesting hardwood timber from wetland soils and swamps. Farming the cleared timber soils was difficult, and the land remained relatively cheap until farmers developed strategies for draining it. The draining of the Tywappity bottomlands (map 7.1) was accomplished by cutting and burning trees, removing stumps using steam power, constructing drainage ditches, aerating the

FIGURE 7.2 When open, the New Madrid Floodway setback levee gate drains the 200,000-acre St. Johns Bayou basin to Main Ditch and into the Mississippi River. It is closed under flood conditions to prevent Mississippi River backup water from flowing into the basin.

soil, and then cultivating the soils. By 1975 only 100,000 acres of original forest remained in both the St. Johns Bayou and the adjacent Little River Drainage District (see chapters 5 and 6) located to the north, west, and south.

Commerce Farmer Levee Breach

During the Great Flood of 2011, the Commerce farmer levee, located between the USACE Commerce to Birds Point levee and the Mississippi River, breached about two miles south of Commerce (map 7.1 and figure 7.4). This breach occurred when a sand boil undermined the earthen levee shortly after the New Madrid Floodway was opened on May 2, 2011, to relieve the pressure on the floodwalls and levees near the confluence of the Ohio and Mississippi rivers. Floodwaters poured through the breach and covered the entire 5,100 acres of agricultural lands within the Commerce Farmer Levee and Drainage District with water. The entire 2010 winter wheat crop drowned [3, 4]. This area drained by the middle of June of 2011, leaving behind a thin organic-clay coating, and soybeans were planted. The levee was temporarily repaired in 2011 and fully repaired in the drought year of 2012. The cost of the levee repair raised the Commerce Farmer Levee and Drainage District assessment from $5 per acre to $25 per acre in 2013. The Commerce to Birds Point federal levee, the western boundary of the Commerce Farmer Levee and Drainage District, did not breach and protected 2.5 million acres of Missouri and Arkansas bottomland [3, 4], including the agricultural lands created from the drained Tywappity bottomlands.

The economic value of the Commerce farmer levee system was demonstrated by recent farmland sales (figure 7.5). Approximately 500 acres of the 5,100 acres located between the Commerce farmer levee and the federal Commerce to Birds Point levee (map 7.1) were sold as three separate tracts at an auction on October 21, 2013. The winning bids averaged $6,100 per acre plus a $360 per acre broker fee. At the same time, 105 acres of unprotected bottomlands between the Commerce farmer levee and the Mississippi River did not sell. A $2,000 per acre bid was not accepted. The farmer levee-protected land was valued at least $4,000 per acre higher than unprotected land.

The Gap in the Birds Point–New Madrid Floodway

The Birds Point–New Madrid Floodway was built between 1928 and 1932 with a 1,500-foot gap between the frontline and setback levees just northeast of New Madrid, Missouri, to provide an outlet (map 7.1) to the Mississippi River when the emergency floodway is opened by the USACE. The creation of the floodway, by the addition of a setback levee 4 to 10 miles to the west of the frontline levee, had the effect of blocking the St. Johns Bayou basin runoff water from draining freely into the Mississippi River. A gate constructed (figure 7.2) in the setback levee allowed closure during river flood stage and opening to permit drainage through Main Ditch when the Mississippi River was low (figure 7.1). Main Ditch currently drains approximately 300,000 acres including 100,000 acres of the floodway and 200,000 acres of St. Johns Bayou basin through the gate in the setback levee.

In 1954 Congress first authorized a plan to close the New Madrid Floodway gap between the frontline levee near the Mississippi River at the eastern edge of Birds Point–New Madrid Floodway and the setback levee. Backwater flooding from lower magnitude high water river events plagued over 80,000 acres in the southern portion of the floodway. The Flood Control Act of 1954 attempted to address this issue by authorizing construction of a new levee across the 1,500-foot gap near New Madrid and creating a gravity drainage structure. Up to 32,000 acres of lowlands near the control structure would form a wetland ponding area. The legislation mandated that state and local partners furnish all lands, easements, rights-of-way, and flowage rights.

In 1959, the St. Johns Levee and Drainage District began acquiring easements within the backwater area from private landowners and obtained easements for 57,000 acres. However, landowners closer to the gap did not want their property used as a ponding area and resisted participation in the acquisition effort. These landowners preferred the construction of a pumping sta-

Date	Mississippi River Height (Feet)	Flooding in New Madrid Floodway Basin (Acres)	St. Johns Bayou Basin Height (Feet)	Flooding in St. Johns Bayou Basin (Acres)
March 16, 1997	298.4	66,950	296.9	25,265
April 15, 2008	296.9	58,355	296.4	35,265
May 1, 2011	299.4	70,000	297.4	40,000

FIGURE 7.3 Floodwater heights in 1997, 2008, and 2011, and corresponding flooded areas in the New Madrid Floodway and the St. Johns Bayou basin.

tion that would remove water from the land. According to Charles Camillo of the Mississippi River Commission, the levee district was unable to acquire the remaining 21,500 acres necessary to meet the easement requirement without the cooperation of these key landowners.

The primary goal of the 1954 plan was to reduce river backwater flooding at the lower end of the floodway to provide year-round access to agricultural roads and fields and protect against lost crops and residential damage when seasonal flooding occurred. However, closing the frontline levee gap would effectively further disconnect the hydrology of the floodway wetlands, including Big Oak Tree State Park, from the river and impact wetland habitat, waterfowl, shorebirds, fish, and other riverine species that move into flooded areas during spring flooding. A 2006 lawsuit by the Environmental Defense Fund and the National Wildlife Federation resulted in a 2007 injunction that halted the work, and the gap remained open.

The aftermath of early snowmelt and heavy spring rains in the upper Mississippi and Ohio River basins resulted in the flood of 2011 and the opening of the floodway for the first time in 74 years [5, 6]. It also reignited public conversations about the urgency for internal flood mitigation in both of these basins. During the spring of 2011, the St. Johns Bayou gates were closed to protect against river backwater. However, this caused internal drainage water within the basin to reach 300 feet above sea level and to back up all the way to Interstate 55. Sandbags were placed on the east side of the interstate to keep floodwaters from blocking traffic on Interstate 55 (map 7.1) and disrupting commerce, but the flooding caused considerable damage to residents and agricultural lands.

FIGURE 7.5 A land auction sign is posted on land protected by the Commerce farmer levee, which was breached in May of 2011.

United States Army Corps of Engineers Flood Mitigation Efforts

Despite an extensive system of levees and internal drainage channels, flooding continues to plague the two basins and limit the agricultural productivity of the region. The USACE resumed efforts late in 2006 to close the 1,500-foot frontline levee gap and address the frequent seasonal flooding problems as authorized by

FIGURE 7.4 The Commerce Farmer Levee and Drainage District repaired a 2011 levee breach located two miles south of Commerce, Missouri. Note the Mississippi River on the right and the crater lake on the left created by the force of water pouring through the breach.

1954 and 1986 legislation at a projected cost of $7 million. However the levee construction was halted after the National Wildlife Federation and the Environmental Defense Fund filed a federal suit, alleging serious flaws in the USACE analysis of the project's environmental impact [7]. Environmentalists claimed that the gap was the last remaining area in Missouri where the Mississippi could reach its floodplain. Further they charged that it would have the largest impact on wetlands of any project in the region that included Missouri, Iowa, Nebraska, and Kansas. In 2007, US District Judge James Robertson ruled that the USACE had acted "arbitrarily and capriciously" by claiming falsely that its plan "would fully mitigate impacts" to the fisheries habitat [7]. This ruling blocked the USACE from proceeding with the project and included an order to "deconstruct that portion of the project which was already built." The removal cost an additional $10 million. After three years of attempting to close the gap and an expenditure of $17 million, the gap remained open, and Pinhook, Missouri (map 7.1), and the lower portion of the floodway continued to be flooded annually when river water pushed back up the internal drainage system [8].

For many years congressional leaders representing Missouri suggested that environmental groups were exaggerating the project's impact on wetlands and that every backup event from the river had a high economic cost to the region. Bills were proposed to build the levee across the gap and construct a pumping station to keep Mississippi River backwater out of the floodway. The St. Johns Levee and Drainage District, headquartered in East Prairie, Missouri, is a key local player in attempts to find solutions to these drainage issues. In 2012, two floodway farmers and members of the drainage district's board of supervisors charged that environmental groups were trying to block the project by conducting a distortion campaign to confuse the public. Taking issue with environmental critics' assertion that closing the levee gap would impede the floodway's future operation and damage the environment, these farmers contended the project would "increase forested lands in St. Johns Bayou basin by 35%, the Birds Point–New Madrid Floodway by 58%, and triple the size of Big Oak Tree State Park" [7]. The project would also result in a net loss of agricultural lands.

The National Wildlife Federation's Water Protection Network analyzed the landownership in the southern part of the floodway, the section in New Madrid County where the backwater problems occur. They reported that approximately 53 landowners or firms as well as the village of Pinhook, which was so badly damaged in the 2011 flood that it was vacated, would likely benefit from closing the levee gap [9]. Big Oak Tree State Park, one of the last forested wetland remnants located in the lower portion of the floodway, provides a small sanctuary for many local species of flora and fauna, including several state and national trees of record (black willow [*Salix nigra*], pumpkin ash [*Faxinus profunda*], persimmon tree [*Dispyros virginiana*], burr oak [*Quercus macrocarpa*], and swamp chestnut oak [*Quercus michauxii*]). There is some evidence that this park is already threatened by drainage channels [10], which can bring pesticides from surrounding cropland into the park and weaken the hydrologic connections to the river. In 2011 floodwater from the induced breach and use of the floodway covered Big Oak Tree State Park and pond (figure 7.6). This flooding may have resulted in transporting weeping willow (*Salix babylonica*) seedlings, an invasive species, into the pond; the seeds germinated and grew, and seemed to be crowding out native species during the drought of years of 2012 and 2013. Also of concern is the bald cypress (*Taxodium distichium*), which historically formed dense forests in these swamplands and continues to show stress after the drought of 2012.

Environmental Impact Statement and Public Meetings

Changes in land use and drainage modifications over time have altered the hydrology of the St. Johns Bayou and New Madrid Floodway basins. The USACE estimates that 86% of the historical bottomland hardwood forest in the region has been converted to cropland. In 2013 the USACE prepared a draft environmental impact statement analyzing several flood mitigation scenarios to address the frequent flooding of agricultural, commercial, and residential areas within the basins. Alternative scenarios proposed a variety of modifications to the current infrastructure (e.g., levee gap closure, pumping stations, ditch modifications, ring levee around town of East Prairie, or the raising of road surface elevations) and changes in land use (e.g., relocate the town of East Prairie, expand or modify existing wildlife refuges, convert nonflood-tolerant crops to flood-tolerant crops, or convert agricultural crops to silviculture). The USACE evaluated these alternatives based on their ability to achieve three key planning objectives: (1) reduce the number of days that communities are isolated by floodwaters, (2) reduce crop and non-crop agricultural damage, and (3) reduce critical infrastructure damages to streets and roads.

Environmental impact analyses examined simulated interior water surface elevations from 1943 to 2009;

FIGURE 7.6 The pond at Big Oak Tree State Park on October 24, 2013, after the flood of 2011 and drought of 2012.

riverine and permanent fish species; resident, migrating, and overwintering waterfowl; shorebirds that forage in areas of sparse vegetation such as harvested agricultural lands; and wetlands. The wetlands have three classifications: (1) low gradient riverine backwater wetlands frequently flooded by low velocity slack water, (2) low gradient riverine overbank wetlands that frequently flood along or near stream banks or islands within the river channel, and (3) connected depression wetlands that are remnants of abandoned stream channels or swales left behind by migrating channels.

The USACE [10] held two public meetings (on August 27, 2013, in East Prairie, Missouri, and on August 28, 2013, in Cairo, Illinois) to present their findings and obtain citizen input on the proposed $170 million flood control project with mitigation elements that compensated for multiple impacts. In the St. Johns Bayou basin, channel enlargement and vegetative clearing extending from the gate in the setback levee outside New Madrid up the St. James Ditch to the East Prairie area and a 1,000-cubic-feet-per-second pump to move water over the levee and into the river when the river gate was closed were proposed. At the bottom of the floodway, the USACE proposed a closer levee with four 10-by-10-foot gated structures within the 1,500-foot gap that could be closed and opened seasonally to allow the river to come into the floodway at the lowest elevations in order to retain hydrologic connectivity to the river and wildlife pools while reducing flooding of 40,597 acres of agricultural lands. When the proposed gate is closed, a 1,500-cubic-feet-per-second pump would manage the interior drainage elevations until the Mississippi River dropped sufficiently to allow for gravity drainage.

Public testimony at the Cairo, Illinois, and East Prairie, Missouri, hearing sites revealed strident differences among rural leaders and landowners of the two basins; outside environmental interests; and upstream Cairo, Illinois, residents and rural Illinois counterparts who experienced historical and recent levee breaching and agricultural land flooding. Underlying these differences were the geographical impacts experienced from the flood of 2011 and actions by the USACE to induce levee breaching of the Birds Point–New Madrid frontline levee, placing in use the floodway to prevent massive levee and floodwall failures along the Mississippi and Ohio rivers [6].

The channel modifications and addition of a pump in St. Johns Bayou were generally supported by local Missouri landowners, elected officials, and environmentalists. A local leader representing St. Johns Bayou basin framed the human and agricultural economic concerns, "[the basin] has about 481 square miles….34,000 residents, 1,046 drinking wells, and produces almost $600 million a year income…[we] agree with the drainage ditch proposal and of course the pump station idea…"

One environmentalist began his testimony in East Prairie by claiming common ground, "...we support fixing the drainage problems that have been plaguing your community [St. Johns Bayou] for so long. [It's] a problem that affects people, it affect roads, it affects infrastructure; and it needs to be fixed."

However, there was strong opposition from local Missouri landowners in both basins to converting existing farmland into wetland reserves or allowing farmland to be seasonally inundated to provide wildlife habitat. Environmentalists representing a variety of state and national groups asked the USACE to decouple the flood mitigation strategies in St. Johns Bayou from the floodway levee gap closure. Speaker after speaker in the East Prairie public hearing reminded the USACE that this was a flood control project, not an ecological restoration project. Further, there was strong sentiment from local speakers and the audience that the two basin proposals not be decoupled. One speaker received prolonged applause when he said, "...1950s, our communities since then have been standing together, trying to get this project done. We refuse to accept that this project can't be done... this project is about neighbors...providing flood protection on both sides of that setback levee..."

Illinois landowners, Mayor Colman of Cairo, and environmental groups testified at East Prairie and Cairo locations and were strongly opposed to closing the floodway gap near New Madrid. The groups were concerned that building a new levee would, in the words of one speaker, increase "...opportunities for more intensified agriculture in the area." Illinois residents were concerned this would increase political pressure to not open the floodway, when needed, in the future and put the Cairo floodwall and levee system at a greater risk of failure. The Illinois landowner opposition was led by homeowners and farmers who had put sandbags around their residences during the 2011 flood. The homeowners were convinced that had the New Madrid Floodway been opened according to the 1986 New Madrid Floodway operational plan, they would have been able to save their homes from flood damage. When the forecast peak issued on April 26, 2011, reached 60 feet on the Cairo gage, the USACE could have begun the 15- to 24-hour preparation of the fuse plug to be opened using trinitrotoluene (TNT). The operation plan, however, did not call for opening of the fuse plug prior 60 feet. It called for the floodway to be prepped for operation by the time the Ohio River reached 60 feet, which actually occurred on April 30, but the site preparations had not been ordered by the Mississippi River Commission; the Birds Point levee fuse plug was not activated (no TNT had been loaded into the fuse plug), and could not be opened at that time. However, the Birds Point fuse plug was filled with TNT (a 15-hour operation) on May 1 and 2 in the middle of a 7.5-inch rain event, and the floodway was opened at about 10:00 PM on May 2. The Len Small levee breach on the Mississippi River occurred on the morning of May 2, 2011, and flooded homes just hours before the floodway was opened to relieve the pressure on the confluence area levees and floodwall. At the time of the Birds Point fuse plug opening, the Ohio River was at 61.7 feet (a record) on the Cairo gage, or 1.7 feet higher than the 1986 New Madrid operational plan depth of 60 feet.

Environmentalists worried that closing the last quarter-mile opening connecting the Mississippi River to its floodplain (figure 7.1) would "result in loss of critical wetlands for fisheries and wildlife whose unique value is their dynamic relationship with the river." They further claimed that agriculture in the floodway was "already profitable and reliable" and did not need taxpayers to pay for the costs of pumps and levee fortification.

Federal Agency and Local Leadership Challenges

The construction of levees and the Birds Point–New Madrid Floodway separated the land in the floodway and St. Johns Bayou basins from their natural drainage pathways. The St. Johns Levee and Drainage District has tried for 83 years to regain access to drain local basin internal floodwater directly to the Mississippi River. Once the floodway setback levee was built, local landowners within the floodway had to sign easements in 1930s and 1960s giving the USACE the right to pass floodwater over their land. If the floodway had never been built, both St. Johns Bayou and New Madrid Floodway basin farmers would still have been affected by Mississippi River floodwaters every time the river reached flood stage. However, when the setback levee gate is closed, there is no effective way for the local St. Johns Bayou basin water runoff to drain to the Mississippi River. If the St. Johns Bayou phase of the USACE project is built, it would appear to allow the local St. Johns Bayou basin floodwater to be pumped over the setback levee during the times when the gate is closed. This should reduce internal flooding in the St. Johns Bayou basin, which has adversely affected agricultural production and constrained the intensification of agricultural land use.

A primary USACE goal of the $170 million project is to reduce river backwater flooding at the lower end of the floodway to provide year-round access to agricultural roads and fields and protect against lost crops and residential damage when seasonal flooding occurs. However, closing the floodway frontline levee gap would effectively further disconnect the hydrology of the floodway wetlands from the river and impact wetland habitat. The USACE project attempts to mitigate the internal flooding experienced in St. Johns Bayou basin and the impact on the hydrology of the floodway ecosystem. The project mitigations would not be met by funding the St. Johns Bayou basin phase only. Both mitigation goals could be met if both St. Johns Bayou and floodway gap-closing phases of the project are implemented including land use change, drainage ditch realignment, and pump station construction in both basins.

One of the USACE's greatest challenges is to manage variable river conditions, the uncertainties associated with the concentration and flow of water, and unpredictable weather and changing climate conditions while balancing diverse and competing river commerce, agricultural, residential, and environmental interests. The building of the floodway introduced a new era in engineering design, moving from the confinement of levees only [11] to a dispersion strategy that allowed the river to temporarily spill into its natural bottomlands to relieve flooding pressures on urban settlements and downstream levees [12]. Paradoxically, past infrastructure investments intended to reduce direct risks of flooding have led to interior flooding problems and unexpected consequences to the larger ecosystem. Levees have been a critical infrastructure in opening new lands to agricultural production; however they may be inadequate as the distribution, seasonality, and intensity of precipitation patterns change [13]. Given the economic and social constraints, the Mississippi River floodplains are not likely to be fully restored as wetlands to mitigate flood hazards. This suggests a need for a new kind of engineering, one that offers greater resilience to the floodplain system [14]. Resilience engineering goes beyond the levees and floodway structures to strategically reconnect the hydrology of levee-protected lands, portions of former wetlands, and the river. This approach acknowledges ecological functions, such as floodwater storage and wildlife habitat, and utilizes these functions to absorb future uncertainties associated with flooding.

---

[1] Olson, K.R., and L.W. Morton. 2014. Ohio River flooding of the Cache River Valley in southern Illinois. Journal of Soil and Water Conservation 69(1):5A-10A, doi:10.2489/jswc.69.1.5A.

[2] Olson, K.R., and L.W. Morton. 2013. Impacts of 2011 Len Small levee breach on private and public Illinois lands. Journal of Soil and Water Conservation 68(4):89A-96A, doi:10.2489/jswc.68.4.89A.

[3] Olson, K.R., and L.W Morton. 2013. Soil and crop damages as a result of levee breaches on Ohio and Mississippi rivers. Journal of Earth Sciences and Engineering 3:139-158.

[4] Olson, K.R., and L.W. Morton. 2014. Dredging of the fracture bedrock lined Mississippi River channel at Thebes, Illinois. Journal of Soil and Water Conservation 69(2):31A-35A, doi:10.2489/jswc.69.2.31A.

[5] Olson, K.R., and L.W. Morton. 2012. The impacts of 2011 induced levee breaches on agricultural lands of Mississippi River Valley. Journal of Soil and Water Conservation 67(1):5A-10A, doi:10.2489/jswc.67.1.5A.

[6] Olson, K.R., and L.W. Morton. 2012. The effects of 2011 Ohio and Mississippi River valley flooding on Cairo, Illinois, area. Journal of Soil and Water Conservation 67(2):42A-46A, doi:10.2489/jswc.67.2.42A.

[7] Koenig, R. 2012. Corps balancing levee repairs on Missouri, Illinois sides of Mississippi. St. Louis Beacon. July 30, 2012.

[8] Koenig, R. 2013. Environmental groups bash new impact statement on St. John's Bayou project. St. Louis Beacon. July 19, 2013.

[9] Morton, L.W., and K.R. Olson. 2013. Birds Point–New Madrid Floodway: Redesign, reconstruction and restoration. Journal of Soil and Water Conservation 69(2):35A-40A, doi:10.2489/jswc.69.2.35A.

[10] US Army Corps of Engineers Memphis District. 2013. Public Hearing for the Draft Environmental Impact Statement on St. Johns Bayou-New Madrid Floodway Project. East Prairie, Missouri, August 27, 2013, and Cairo, Illinois, August 28, 2013.

[11] Barry, J. M. 1997. Rising Tide: The Great Mississippi Flood of 1927 and How it Changed America. New York: Simon and Schuster.

[12] Camillo, C.A. 2012. Divine Providence: The 2011 Flood in Mississippi River and Tributaries Project. Vicksburg, MS: The Mississippi River Commission.

[13] Morton, L.W., and K.R. Olson. 2014. Addressing soil degradation and flood risk decision making in levee protected agricultural lands under increasingly variable climate conditions. Journal of Environmental Protection 5:1220-1234.

[14] Park, J.T., P. Seager, S.C. Rao, N. Convertino, and I. Linkov. 2012. Integrating risk and resilience approaches to catastrophe management in engineering systems. Risk Analysis 33(3):356-367.

Flooding and Levee Breach Impacts on Protected Agricultural Lands

Alluvial soils are developed from fine-textured clay and silt sediments deposited in floodplains when rivers overflow their natural banks and flood into adjacent bottomlands. These water-saturated lands experience annual flooding for many months each year as the river levels vary with local and upstream precipitation and snowmelt. Fast-moving floodwaters also can transport and deposit sand and gravel onto alluvial bottomlands. When these lands are drained and leveed to protect from river flooding, they are some of the most fertile and productive agricultural soils in the world. Whenever levees on rivers are breached, there are soil and crop damages in the flooded bottomland areas that impact agricultural management capacities and crop productivity (figure 8.1). Earthen levees and floodwalls can be undermined by sand boils, fail after weeks of high floodwater pressure and soil saturation, or even be topped. When a levee fails, there is often a crater lake created adjacent to the levee breach (figure 8.2), with gullies and land scouring extending into the previously protected lands. As the water spreads out and slows down, sand and sediments are deposited on the bottomlands and in road ditches and drainage ditches. Floodwaters may drown crops and coat the entire flooded land surface with sediments containing a variety of pollutants, nutrients, and contaminants. Restoration of craters, gullies, land scoured areas, and sediment depositional sites is necessary if agricultural lands are to be returned to some level of productivity.

River Bottomlands, Agriculture, and Levees

In the United States, the Mississippi River and tributaries drain 41% of the continental land mass including millions of acres of agricultural lands protected by thousands of miles of earthen levees (figure 8.3) and floodwalls. By 1926, an extensive system of private and public levees along the Mississippi River was engineered to secure agricultural lands and river cities against flooding [1] from the confluence of the Ohio and Mississippi rivers at Cairo, Illinois, all the way south to the Gulf of Mexico. Today the US Army Corps of Engineers (USACE) and local levee and drainage districts continue to actively manage river levels to maintain navigation and protect against flooding. Prior to the construction of these levees, most river bottomlands experienced crop loss every time the

FIGURE 8.1 A sand delta created by a breach covers a corn field with tree residue transported from the river and levee into an agricultural field.

FIGURE 8.2 Water pours through a breached levee on the Embarras River in June of 2008.

river flooded, but the flooding caused little if any soil damage. Although these massive fortress-like structures seem impermeable, levees do fail for a variety of reasons and allow floodwater to flow through breaches with an intensity that can do substantive damage to agricultural crops. Not only does the crop drown if flooded during growing season, but considerable soil damage can occur as a result of the levee breach and lead to the creation of crater lakes, gullies that extend into agricultural lands, land scouring, and sand and sediment deposition on bottomlands as well as in drainage and road ditches. Further, as these fast-moving waters slow down, the drowned crop and land surface are coated with silt, clay, organic matter, and other chemicals that the water carried. This flooding of crops and soils, and the maintenance, repair, reconstruction (figure 8.4), and strengthening of degraded levees cost millions of private and public dollars after every levee breach.

Changes in climate [2], such as shifts in the long-term seasonality and frequency of extreme weather events, can result in record flooding and droughts and increase the vulnerability and risks associated with managing levee-protected agricultural lands. Those who live, work, and farm in levee-protected areas have a distinct language and set of terms they use when talking about their systems of levees, the threat of flooding and potential levee breaching, and different engineered structures used to manage the river. This chapter details some of these terms to help explain the impacts of natural and induced levee breaching during high water and flood events, how the levee system works, the processes and mechanisms by which land scouring and sediment deposition occur when bottomlands are flooded, and the remedial activities that are used to repair and guard against future levee breaching.

Earthen Levees and Floodwalls

The levee is a massive earthworks designed to contain and channelize the river at flood stage and protect agriculture and other land uses against flooding. It has a flat crown with three-to-one sloped sides. The texture of the materials used in earthen levee construction can vary from silty clay to sandy. Grasses with thick, dense roots are planted on the levee to hold the soil in place and reduce the erosive effect of water. Breaks in the levee, called breaches, can occur when a portion of the levee is eroded or breaks from a subsurface weakness. The higher the levee, the greater the force of the river on the protected land when a breach occurs. A levee as high as a three- or four-story building can explode with the same power and suddenness of a dam bursting. Silty clay levees with a sand core can be affected by vegetation and animal burrowing, which in turn influences the susceptibility of the levee to erosion and natural

FIGURE 8.3 The Commerce farmer levee located south of Commerce, Missouri, protects a soybean field ready for harvest.

FIGURE 8.4 A levee at Birds Point, Missouri, is reconstructed after the induced breach of 2011.

breaching. However, the greatest danger to levee failure is constant water pressure against the levee [1, 3]. The weight of the river can push water underneath the levee and create sand boils, which undermine the strength of the levee and its capacity to hold back water. Another type of levee is a floodwall constructed of concrete (figure 8.5). These are often built in urban areas at the most erosive points in the river, usually the bend where strong currents and constant pressure of flowing water can erode an earthen levee. Floodwalls may also replace earthen levees when desired soil materials are not available or are too costly to transport.

Levee Saturation and Topping

The water-holding capacities of the soil in earthen levees affect its strength to withstand the pressure of the river. During record flooding, the levee can be saturated for a prolonged period and fail, or the floodwater can be higher than the levee and top (run over) the levee crest. When floodwater starts over the top of the levee (see figure 15.1), it can cut an erosion channel into and through the levee. Once the floodwater flows over or through a break in the levee, the river water drops with great force and cuts a crater lake on the inside of the levee. As more water flows through the eroding crack, the floodwaters pick up speed and widen the breach by removing sections of the adjacent levee [4]. The breach can end up 300 feet or more in width and result in a deep crater lake.

Sand Boils

When floodwaters put significant pressure on floodwalls and levee systems, seepage and sand boils can occur, especially if there are sandy soils underneath the floodwall (figure 8.6) or earthen levee [3]. Sand boils, including a mega sand boil, occur when the river is at or above flood stage and is putting enormous pressure

FIGURE 8.5 The concrete floodwall on the east side of Cairo, Illinois, protects the city from Ohio River flooding.

on the levee system. The bottomland inside the levee acts like an empty sunken basin with the higher floodwater outside the basin creating a hydraulic gradient that can cause an internal erosion of the embankment. As the water seeks to equalize the pressure on both sides of the levee, a stream of water (called piping) can force its way through even a tiny opening in the river side of the levee or floodwall [5]. Once the water is piped through the soil into the basin, sand and water can burst through the ground and release the pressure. With the release of pressure, water shoots up through the porous earth or sand in a churning or boiling action from which the name "sand boil" is derived [6].

Uncontrolled seepage, a major cause of levee failure, creates instability when high water pressure and soil saturation cause the earth materials to lose strength. Most small sand boils are treated with a 5-foot-high sandbag ring and filled with water [4]. The sandbag dike is normally a temporary measure to increase the depth and weight of water over the sand boil in order to decrease the hydraulic gradient across the seepage path and reduce the potential for erosion of earth materials along the piping path [6]. A sand boil can start small and quickly turn into a high-energy sand boil, which is difficult to stop (figure 8.6). In a few hours, it can enlarge dramatically from a few inches to 2 feet in diameter. In the case of a mega sand boil, the crew often has to construct a 50-foot ring berm to a height of 6.5 feet or more. When the counterweight of water alone is insufficient to control a mega sand boil, it is covered with a few yards of fly ash cinders or other available materials. The treatment of a mega sand boil can require big earth-moving equipment, such as bulldozers, backhoes, loaders, excavators, and dump trucks, and a large crew to contain the mega sand boil and keep watch until flood stage recedes.

Crater Lakes, Adjacent Gullies, and Thick Sand Deposits

As the floodwater tops or breaks through an earthen levee, it often drops many feet to the bottomland inside the levee and causes deep erosion of the soil and underlying geologic parent material (figure 8.7). Most crater lakes are many feet deep [3]. As the floodwater flows rapidly into the previously protected bottomland, gullies originating adjacent to the crater lake are cut into the soil and can extend 30 feet to more than 300 feet beyond the lake. Geologic materials, soil, sandbars, and sediment carried downriver and from the degraded earthen levee are deposited in the bottomland after the floodwaters slow down (figure 8.7) [4]. The thick sand can be deposited on agricultural crops and other vegetation on the land surface as well as in drainage ditches and in road ditches.

Land Scouring and Gully Fields

After a levee breach occurs and the fast current of the water has created crater lakes, extended gullies into the bottomland (figure 8.8), and scoured the land surface, the speed of the advancing floodwater begins to slow and deposit sand particles on the bottomland in large sand deltas. As the floodwaters continue to slow, the silts drop and then the clay and organic particles settle out and coat plant residues and the land surface [7, 8]. Significant land scouring can result in many hundreds of acres of land losing half an inch or more of topsoil after each levee breach.

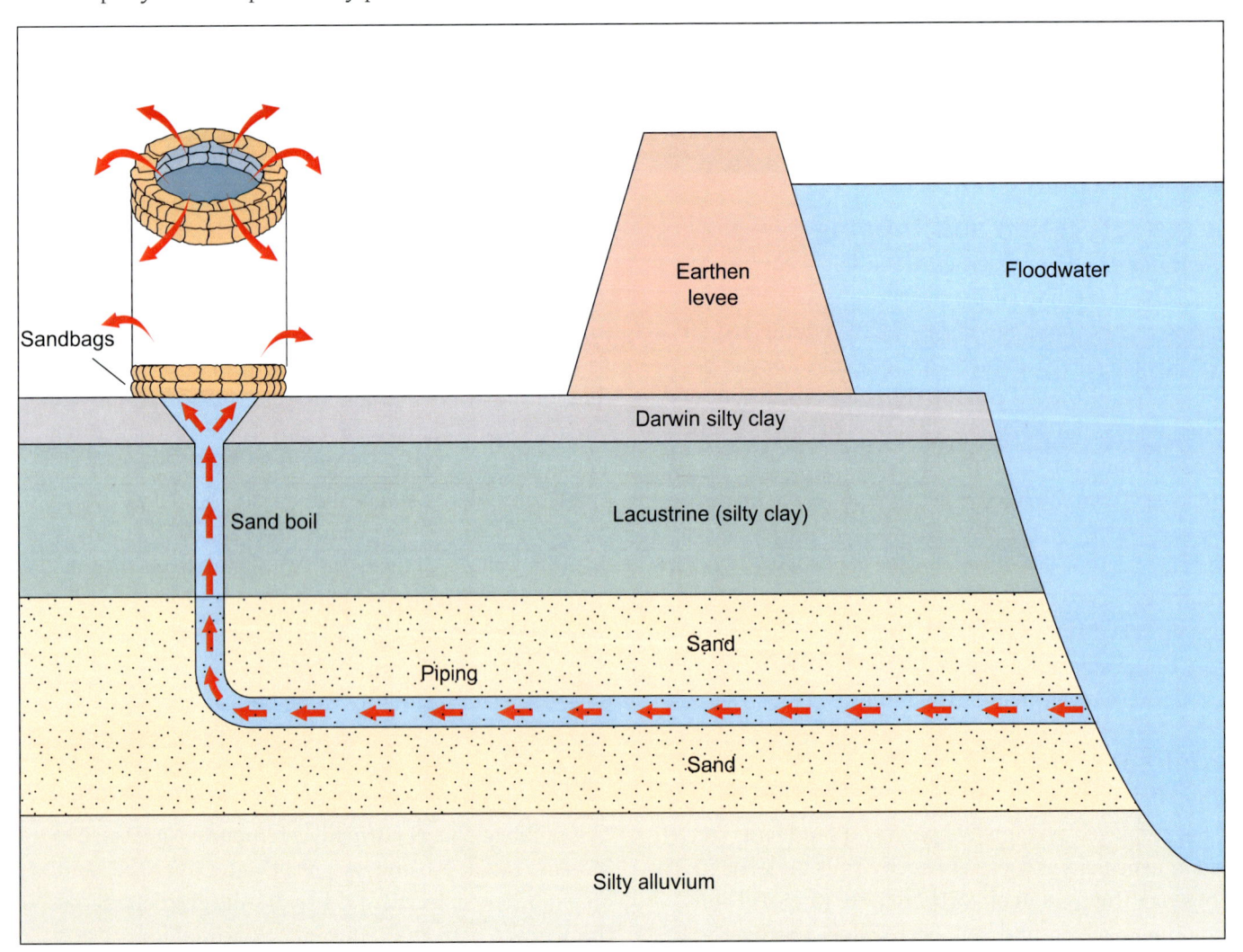

FIGURE 8.6 Anatomy of a sand boil.

FIGURE 8.7 A crater lake extends 300 feet through the levee breach.

Levee-protected bottomlands normally have very little slope and are almost flat but can contain higher natural levees formed from old oxbows cut off from the river. An oxbow is the wide curve portion of a meandering river channel. The floodwaters will pond in front of these meander scars or natural levees. When water flows over a natural levee ridge and drops down to another alluvial bottom, it concentrates and creates an eroded channel or waterway. This erosion process is called hydraulic jumping. As high-energy floodwater flows into the newly created waterway, it can cut into the ridge and carve additional new channels and gullies.

These deeply eroded soils, or gully fields (figure 8.8), are extremely difficult and costly to reclaim [7]. Often bulldozers are used to push in the tops of the vertical gully walls to fill in the bottom and then grade the side slopes. The pushing of topsoil into the gullies (figure 8.9) places the soil material on slopes that are highly erodible, and the exposed subsoil and parent material of the scrapped areas lower the productivity of the original soils. Topsoil must be hauled in to raise the soil organic carbon content of the soil if the land is to be returned to the previous level of agricultural productivity. Terracing is another option for reshaping the side slopes above the gully bottom. This approach will still result in the loss of long-term soil productivity and crop yields since the newly created soils will be less productive than the original soils as a result of mixing topsoil with parent material low in soil organic carbon. Reclamation efforts to restore these land scoured ridges and gully fields can cost hundreds of thousands of dollars and are likely to still result in lowered soil productivity and crop yields when compared to original crop yields [9]. When gully fields are created by levee breaching, the land use may change. Gullies that are not reclaimed will collect water and become wetlands. Loss of agricultural land to ponds and wetlands can result in a net loss of agricultural productivity [9].

FIGURE 8.8 An O'Bryan Ridge gully created from an induced breach in May of 2011 is filled with water and removed land from agricultural production.

FIGURE 8.9 Bulldozers push soil into wetlands and ponds at the bottom of the gullies in O'Bryan Ridge.

Effect of Growing Crops and Crop Residue on Erosion and Deposition

Crops grown in the Mississippi and Ohio river bottomlands are primarily wheat, corn, soybean, and forages. Depending on the time of year, the land cover may include these crops in various stages of growth or only plant residues remaining from the previous year's crop, such as soybean stubble, corn stalks, and wheat stubble. When spring floods occur, winter wheat and forages are growing and are likely to drown. However, these fully developed plants can hold the soil in place and prevent land scouring. Sediment carried by floodwaters is caught by the wheat and forage vegetation and deposited on the crop. If levee breaching and flooding occur in the spring before corn and soybeans are planted, only previous crop plant residues are protecting the bottomland soils. These plant residues are often picked up, carried along by floodwaters, and lose their protective capacity to prevent land scouring. When flooding and levee breaching occur later in the growing season, the soybean and corn plants help slow the speed of the floodwaters and anchor the soil.

After the floodwaters recede and the land drains, the soil can either be dried and planted or left idle but tilled to eventually mix in the sediments to help dry out the fields depending on the timing of the flooding. Thin layers of silt and clay deposits can be treated by sunlight, drying, and tillage to incorporate into the plow layer. Tillage equipment, such as chisel plows and moldboard plows, can be used on the areas with thin sand deposits (less than six inches) in an attempt to mix the sand into the underlying bottomland with silty and clayey topsoil (figure 8.10). Crop damages depend on the type of crops commonly grown in the area and the timing of the levee breach and subsequent flooding. If flooding occurs in the growing season of a crop and the plants are under floodwater for 24 to 36 hours, the submerged crop will drown, and this can result in a total crop failure for that year. If flooded early in the growing season, the crop can be replanted, but usually lower crop yields result. Crop insurance can provide replacement for a portion of the income for farmers who have purchased it [8, 10].

Sediment Deposition in Road and Drainage Ditches

After a levee breach and flood event, road and drainage ditches in the area are filled with sediments and sand, sometimes as much as three to seven feet deep.

FIGURE 8.10 Tillage equipment incorporates sediment left behind from flooding into the topsoil.

FIGURE 8.11 Trees on the river side of a levee are transported by floodwaters through the levee breach into adjacent fields.

Excavators are usually brought in by either the county, drainage districts, or the Natural Resources Conservation Service (NRCS) to remove debris and sediment that block drainageways and ditches to speed up the drainage process and to accelerate the drying out of low-lying areas [7, 8]. Sediment removal from private drainageways of most qualified landowners can sometimes be partially financed by the United States Department of Agriculture (USDA) Emergency Services Agency's Conservation Program. The local drainage district often provides additional matching funds for these projects.

Levee Repair, Sand Delta Removal, and Crater Lake Filling

If the funds are available, the USACE or the farmer levee and drainage districts usually begin reconstructing levees immediately after the floodwaters have drained. Sometimes the levee is repaired in stages if the funds are insufficient for the entire reconstruction of the levee. Restoration and repair of the levee to the original height or higher is usually a concern of the landowners flooded by the levee breach [8]. The sand deposits are often collected and used to fill in the crater lake and then topsoil is trucked in and spread over the crater lake to restore the previous land use. The thick deltaic sand deposits can be 8 to 50 acres in size and between 4 inches and 3 feet deep. Both the crater lakes and the thick sand deposits can result in a permanent loss of agricultural productivity [9, 11] if they are not filled in or removed. The transported topsoil may come from other levees on smaller tributaries and drainage ditches or other adjacent soil deposits. The reconstructed soil profile will still be less productive than the original soils.

Damage to Farm Buildings and Homes

When a levee breach occurs there is always the risk that lives and property may be lost or seriously harmed [1]. When a levee breach is imminent, the US National Guard usually sweeps the area to make sure the people living and working in levee-protected bottomlands are evacuated. There is also a high risk of damage to the homes (see figure 12.2), barns, and other structures on these flooded bottomlands [12]. Buildings can be impacted by the force of the flowing floodwater and become fully or partially submerged in the floodwater. Water pressure can result in loss of the lower half of entire walls, damage wooden floors, or destroy structures completely. In addition, farm structures (sheds, barns, and grain bins) can be damaged, and depending on insurance coverage, only a few of these structures may be repaired.

Protected and Unprotected Agricultural Lands

River bottomland areas that are not protected by levees usually receive a thin layer of silt, clay, and organic matter during flood events. The crop is lost if flooding occurs in the growing season, but soils do not usually suffer permanent damage [13]. This is not the case where levees fail. Levees can fail as a result of a sand boil or by levee topping. Blow-out holes or craters between 3 and 25 acres in size can be created and hold water. Fast-flowing water can remove hundreds of feet of levee embankments and erode thousands of cubic feet of soils and underlying outwash parent material to depths of many feet below the base of the earthen levee when the levee breaks [4]. The force of the rushing water can uproot trees growing between the river and the levee and deposit them, as residue, on the previously protected floodplain (figure 8.11). Deep gullies can extend many hundred feet into the bottomland, and hundreds of mature trees can be transported hundreds to thousands of feet. Deltaic sand deposits up to three feet thick can cover many acres on the floodplain at each breach site with additional acres covered with a thin layer of sand. The remaining hundreds of acres of previously protected floodplain soils receive a thin coating of silt and clay and can remain under floodwaters long enough to drown the current year's crop if it was planted and not already removed by the wall of advancing floodwater. After a few weeks, the floodwater usually drains from the bottomland and back into the river or slowly evaporates and infiltrates bottomland soils sufficiently for local landowners to begin the task of moving the trees from near the blow-out holes and floodplain and begin filling in the craters and gullies [4].

Managing Soil Damages from Levee Breaching

A changing climate can amplify the risks associated with snowmelt, rainfall, runoff patterns, and river flooding. As the odds for certain types of weather extremes increase in a warming climate, farmers, rural residents, and supporting institutions as well as public and private levee districts will need short- and long-term strategies to sustain their system of levees, address breaching events and reclamation of agricultural lands, and put in place adaptive management plans that anticipate future events. Levees are complex engineered systems coupled with natural and social systems. Future levee redesigns must not only account for risks to the engineered system but also consider how to make the social, economic, and environmental systems more resilient. One need is to better understand how the soils are affected by flooding and levee breaching and their capacities to support future agricultural productivity.

Three recommendations are suggested to provide the critical data needed to assess soil degradation and to guide restoration in making levee-protected agricultural bottomlands more resilient: (1) improve characterization and measurement of eroded soils and distribution of sediment contaminants after levee breaching, (2) assess contamination effects on soil productivity and long-term agricultural production in order to understand the impacts of flooding on agricultural soils, and (3) evaluate reconstruction investments needed to repair levees based on return of the land to productivity and on reduction of vulnerability to future flooding and levee breaching stress.

[1] Barry, J.M. 1997. Rising Tide: The Great Mississippi Flood of 1927 and How It Changed America. New York: Simon and Schuster.

[2] Palmer, M.A., D.P. Lettenmaier, N.L. Poff, S.L. Postel, B. Richter, and R. Warner. 2009. Climate change and river ecosystems: Protection and adaptation options. Environmental Management 44(6):1053-1068, Epub. PubMed PMID: 19597873.

[3] Olson, K.R., and L.W. Morton. 2012. The effects of 2011 Ohio and Mississippi River valley flooding on Cairo, Illinois area. Journal of Soil of Water Conservation 67(2):42A-46A, doi:10.2489/jswc.67.2.42A.

[4] Olson, K.R. 2009. Impacts of 2008 flooding on agricultural lands in Illinois, Missouri, and Indiana. Journal of Soil and Water Conservation 64(6):167A-171A, doi:10.2489/jswc.64.6.167A.

[5] Lansden, J.M. [1910] 2009. A History of the City of Cairo, Illinois. Reprint, Carbondale: IL: Southern Illinois University Press.

[6] Veesaert, C.J. 1990. Inspection of Embankment Dams. Session X in Embankment Dams. Bureau of Reclamation. http://www.michigan.gov/documents/deq/deq-p2ca-embankmentdaminspection_281088_7.pdf.

[7] Olson, K.R., and L.W. Morton. 2012. The impacts of 2011 induced levee breaches on agricultural lands of Mississippi River Valley. Journal of Soil of Water Conservation 67(1):5A-10A, doi:10.2489/jswc.67.1.5A.

[8] Olson, K.R., and L.W. Morton. 2013. Restoration of 2011 flood-damaged Birds Point-New Madrid Floodway. Journal of Soil of Water Conservation 68(1):13A-18A, doi:10.2489/jswc.68.1.13A.

[9] Olson, K.R., J. Matthews, L.W. Morton, and J. Sloan. 2015. Impact of levee breaches, flooding, and land scouring on soil productivity. Journal of Soil and Water Conservation 70(1):5A-10A, doi:2489/jswc.70.1.5A.

[10] Morton, L.W., and K.R. Olson. 2013. Birds Point-New Madrid Floodway: Redesign, reconstruction, and restoration. Journal of Soil and Water Conservation 68(2):35A-40A, doi:10.2489/jswc.68.2.35A.

[11] Lowery, B., C. Cox, D. Lemke, D.P. Nowak, K.R. Olson, and J. Strock. 2009. The 2008 Midwest flooding impact on soil erosion and water quality: Implications for soil erosion control practices. Journal of Soil and Water Conservation 64(6):61A-66A, doi:10.2489/jswc.64.6.166A.

[12] Camillo, C.A. 2012. Divine Providence: The 2011 Flood in Mississippi River and Tributaries Project. Vicksburg, MS: Mississippi River Commission.

[13] Olson, K.R., and L.W. Morton. 2013. Soil and crop damages as a result of levee breaches on Ohio and Mississippi rivers. Journal of Earth Science and Engineering 3(3):139-158.

Impacts of 2008 Flooding on Agricultural Lands in Illinois and Indiana

Eight to twelve inches of rain fell across the upper Midwest from May 30 to June 12, 2008 (map 9.1). Iowa, east-central Illinois, and Indiana were the hardest hit, and the governor of Indiana declared a state of emergency in 23 counties. After months of heavy precipitation, previously saturated soils could not hold any more rainfall, and the wetlands, potholes, and depressions across the landscape filled with water (figure 9.1) and then began to run off through waterways into small streams (figure 9.2) and rivers. Levees broke, unable to withstand the pressure of rising floodwaters. The Mississippi, Missouri, and Wabash rivers and their tributaries flooded cities and farmland as runoff accumulated in main stem rivers and crested downstream.

The Mississippi River peaked at St. Louis, Missouri, on July 1, 2008, but at a lower height than the 1993 flood. Although there were no levee breaks on the Mississippi River south of St. Louis in 2008, there was substantial flooding of agricultural lands and roads throughout the floodplain. Rising floodwaters caused evacuation of residents in towns such as Winfield, Missouri; Meyers, Illinois; and Keithsburg, Illinois; closed many local roads and bridges; and flooded adjacent agricultural lands. National news coverage focused on Iowa where the Cedar and Iowa rivers flooded the cities of Cedar Rapids and Iowa City and destroyed portions of their downtowns. Lands along the rain-swollen rivers in the Wabash watershed in Indiana and Illinois (map 9.2) also experienced disaster as towns were evacuated, levees failed, and agricultural fields flooded.

Many Indiana and Illinois farmers delayed planting, and corn and soybean fields already planted drowned. As much as 30% of the upland soils in south-central Illinois and southern Indiana were affected by ponding. Approximately one-third of that ponded acreage was not replanted in 2008. As overland flow started to occur, so did sheet, rill, and gully erosion causing loss of topsoil. Fields with significant topsoil loss are at risk of eventually moving into an erosion phase change of the soil. Any soil erosion phase change from slightly to moderately or severely eroded can reduce the crop yield potential from 5 to 15 bushels per acre, depending on whether the soils have favorable or unfavorable subsoils for rooting. One year's erosion events do not change the erosion phase of the soil unless gullying occurs. However, the 2008 soil

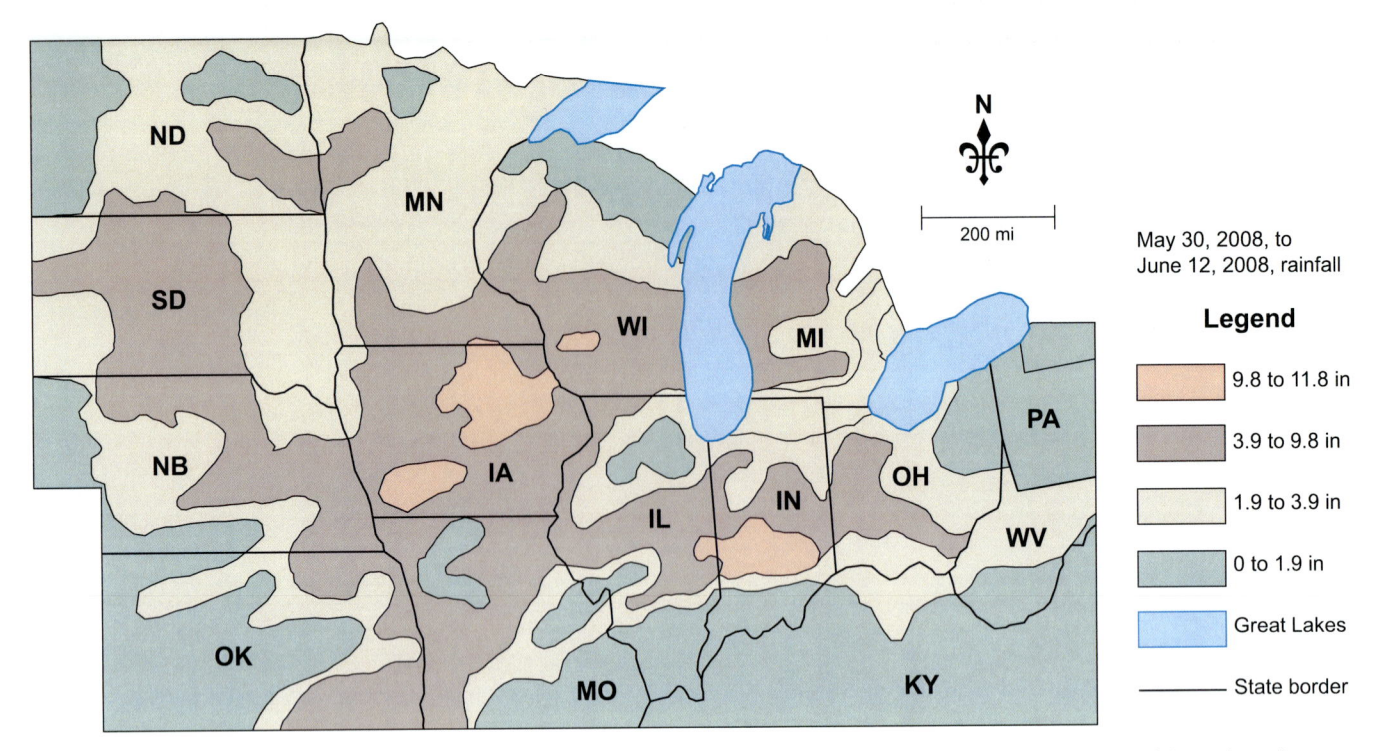

MAP 9.1 Total rainfall between May 30 and June 12, 2008, in the north-central United States. Modified from a National Oceanic and Atmospheric Administration weather map for the same time period.

loss in many fields, when added to the soil loss from erosion in previous years, had the potential to result in a soil erosion phase change.

Wabash Watershed and the Embarras River

The Wabash watershed (map 9.2) drains almost three-quarters of Indiana and a portion of east-central Illinois. Its main stem river, the Wabash, runs 503 miles from headwaters in Ohio southwest across the state of Indiana to Terre Haute where it forms the Indiana-Illinois border as it flows south to confluence with the Ohio River. During much of the nineteenth century, large ships traveled the Wabash River between Terre Haute and the Ohio River with steamships frequently stopping in Terre Haute. However, by the late 1800s, farmland erosion and sediment deposits from runoff filled the Wabash River making it impassable.

The Embarras River in eastern Illinois is a 195-mile tributary of the Wabash River (map 9.2). It drains more 1.5 million acres as it flows south from Champaign County to Jasper County where it turns southeast and joins the Wabash River southwest of Vincennes, Indiana. Jasper County is part of Springfield Plain, which lies within the Till Plains section of the central Interior Lowland province [1]. Springfield Plain is a nearly level plain formed of till deposited by the Illinoian glacier (see map 2.5). This till overlies till deposited by the Kansan glacier. The total depth of the two glacial tills average about 35 feet. The till plain is covered by 2 to 4 feet of highly erodible loess [2] and other sediments at the surface of the Illinoian till.

The primary floodplain soils of the Embarras River were formed in silty and loamy alluvium including the somewhat poorly drained Wakeland silt loam, the poorly drained Petrolia silty clay loam, the well-drained Landes fine sandy loam, and the well-drained Haymond silt loam [3]. Historically, all four alluvial soils in lands adjacent to the Embarras River frequently experienced seasonal flooding. Today, earthen levees built along the

FIGURE 9.1 Aerial view of ponded depressions and potholes on the uplands in central Iowa after heavy rain in June of 2008. Photo credit: Eddie Miller, Gilbert, Iowa.

Embarras protect many agricultural fields, and the soils are rarely flooded. However, prolonged periods of rain and saturated soils increase the runoff along the river and its tributaries, and extreme flooding can become a threat to this levee system. In 2008, two weeks of steady rainfall on top of saturated soils accelerated runoff and the pressure of the flooded Embarras began to weaken the levees. Farmer levees along oxbows in the river were particularly vulnerable.

2008 Flood on the Embarras River and Tributaries

The 2008 spring rains in the Wabash watershed delayed planting, drowned corn and soybean plants, and resulted in significant replanting. Corn and soybean planting in Illinois and Indiana was more than three weeks behind schedule by May 30, 2008, due to wet and cool weather conditions. Many of the soils remained near saturated conditions, and the soil could not store the additional 8 to 12 inches the area received (map 9.1). By mid-June local flooding was substantial, and levees broke on the Embarras (figure 9.3), White, and Wabash rivers. Thousands of acres of agricultural lands were impacted. Much of the 2008 corn crop planted by June 8, 2008, on floodplain soils was lost due to flooding, and many areas did not dry out sufficiently for crop planting until after July 15, 2008, making it too late to replant.

The areas that were not protected by levees and flooded received a thin layer of silt and clay, but the soils did not suffer permanent damage although the 2008 crop was lost. This was not the case where levees failed. Water removed hundreds of feet of the levee embankments (figure 9.4) and eroded thousands of cubic feet of soils and underlying outwash parent material to depths of 10 to 20 feet below the base of the earthen levee when the levees broke (figure 9.5). The force of the rushing water uprooted trees growing between the river and the levee and, after the breach, deposited them on the previously protected floodplain (figure 9.6). The 2008 crop on the floodplain soils behind the broken levees was a total loss, and permanent soil damage was great.

Levee breaches often occur where a levee is placed across a filled-in meander channel as a result of piping (see figure 8.7) and sand boils, which undermine the earthen levee. This situation happened at two levee breaks southeast of Sainte Marie, Illinois (map 9.2). About 300 feet of levee was lost at each break. Blowout holes or craters were created that were 1 to 3 acres in size and retained water. Three- to ten-foot-deep gullies extended a few hundred feet from the blowout holes, and hundreds of 65-foot high trees were transported

FIGURE 9.2 Rapid runoff after 5.25 inches of rain in 24 hours on May 30, 2008, caused a small headwater creek in central Iowa to flood and overflow its banks into surrounding pasture land.

FIGURE 9.3 Full temporary storage ponds surrounded by flooding from the Embarras River near Sainte Marie, Illinois, in June of 2008. Photo credit: Ken Flexter, Jasper County Natural Resources Conservation Service Field Office, Newton, Illinois.

FIGURE 9.4 The Embarras River cuts through a levee near Sainte Marie, Illinois, in June of 2008.

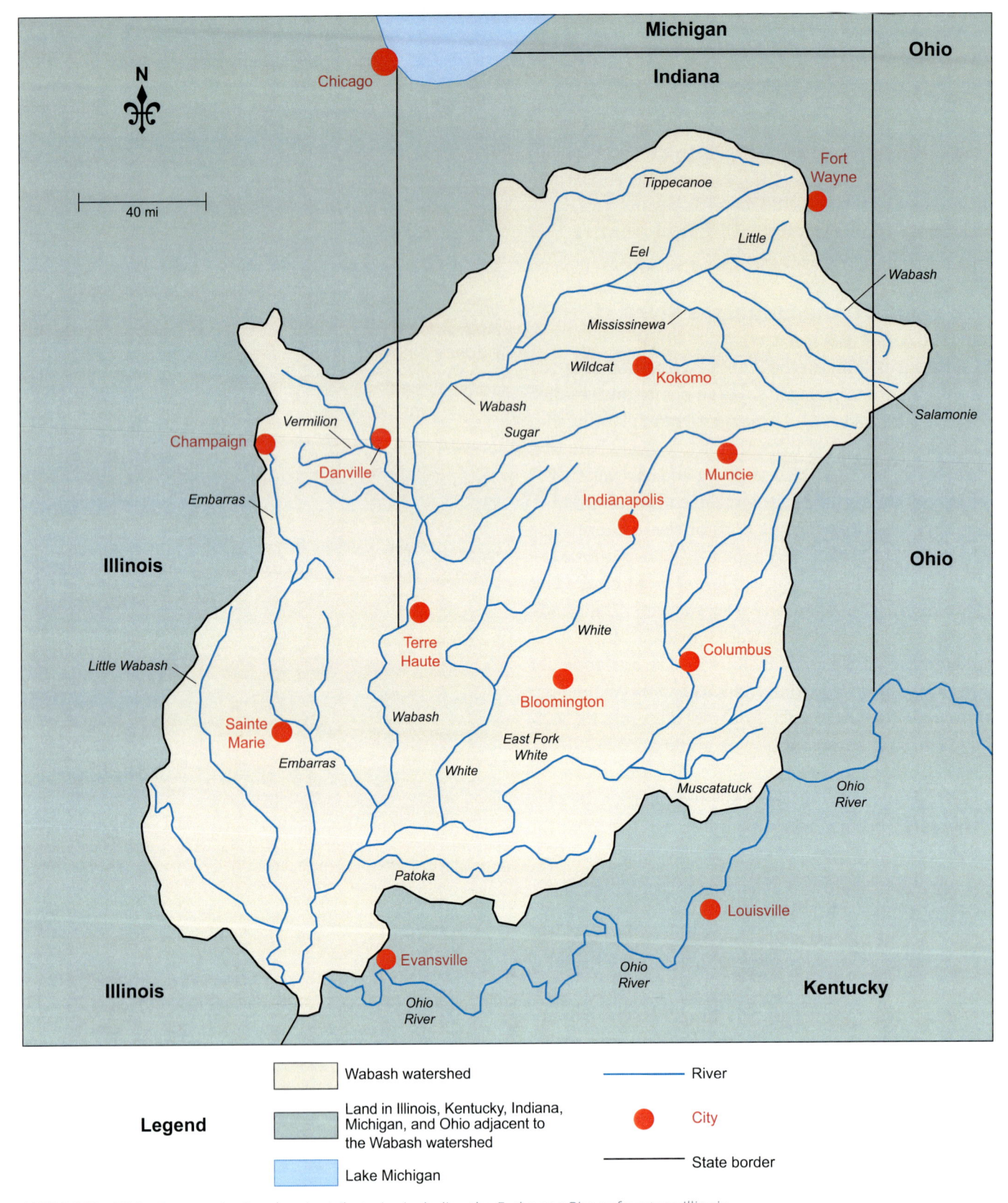

MAP 9.2 The Wabash watershed and major tributaries including the Embarras River of eastern Illinois.

hundreds of feet onto the previously protected floodplain. Deltaic sand deposits up to 1.6 feet thick covered 74 acres or more on the floodplain at each site, with an additional 198 acres covered with a few inches of sand. The remaining hundreds of acres of previously protected floodplain soils received a thin coating of silt and clay and remained under floodwater long enough to drown out the year's crop if it was planted.

Road and drainage ditches on the previously protected floodplain were also filled with sand more

FIGURE 9.5 In the background is the Embarras River, Illinois, adjacent to a missing 300-foot section of the levee and the crater lake that was created by rushing waters.

than 1.2 miles from the levee break. By June 23, 2008, the water had drained from the floodplains and back into the Embarras River sufficiently for the local farmers to hire contractors with bulldozers, pans, graders, backhoes, and buckets to begin the task of moving the trees from near the blowout holes and floodplains and to begin filling in the craters and gullies (figure 9.7). In addition, temporary levee embankments were constructed either around the blowout holes or across them to prevent any future flooding. The material for the temporary levee was obtained from the thick sand deposit beyond the blowout holes, transported to the edge of the blowout hole or crater, and then compacted by a bulldozer. Other equipment was used to scrape and pile the sand either for use in the temporary dam or to fill in parts of the craters and gullies. Tillage equipment such as chisel plows and moldboard plows were used on the areas with thin sand deposits (less than six inches) in an attempt to mix the sand into the underlying silty and clayey topsoil.

Post-Breach Sand Deposits

One might wonder where all the sand came from. Most of the surrounding Illinois soils are low in sand content. Most of the upland soils formed in loess that has less than 3% sand, and even the bottomland soils in the area have less than 15% sand. It is likely that the sand came from sandbars that may have developed in the river as the fine silts and clays were carried beyond where the sand dropped onto the riverbed. Also, this region is just south of the Wisconsinan terminal moraine that was also topped thousands of years ago by rising water moving rapidly, and sandy outwash would have deposited in the existing Embarras River valley. This underlying outwash is higher in sand and may have been used to create the levee embankment in the 1920s and 1930s. It is the parent material under the current alluvial soils at depths

FIGURE 9.6 Agricultural fields are covered by deltaic sand deposits, water, and trees moved and deposited by rushing water through a levee break on the Embarras River, Illinois, in June of 2008. Photo credit: Ken Flexter, Jasper County Natural Resources Conservation Service Field Office, Newton, Illinois.

FIGURE 9.7 Sand is piled for transport off the agricultural lands previously protected by a levee.

from 4 to 26 feet. The levee itself contained a higher sand content than the soils in the area, and significant sections of the levee were removed in addition to soil from the deep blowout hole. When the current Embarras River cut through the levee and dissipated sufficient energy to slow the water flow down, the sand dropped out and the remainder of the finer soil materials were carried further out into the valley floodplain.

Landowners often ask if soils in the floodplain are damaged by flooding, and if they are, how they can be restored. When water flows over alluvial soils not protected by a levee on the upper Mississippi and Ohio rivers, the rising water level does not normally cause significant soil erosion damage in the form of sheet, rill, or gully erosion. The unprotected alluvial soils often receive a thin layer of sediment, which can usually be mixed into the underlying topsoil with tillage equipment. However, when a levee fails, a several-acre blowout hole becomes a pond, resulting in the permanent loss of floodplain soils and agricultural land. Additional damage can be caused by thick sand deposits adjacent to the crater. These sand deposits bury the underlying soils with up to 1.6 feet of sand and create a delta which can cover 74 acres or more. This sand deposit has to be removed or the soils will remain too droughty for growing row crops in future years. The soils in areas that receive less than 6 inches of sand can often be mixed with the underlying silty and clayey soils and farmed in future years. Future crop yields may or may not be affected depending on the success of the mixing.

Effects of Conservation Practices on Flooded Soil

Some soil conservation practices can protect soils when they are flooded. During the last 30 years, soil and water conservation practices and structures have changed with soil erosion standards met using conservation tillage and no-till systems. In the Embarras watershed, most remaining terraces, contour farming, strip cropping, and waterways were effective. However, many waterways were filled above capacity and were eroded by fast-moving water or had significant sediment deposition. Culverts and other drainage structures were often plugged by soybean and corn residue (figure 9.8), primarily from no-till systems. No-till systems reduce raindrop impact and erosion, but once overland flow started on sloping lands, the residue was transported into the streams, blocked drainageways and structures, and resulted in local flooding of fields and even highways. Water storage structures, such as retention ponds, filled quickly with water and in some cases were covered by floodwaters. Risers and tile outlets were often insufficient to drain crop areas within 24 or 48 hours, resulting in significant numbers of corn and soybean plants lost. Some areas were eventually replanted to corn or soybeans.

Soil Drainage

Watersheds with a high slope gradient have even greater runoff potential than flatter lowlands. The hydrologic soil groupings in some watersheds also affect the runoff rate as does the type of vegetation and crops planted. The crop rotation in Illinois and Indiana is up to 90% corn and soybeans, with limited acreage in small grains and forages. Further, urban and highway development in floodplains within the Mississippi and Ohio river watersheds contribute to flooding problems. In 2008, drainage systems in the upland designed to remove excess water to open outlets in 24 hours reduced crop plant loss but contributed to higher flooding levels on floodplain soils.

It is important to separate watersheds with well-drained soils and high-slope gradients from watersheds

FIGURE 9.8 Corn stalks clog drainageways, waterways, and culverts after flood events.

that are relatively flat with poorly drained soils to understand how to manage flooding impacts on soils. In flat watersheds with poorly drained soils, such as the Embarras River watershed (Illinois), most soil and water conservation practices only include waterways, conservation tillage, and no-till systems. Historically, the biggest management problem was soil drainage. Many of the soils were too wet to grow row crops. This was addressed with the 1879 Illinois drainage law, which permitted digging drainage ditches, channelization, surface drainage, and subsurface tile drainage with the goal of removing surface ponding within 24 hours to prevent corn plants (soybeans had not yet been introduced) from drowning and to permit cropping. The only measures taken to reduce flooding in wet years were to build private levees near the banks of the streams and rivers, which reduced the width of the floodplains but protected, in most years, the adjacent agricultural cropland on the floodplain soils from crop loss. Thus, significant crop loss and damage to floodplain soils only occurs when levees break.

In more sloping watersheds, such as the Iowa and Cedar river watersheds (Iowa) or similar ones in Missouri, southern Indiana, and southern Illinois, most of the soil and water conservation practices have focused on soil erosion control and have included terraces, waterways, contour farming, strip cropping, grass waterways, and upland water storage dams and settling basins [4]. Soil drainage is not required for most well-drained soils in the upland, and drainage systems (drainage tile and waterways) have been used to safely remove water from the upland without soil erosion. Terraces usually drain to tile outlets or waterways. In sloping watersheds, heavy rains create both soil erosion and flooding problems.

There are a number of potential solutions to reduce the flooding impact on agricultural lands in flat watersheds with poorly drained soils: (1) slow the runoff and land drainage rate; (2) temporarily store more water in the uplands; (3) change the upland crop rotation to include more forages rather than row crops; (4) convert additional agricultural land to pasture, timber, or forages so the soils can infiltrate, use, or store more water; (5) build stronger and higher levees on the floodplains but further away from the streams and rivers to widen the stream or river channel; and (6) stop farming the floodplain soils and change to land uses that do not sustain damage during periodic flooding.

Leveed Agricultural Lands

Floodwaters in 2008 on floodplains without levees resulted in 100% crop loss, and these soils received thin silt, sand, or clay deposition, which could be mixed with tillage equipment into the topsoil prior to planting the 2009 crops. Floodplains with levees that did not fail had little 2008 crop lost except where tributary streams ponded water behind the levees. However, floodplains with levees that broke lost both the 2008 crops and agricultural land. Lands adjacent to levee breaks were subject to rushing water that often created blowout holes or craters, resulting in total loss of soil profiles as the area became a pond. Thick sand deposits often tens of acres in size occurred adjacent to the blowout holes in the form of sand deltas. These deposits resulted in up to 100% crop loss in 2008 and could lower future yields of buried alluvial soils.

Flooding severity depends on geography, the climate, and soils where the watershed is located. In southern Wisconsin or Iowa, the 2008 flood was the worst flooding in 500 years; in the Embarras River watershed (Illinois), it was the worst flooding in 100 years; and near St. Louis, Missouri, on the Mississippi River, it was the most severe flooding in past 15 years. The National Oceanic and Atmospheric Administration (NOAA) weather map (map 9.1) of the May 30 to June 14, 2008, time period clearly shows that where 10 inches of rain fell on already saturated soils, the result in the upper Mississippi River basin and tributaries was flooding, crop loss, and erosional damage to levees and adjacent agricultural bottomlands. However, as these waters moved southward during the month of July, it was a rather uneventful period for the lower Mississippi River south of confluence of the Ohio and Mississippi rivers near Cairo, Illinois.

---

[1] Leighton, M.M., G.E. Ekblaw, and L. Horberg. 1948. Physiographic divisions of Illinois. Report of Invest. No. 129. Champaign, IL: Illinois Geological Survey.

[2] Bramstedt, M.W. 1992. Soil Survey of Jasper County, Illinois. Washington, DC: USDA Soil Conservation Service.

[3] Fehrenbacher, J.B., K.R. Olson, and I.J. Jansen. 1986. Loess thickness in Illinois. Soil Science 141:423-431.

[4] Lowery, B., C. Cox, D. Lemke, P. Nowak, K.R. Olson, and J. Strock. 2009. The 2008 Midwest flooding impact on soil erosion and water quality: Implications for soil erosion control practices. Journal of Soil and Water Conservation 64(6):166A, doi:10.2489/jswc.64.6.166A.

Impacts of 2011 Induced Levee Breaches on Agricultural Lands of the Mississippi River Valley

To control the Mississippi River is a mighty task [1], and the 2011 flooding of its alluvial valley was a reminder of just how difficult this task can be. Heavy snowmelt and rainfall 10 times greater than average across the eastern half of the 200,000-square mile Mississippi watershed in spring and early summer of 2011 produced one of the most powerful floods in the river's known history [2]. The water from the Mississippi and Ohio rivers arrived in the Cairo, Illinois, area (map 10.1) at about the same time, straining the levees and floodwall system designed to confine the rivers and protect cities and farmlands.

The deliberate breaching of the levees in the New Madrid Floodway below Cairo in May of 2011 was a planned strategy to reduce water pressure and prevent levee failures where harm to human life might occur. The induced breach and the flooding of 133,000 acres of Missouri farmland resulted in the loss of 2011 crops and damage to future soil productivity. The strong current and sweep of water through the Missouri Birds Point levee created deep gullies; displaced tons of soil; and damaged irrigation equipment, farms, and home buildings.

The starting point of the lower Mississippi River is the confluence of the Mississippi and the Ohio rivers at Cairo, Illinois, 279 feet above sea level. In the aftermath of the deadly 1927 flood, the US Army Corps of Engineers (USACE) designed the New Madrid Floodway project as part of a larger Mississippi River basin plan to manage the river and control flooding when the Ohio and Mississippi rivers converge and threaten to overflow the frontline levees and floodwalls that contain these rivers. The New Madrid Floodway (map 10.1) and border levees (figure 10.1) were built under the authority of the Flood Control Act of 1928. The floodway is approximately 33 miles long and is between 4 and 10 miles wide. The floodway is enclosed by frontline and setback levees, except for a 1,500-foot gap at the lower end that serves as a drainage outlet and allows flood backwaters to enter the Mississippi River. The frontline levee includes an 11-mile-long upper fuse plug section, a 5-mile-long lower fuse plug section, and a 16-mile-long frontline levee section connecting the two plugs. The frontline levee, which forms the eastern boundary of the floodway, was constructed to protect the floodway until the Mississippi River reached the 55-foot stage, at which time the floodwater would naturally

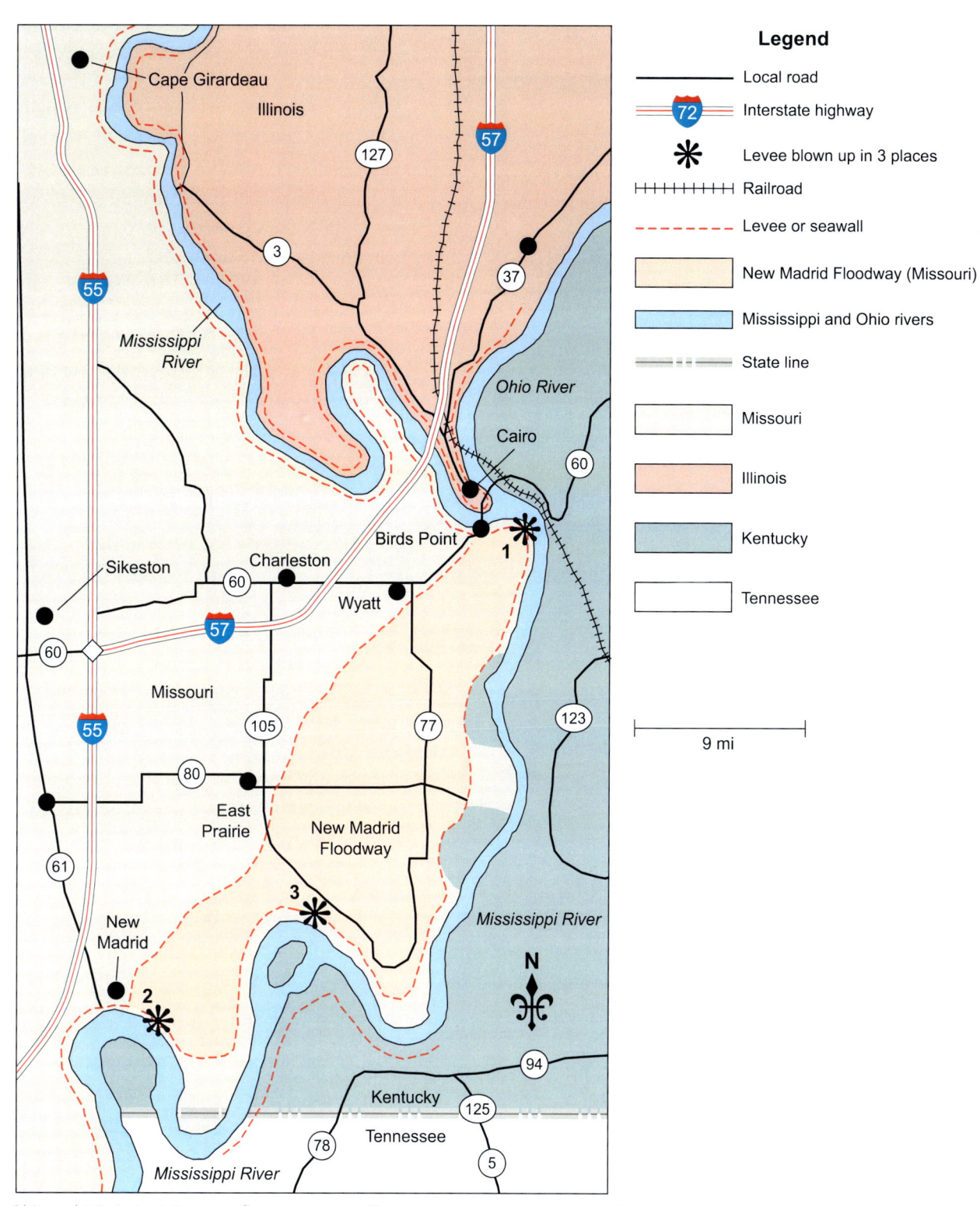

MAP 10.1 The Ohio and Mississippi rivers confluence at Cairo, Illinois, to become the Mississippi River. The floodplain to the west of the Mississippi River as it flows south from Cairo to New Madrid, Missouri, is a natural overflow spillway or floodway to store excess water when the rivers are above flood stage.

overtop the frontline levee. The setback levee is 36 miles long and runs from Birds Point, Missouri, to New Madrid, Missouri (map 10.1). The USACE obtained easements in 1928 from the landowners of the 133,000 acres of the New Madrid Floodway, which gave them the right to pass floodwater into and through the New Madrid Floodway and temporarily store floodwater in the basin.

Congress authorized modification of the New Madrid Floodway operational plan in the Flood Control Act of 1965 (Public Law 89-298). The plan called for raising the upper and lower fuse plugs to 60.5 feet, the frontline levee to 62.5 feet, and the setback levee to 65.5 feet. The floodway plan also called for artificial crevassing (breaking or breaching) of the levee by means of

FIGURE 10.1 The New Madrid Floodway and basin still had four to six feet of floodwater on May 20, 2011, 18 days after the US Army Corps of Engineers induced a breach at Birds Point, Missouri.

explosives on only the upper fuse plug section of the frontline levee when river stages were at or above 60 feet on the Cairo gage. The USACE could activate (load with TNT) the New Madrid levee fuse plug site when the weather forecast predicted a peak of 60 feet or more. The plan envisioned that natural breaching might occur in the lower fuse plug section in the event of a flood of that magnitude (such as the flood of 1937). After approval of the 1966 plan, the levee system's construction was significantly improved by raising and strengthening the frontline levee. The 1965 Flood Control Act gave the USACE the authority to pass floodwaters through the New Madrid Floodway. The USACE, armed with the power of eminent domain, obtained modified flowage easements within the floodway lands to permit the artificial breaching. The Birds Point upper fuse plug levee was designed to be blown up with explosives in the event of a great flood. No such event happened after 1966, not even during the floods of 1975 and 1979.

The USACE adopted a modified plan in 1983, calling for the artificial breaching of both the upper (near Birds Point) and lower (near New Madrid) fuse plugs and the middle section of the frontline levee south of Big Oak Tree State Park, which are in Mississippi and New Madrid counties, Missouri. Landowners in the New Madrid Floodway filed suit against the USACE (Story vs. Marsh) in opposition to the 1983 artificial breaching plan of the frontline levee between the two fuse plug sections near Big Oak Tree State Park. The Eastern District Court of Missouri ruled in favor of the landowners and enjoined the Secretary of US Army and subordinates from artificially crevassing (breaching) the frontline levee. The district court determined that the 1983 USACE new plan was not authorized by Congress and instead relied on the 1965 Flood Control Act for the authority to operate and maintain the floodway. The district court ruling denied condemnation, and immediate possession was reversed. Further the court required that the federal government escrow anticipated levee repairs with a deposit of $6,400,000 for Levee District Number 3 and $4,000,000 for St. John's Levee and Drainage District. This "just compensation" estimate was based on the amount determined necessary to restore the levees and the operation of the floodway.

However, on April 13, 1984, the US Court of Appeals for the Eighth Circuit Court reversed all of the district court rulings and permitted the USACE to artificially crevasse (breach) the frontline and fuse plug levees and negated the need for the escrow deposit. The Eighth Circuit Court determined that the 1965 Flood Control Act gave the authority to the USACE to operate the New Madrid Floodway and as such already required them to restore the levees after any artificial breaching events. This case law became the legal precedent used by the federal court system in 2011 to decide a last-minute appeal by the Missouri attorney general on behalf of floodway landowners to prevent the USACE from blowing up the frontline levee once the Cairo gage floodwater forecast peak of 60 feet or higher was issued. Such a National Oceanic and Atmospheric Administration (NOAA) forecast was issued April 28, 2011, but the seven-member Mississippi River Commission (MRC) chaired by Major General Walsh, delayed, for many reasons, the decision to open the floodway. The drama surrounding this MRC decision-making process and the timing of the decision was later detailed in *Divine Providence: The 2011 Flood in the Mississippi River and Tributaries Project* by Charles Camillo [3].

Consequently, floodwaters continued to rise through April and until May 1, 2011 (a Sunday night), when the Supreme Court affirmed, by not accepting the Missouri Attorney General's lawsuit on behalf of the citizens of Missouri, the USACE right to activate the New Madrid Floodway fuse plug after a weather forecast of a 60 foot or higher peak on the Cairo gage. The New Madrid Floodway fuse plug could then be opened any time after the Ohio River peak reached 60 feet. In the middle of a heavy rainstorm on May 2, 2011, the USACE loaded 265 tons of TNT from boats into the Birds Point fuse plug in the frontline levee. The loading of the TNT took 15 hours and was life-threating work since it was done in the middle of a 7.5-inch rainstorm while working on 2-foot-high slippery soil platforms surrounding each entry pipe as the floodwaters were already starting to top the 2-mile-long, 60.5-foot fuse plug levee. A 1.2-mile section of the Birds Point fuse plug was blown up simultaneously at six points at 10:00 PM on May 2, 2011, and one-fourth of the lower Mississippi River water passed into and through the New Madrid Floodway and onto bottomland soils (Caruthersville very fine sandy loam, Commerce silty clay loam, Dundee silt loam, and Forestdale silt loam) [4, 5]. The amount of temporary water storage (at initial depths of about 3 to 4 feet) and pass-through water in the New Madrid Floodway was 25 to 28 times greater than what could have been stored in the Cairo and Future City, Illinois, areas and adjacent agricultural areas of Illinois if the Cairo and Future City floodwalls or levee system were naturally breached. The impacts of the 2011 Mississippi and Ohio rivers' flooding on Cairo, Illinois, are detailed in chapters 16 and 17.

The induced levee breach at Birds Point, Missouri, the first explosion site (map 10.1), and flooding of the New Madrid Floodway and basin on May 2, 2011, resulted in no loss of life thanks in part to the US National Guard sweep of the area to make sure the people living and working in the floodway were evacuated. The force and impact of the floodwater on the 208 square miles of the floodway may have been greater than projected in the 1983 USACE operational plan. The floodway was subjected to rapidly moving floodwaters through the 1.2-mile-wide breach, and on May 2 the waters were 1.7 feet higher (13 feet vs. 11.3 feet) than if the fuse plug had been activated on the peak forecast date, April 28. There was severe land scouring damage near all breaches and on O'Bryan Ridge. Property damages included most of the 200 buildings, including homes, which were exposed to the rapidly advancing floodwaters and then partially submerged in 4 feet of ponded floodwater. Water pressure caused loss of the lower one-third to one-half of entire building walls, damaged most wooden floors, and in some cases completely destroyed structures.

On May 3, 2011, a second explosion site (map 10.1) in the levee near New Madrid, Missouri, blew up the lower plug to accelerate the return of the stored floodwater back into the Mississippi River. However, the Mississippi River was still too high to allow a quick drop in the floodwater in the New Madrid Floodway. The USACE planned to blow up the frontline levee near Big Oak Tree State Park in Missouri, the third explosion site (map 10.1) on May 4, 2011, but weather and shortage of TNT delayed the third explosion until May 5, 2011. It appears that for the next few days the Mississippi River floodwater at the third breached levee site may or may not have flowed out before it flowed back in since the Mississippi River remained above flood stage. Evidence of the inward flow includes a crater lake adjacent to the west side of the frontline levee breach and a sand deposit [7]. Local farmers reported this, but it was not confirmed by the USACE that the Mississippi River water flowed out initially on May 5, 2011, but later started to flow rapidly in through this third frontline levee breach, creating a 16-acre lake, gullies, and a thick 250-acre sand deltaic deposit. Over the next few weeks, the New Madrid Floodway continued to drain through the New Madrid levee breach, and the floodwater in the floodway basin dropped to 2 to 6 feet depending on the elevation of bottomland soils. However, 3 to 6 feet of floodwater still remained and covered cropland.

Flooded Agricultural Lands

From an agricultural production point of view, the 133,000 acres of cropland in the New Madrid Floodway were only partially planted to corn and soybeans due to a very wet April. However, there was extensive acreage (20,000 to 30,000 acres) of wheat planted in the fall of 2010 that was 2 feet tall but had not yet reached grain fill stage when the floodway was opened, flooding agricultural fields with 6 to 8 feet of water. There were still 3 to 5 feet of water on much of the farmland as of May 20, and it took a drop in the Mississippi River to permit the draining of the remaining floodwater. By June 6, approximately 90,000 acres had no remaining floodwater on the surface and had dried sufficiently to begin the soybean planting. About 10,000 to 20,000 acres were still covered with water at the south end of the floodway near New Madrid, Missouri, and about 10,000 acres were covered near Big Oak Tree State Park, Missouri, on June 6. By June 15, more than 30 excavators were working to clean out sediment and to remove debris that were blocking drainageways and ditches to speed up the drainage process (figure 10.2). The sediment in private drainageways of most qualified landowners was removed with a 75% cost share from the USDA Farm Services Agency's Emergency Conservation program. It took until October of 2011 for the last 20,000 to 30,000 acres to dry out sufficiently to allow tractor traffic and planting of wheat in the fall of 2011.

Many farmers in the central part of the New Madrid Floodway were able to begin planting soybeans after June 6 on areas with loamy and silty soils (Caruthersville very fine sandy loam, Commerce silty clay loam, and Dundee silt loam) [4]. These soils had good internal drainage and were on higher areas with less initial depth of floodwaters. As many as 30,000 acres of soybeans were planted by June 15, and another 60,000 acres of soybeans were planted before July 15 (last date for planting soybeans in the area) as ponded areas and lakes shrank and more soils dried out. As anticipated, the approximately 30,000 acres that were under water on June 5 remained too wet to plant to soybeans by July 15, and at least half of these areas were planted to wheat in the fall of 2011. Another 15,000 acres on low-lying areas with clayey soils (Sharkey silty clay loam and Alligator silty clay) [4] with low permeability were not planted to either corn or soybeans until the spring of 2012. These soils are not well suited to wheat production, so the crop production loss in 2011 was 20,000 to 30,000 acres of wheat planted in the fall of 2010 and drowned in early May of 2011 and the additional 15,000 acres of low-lying clay soils that remained ponded until late in the summer of 2011.

Almost all of the 2010 winter wheat drowned before filling with grain, and the wheat fields collected more sediment (3.9 to 6 inches) than the fields with corn and soybean residue from the previous year's crop (less than 2 inches). Most of the 2011 soybeans were planted after chisel plowing or, in some cases, using a no-till planting system. Much of the floodway agricultural lands without winter wheat planted in the fall of 2011 were planted to corn and soybeans in the spring of 2012. Some of these areas were double cropped in 2012, with soybeans planted and harvested and then wheat planted in the fall. It appears that about one-fourth of the entire 133,000 acres of highly productive agricultural land was out of production for an entire year, including some areas where the wheat crop drowned and clayey soils that were too wet to plant to soybeans by July 15 or not suited to wheat production. Consequently, there was an adverse effect on farm incomes and some wheat and corn supplies. For insurance purposes, the human-induced flooding was treated as natural flooding (federal declaration) or the same as if the levee breach occurred naturally.

Gully Fields and Sediment Transport

Significant crater lakes existed in June of 2011 at two locations where the explosives were used (Birds Point, Missouri, fuse plug and the frontline levee near Big Oak Tree State Park, Missouri). The extreme force of the rushing water widened the holes in the levees (1.24 miles) at the Birds Point fuse plug and created six crater lakes adjacent to the levee breach. Each Birds Point crater lake was approximately 1.3 acres in size. At a few Birds Point sites, there were gullies extending into the fields from the crater lakes and subsequent creation of a deltaic sand deposit that was 1 to 4.6 feet thick and approximately 7.4 to 49.4 acres in size. Both the crater lakes and the thick sand deposits resulted in a permanent loss of agricultural land until the craters were filled in 2012 and the sand removed in 2013 [7]. The frontline levee near Big Oak Tree State Park had only one 16-acre crater lake extending through the levee and to the west (away from Mississippi River) and a thick, 3.3- to 6.6-foot deltaic sand deposit approximately 74 acres in size with additional sand deposits between 1 to 3.3 feet and more than 173 acres in size. This thick deltaic sand deposit was removed to permit the land to be returned to agricultural use. The road and drainage ditches along Highway 77 in Mississippi County approximately 2.5 to 3.7 miles south of Wyatt, Missouri, were filled with 3 to 6 feet of sediment and sand. Figure 10.2 shows an excavator removing about 2 feet of sediment from the road ditch. Irrigation systems were occasionally overturned by the force of water and wind and then buried by deltaic sand deposits (figure 10.3).

When the thin, organic silt and clay coatings were mixed into the topsoil in 2011 or 2012, little significant loss in future crop yield occurred. By early June, the floodway water had dropped enough for a USDA Natural Resources Conservation Service (NRCS)–led survey team to start determining the extent of the damage to the agricultural land. Preliminary findings suggest that considerable sediment deposition (few inches) occurred in drainage ditches and on the wheat fields. In addition, there were a few hundred acres of cropland with deep gullies (figure 10.4) created on ridges that were land scoured (eroded) as the rapidly moving initial water flowed an extra 1.7 feet. Huge fields with gullies (figure 10.5) were found on higher natural levee or second bottom Dundee silt loam soils [4] or on ridges (such as O'Bryan Ridge south of Wyatt, Missouri) when floodwa-

FIGURE 10.2 Excavator cleaning out sediment in road ditches. Photo credit: Brett Miller, USDA Natural Resources Conservation Service soil conservation technician.

FIGURE 10.3 An irrigation system was overturned by the force of floodwater and wind and then buried by a thick sand deposit. Photo credit: Brett Miller, USDA Natural Resources Conservation Service soil conservation technician.

ters flowed rapidly over the higher flat land surface and then dropped back to lower bottomland soils to the west and south. This rapid dropping of floodwater in existing drainageways created turbulence and eroded the higher Dundee silt loam soils. Once an erosional channel was created, the channel concentrated the water, and up to 0.6-mile-long gullies (channels) were created. These gully fields were primarily located 5 to 10 miles to the south and west of the Birds Point levee breach and were not connected to the crater lakes next to the levee breaches. Some of the gullies were 12 feet deep, 150 feet wide, and 0.6 miles long (figure 10.5). Less than three-fourths of each gully field was planted to soybeans in June of 2011. The force of water, which created these large gullies (approaching small canyons in size), also removed road beds and resulted in the washout of a bridge.

There was a health concern related to any pollutants that might be in floodwaters, such as untreated sewage from plants that were flooded or other chemicals picked up by floodwaters. When the organic and clay particles coating the plants and soils dried, it is likely that sunlight killed any pathogens that were present. Tillage was used to bury or mix this potentially toxic coating into the topsoil layer, which will treat or dilute any toxic chemicals present. It is not known whether the soil organic carbon content of the alluvial soils was permanently increased as a result of carbon-rich sediment deposition and exposure to carbon-rich floodwater. Microbes will decompose carbon deposited with the sediment or in the thin surface coating and release the carbon to the atmosphere as either carbon dioxide or methane gases, depending on whether there are aerobic or anaerobic conditions at the time the microbes are active.

Implications of the Induced Levee Breach

The decision to blow up the frontline levees of the New Madrid levee system (figure 10.6) in three places had significant consequences for rural Missouri landowners, farmers, and residents in the New Madrid Floodway. The impact of the floodwaters on the floodway appears to have

FIGURE 10.4 Gullies extended into cropland. Soybean stems and roots can be seen between the gullies. The gullies were created during the use of the floodway in May of 2011.

FIGURE 10.5 Gullies and channels were created from May 4 to May 16, 2011. The main channel and attached gullies were 12 feet deep and removed cropland from production.

FIGURE 10.6 A crater lake was formed at the Big Oak Tree frontline levee when the fuse plug was blown and floodwaters rushed into the floodway.

been greater than anticipated in part as a result of the delay in opening the floodway. When the USACE opened the floodway, the Mississippi River was 1.7 feet higher than planned for, and the initial additional force and depth of floodwater caused more damage to buildings and more deep land scouring, such as on O'Bryan Ridge, than was probably anticipated. Impacts included the loss of the 2011 wheat crop (20,000 to 30,000 acres) and of crop production from perhaps 20,000 to 30,000 acres of poorly drained clayey soils that were not replanted to soybeans in 2011. Most of the remaining ponded farmland in the floodway dried out sufficiently to permit 2011 fall planting of wheat. The floodway was dry enough by the spring of 2012 to allow the planting of corn and soybeans. It is not clear how much of 2011 farm income replacement came from flood insurance since not all floodway farmers had crop insurance. Over 2.5 million acres of agricultural bottomlands in Missouri and Arkansas were protected by hundreds of miles of levees on the west side of the Mississippi River between Commerce and New Madrid including the floodway setback levee. These levees did not fail before, during, or after the use of the floodway, and the 2011 agricultural production from this region was maintained. There was a one- to three-year loss of some agricultural production on hundreds of acres of land due to large gully fields on those parcels adjacent to the blown levees and on farmland where the new crater lakes occurred.

---

[1] Barry, J.M. 1997. Rising Tide: The Great Mississippi Flood of 1927 and How It Changed America. New York: Simon and Schuster.

[2] USACE (US Army Corps of Engineers). 2011. Great Flood of '11. Our Mississippi. Rock Island, IL: US Army Corps of Engineers. http://www.mvs.usace.army.mil/Our%20Mississippi/ourmississippi_su11_lowres.pdf.

[3] Camillo, C.A. 2012. Divine Providence: The 2011 Flood in Mississippi River and Tributaries Project. Vicksburg, MS: The Mississippi River Commission.

[4] Brown, B.L. 1977. Soil Survey of New Madrid County, Missouri. Washington, DC: USDA Natural Resource Conservation Service.

[5] DeYoung, W. 1924. Soil Survey of Mississippi County, Missouri. Washington, DC: USDA Bureau of Soils.

[6] Olson, K.R., and L.W. Morton. 2012. The effects of 2011 Ohio and Mississippi River valley flooding on Cairo, Illinois, area. Journal of Soil and Water Conservation 67(2):42A-46A, doi:10.2489/jswc.67.2.42A.

[7] Olson, K.R. 2009. Impacts of 2008 flooding on agricultural lands in Illinois, Missouri, and Indiana. Journal of Soil and Water Conservation 64(6):167A-171A, doi:10.2489/jswc.64.6.167A.

Repair of the 2011 Flood-Damaged Birds Point–New Madrid Floodway

During the spring 2011 flooding along the Mississippi River, the strong current and sweep of water through the Birds Point, Missouri, levee breach in May of 2011 created hundreds of acres of deep gullies; scoured hundreds of acres of land; eroded tons of soil; filled ditches with sediment, which blocked drainage; created sand deltas; and damaged irrigation equipment, farm buildings, and homes. Reclamation and restoration of these agricultural lands following the US Army Corps of Engineers (USACE) opening of the New Madrid Floodway to relieve flood pressure on the levee system from the Mississippi River [1, 2, 3] was time consuming and costly to individual landowners and public tax dollars. While levees were rebuilt, ditches cleared of sediment, and many lands in the floodway restored by November of 2012, soil productivity and growing conditions continued to challenge the farmers of this historically highly productive region.

The USACE decision to blow up Birds Point levee along the Lower Mississippi River and flood agricultural lands in the New Madrid Floodway of Missouri (map 11.1) was difficult, with substantive legal challenges as well as social and political trade-offs between human life and property in urban and rural areas [1]. The induced breach and flooding of 133,000 acres of Missouri farmland resulted in partial 2011 crop loss and permanent soil damage because of land scouring, gully fields, and crater lake areas [3]. With the opening of the floodway, 13 feet of floodwater poured through the 1.2-mile hole in the breached levee and onto the floodway bottomlands. There was severe damage to most of the 200 building structures, including 80 homes, that were exposed to flowing floodwater. These buildings were damaged when they were partially submerged in 3 to 6 feet of temporarily stored floodwater [4].

Patching of Fuse Plugs and Reconstruction of the Frontline Levee

In the fall of 2011, the USACE began reconstructing the Birds Point and New Madrid fuse plugs and frontline levee near the Big Oak Tree State Park (figure 11.1). Initially, the fuse plugs at blast sites 1 and 2 (map 11.1), which were at 60.5 feet prior to the explosions, were rebuilt to only 51 feet (figure 11.2) because of insufficient federal funds to repair the levees [5]. The lower levee height was of great concern to floodway landowners who thought their lands were at considerable future

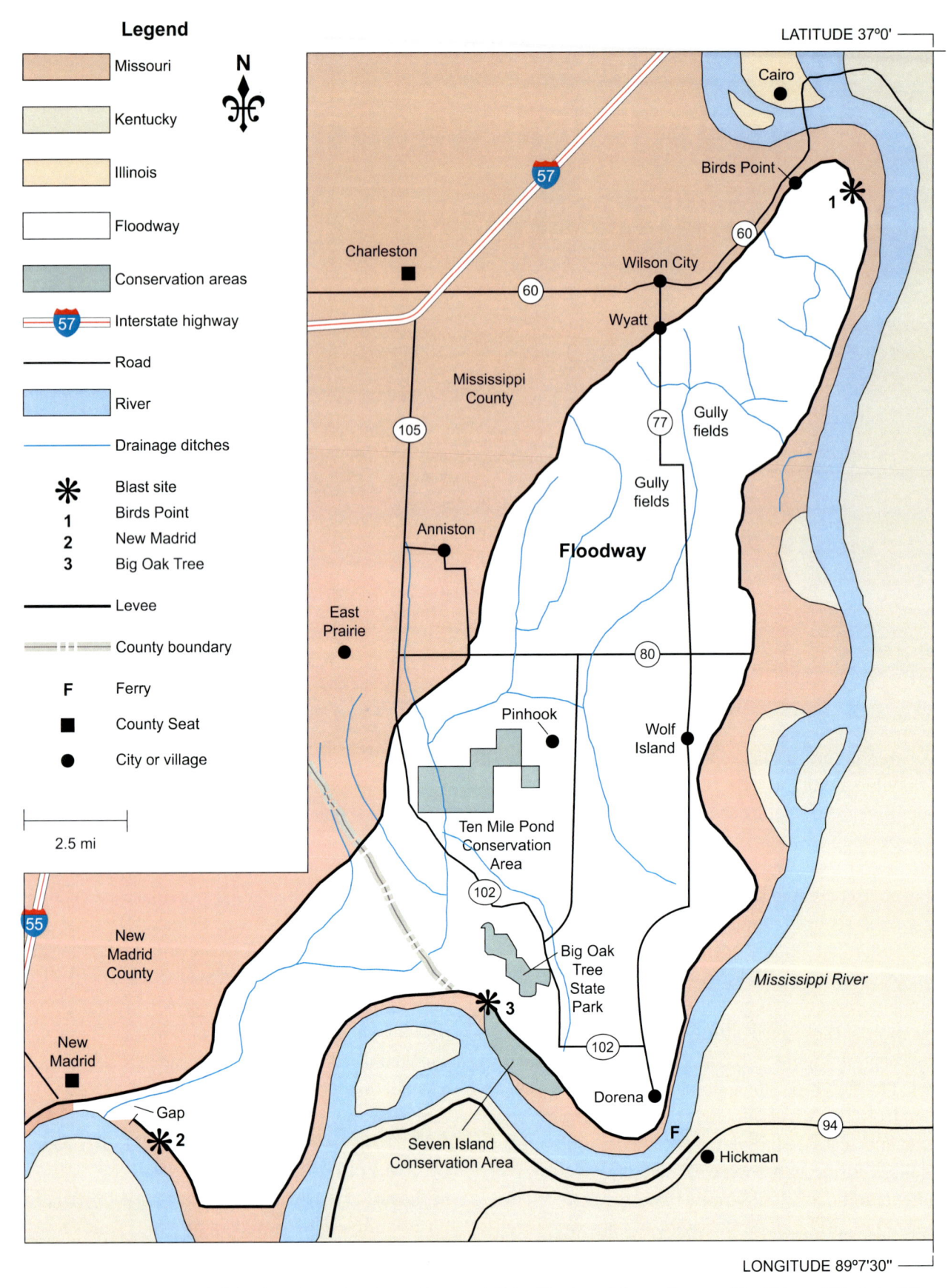

MAP 11.1 The Birds Point–New Madrid Floodway located in Missouri just south Cairo, Illinois, and the confluence of the Ohio and Mississippi Rivers.

risk. When they met with the USACE in the early fall of 2011, the landowners strongly demanded the levee be restored to the original levee height of 60.5 feet (fuse plug) or 62.5 feet (frontline levee) and noted that the proposed 51 feet height would result in flooding every other year based on historic flooding records [6]. Later,

84

FIGURE 11.1 The Big Oak Tree levee blast site was repaired in October of 2011 and reconnected to the 62.5-foot frontline levee.

FIGURE 11.2 The 2011 Birds Point levee patch built to 51 feet rather than the original 60.5 feet.

the USACE obtained additional funds from Congress to rebuild the levee patches (figure 11.2). Since 1884, the Mississippi River flooding exceeded the 55 feet only seven times including the record floods of 1927, 1937, and 2011 [6]. According to Michael Ward, assistant professor of animal ecology and conservation in the Department of Natural Resources and Environmental Sciences at the University of Illinois, the Big Oak Tree frontline levee repair was delayed, until after nesting birds (least tern [*Sternula antillarum*], a state endangered species) left the area in the fall of 2011. The frontline levee near Big Oak Tree State Park (figure 11.3) was then rebuilt to 55 feet in October and November of 2011 to prevent potential flooding in spring of 2012. The top of the frontline levee patch was covered with canvas since it was too late to establish a vegetative cover (figure 11.3).

In March of 2012, the USACE received $22 million to raise the patched floodway levees from 55 to 60.5 feet at the fuse plugs locations and to 62.5 feet at the frontline levee. On May 30, 2012, the USACE released contracts to private contractors to begin raising the three levee patches to original height. A protest letter from a private contractor, who lost the bid due to paperwork problems, delayed the start of the repairs at Birds Point for about seven weeks. Thanks to drier than normal weather, much of the levee work was completed by November of 2012 [7]. The USACE filled in the crater lakes and reestablished washed-out roads at the base of the levee, reconnecting levee roads to the west and south (figure 11.3).

Crater Lake Filling and Soil Replacement

The six crater lakes at the Birds Point fuse plug levee extended into agricultural lands a total of 7.8 acres. They were filled in by the USACE and covered with trucked-in topsoil in October of 2011 to ready the fields for either winter wheat (*Triticum aestivum* L.) in the fall of 2011 or soybeans (*Glycine max* [L.] Merr.) and corn (*Zea mays* L.) in the spring of 2012. There were also deltaic sand deposits 1 to 4.6 feet thick and approximately 7.4 to 49.4 acres in size that needed to be addressed to prevent a permanent loss of what was previously finer textured agricultural land [8, 9]. This reconstructed soil profile will still be less productive than the original soils. Much of the topsoil came from other levees on smaller tributaries and drainage ditches. The hole in the frontline levee, created on May 5, 2011, by the third explosion (map 11.1) let the record-high Mississippi River flow back into the New Madrid Floodway. The inflow created an expanded 5-acre crater lake and a 67-acre sand delta to the west in the floodway on top of prime farmlands near Big Oak Tree State Park (see figure 10.6). The crater lake was partially filled in as the levee was rebuilt to 55 feet in November of 2011. It still remained as a slight depression in May of 2012 and was left idle in the 2012 growing season. Part of the sand deposit was removed in October and November of 2011 and spread on the local lanes and roads. The remaining piles of sand were later sold and hauled away.

Crop and Economic Loss

The total 2011 crop loss included 20,000 to 30,000 acres of low-lying clayey soils [10, 11] (figure 11.4) and 20,000 to 30,000 acres of drowned winter wheat (figure 11.5), which either remained too wet to plant to soybean in 2011 or, if planted to wheat, did not result in a harvested crop in 2011. About 1,500 acres of agricultural lands, near the blown-up levees at Birds Point and Big Oak Tree, had crater lakes and gullies and were not farmed in 2011 since the craters were only partially filled in and the thick sand deposit at the Big Oak Tree breach was only

FIGURE 11.3 The site of the third explosion on the frontline levee is located near Big Oak Tree State Park. The levee is being reconstructed where the crater lake extended into agriculture lands.

FIGURE 11.4 Low-lying soils remained wet on October 24, 2011, more than five months after the floodway opening and could not be cropped in 2011.

partially removed. These areas were not reclaimed in time for spring of 2012 planting. The USDA Risk Management Agency issued a ruling that the induced flooding of the New Madrid Floodway was caused by nature (the heavy rains in Ohio River valley), and the flood damages were covered by claims to crop indemnity insurance as if they had occurred as a result of a natural flooding event. USDA Risk Management Agency 2011 payouts to landowners in Mississippi and New Madrid counties (Missouri) for excess moisture, precipitation, rain, and flooding totaled almost $16.2 million with almost $9.7 million paid for crop damage prior to the May 2011 levee breach [12]. Sixty percent ($8.7 million) of the payout was for loses to the 2011 corn crop; 23.5% or almost $3.4 million was for wheat loses; and 16.5% or almost $4.4 million was for soybean loses [12]. These payouts were only a portion of the total crop losses and restoration costs associated with the 2011 flooding event. The crop damage estimates by landowners in the floodway have been reported to be as much as $42.6 million to $60.6 million in 2011 [13], which were partially covered by $16 million in crop insurance payments.

Sediment Deposits on Land and in Drainageways and Ditches

The 20,000 to 30,000 acres of drowned, fully grown but not grain filled wheat trapped significantly more sediment (few inches; figures 11.5 and 11.6) than fields with soybean stubble. More than 30 excavators worked from late May to November of 2011 to remove debris and sediment that was blocking drainage ditches (figure 11.7) in order to speed up the drainage process and to accelerate the drying out of low-lying areas. The sediment in private drainageways of most qualified landowners was removed with a 75% cost share from the USDA Farm Services Agency Emergency Services Agency's Emergency Conservation program. It took until October of 2011 for half of the last 15,000 acres to dry out sufficiently to allow tractor traffic and planting of wheat in the fall of 2011. The other half of the idle land (figure 11.4) was eventually tilled to mix in the sediment and to help dry out the fields. These remaining fields were planted to soybean or corn in 2012. In some places (especially along Route 77), the road ditches were filled with 3 to 6 feet of fine sediment and sand. In addition, there were hundreds of acres of thick sand deposits at both the Birds Point and the Big Oak Tree levee breaches, which had to be removed in 2011 and 2012.

Thin coatings of silt covered much of the 133,000 acres in the floodway. The sediment deposits were much thicker in the drowned wheat fields with extra plant residue. Most of the thin silt and clay deposits were exposed to sunlight with tillage to speed dry and incorporate deposits into the plow layer (figure 11.6). Through the Emergency Watershed Protection (EWP) program administrated by the USDA Natural Resources Conservation Service (NRCS) and sponsored by county governments or drainage districts, the sediment depositions in large drainage ditches were removed. Sediment deposition on private farm waterways was removed on a cost share basis by a program administered by the Farm Services Agency and NRCS with $3 million in technical support. Excavators in Consolidated Drainage District 1 of Mississippi County scooped out silt from the drainage ditches (figure 11.7) and loaded it onto trucks to create ditch banks further down the ditch. The work was accelerated when the president of Consolidated Drainage District 1 signed an amendment to an existing cooperative agreement to flood recovery work in the Birds Point–New Madrid Floodway utilizing the NRCS EWP program.

This floodway restoration effort was part of a total $35 million project to clean out 900 miles of ditches with 120 excavators and draglines employing more than 200 workers in the Missouri Bootheel bottomlands that extended south to the Arkansas border. Funds were appropriated by Congress through the EWP program. In April of 2012, the NRCS made an additional $2.1 million of federal disaster funds available for flood repair work to supplement the $0.9 million already committed to the Birds Point–New Madrid Floodway as part of the NRCS EWP program. A total of $3 million funded six repair sites with the drainage district providing $335,000 in matching funds. The new project removed sediment and debris from the drainageways and drainage ditches (figure 11.7) in the floodway and disposed of some of the excavated materials. Initially, work in the spring of 2012 focused on Main Ditch, which drains more than 100,000 acres into the Mississippi River just east of the town of New Madrid (map 11.1). Main Ditch goes through the 1,500-foot gap between the frontline levee and the setback levee, which merges with the town of New Madrid frontline levee. Additional excavators worked on other district ditches. Repairs in the Birds Point–New Madrid Floodway were accelerated in April of 2012 to clean out Ditch 29, the main drainage artery for Consolidated Drainage District 1. Sediment depths were found to be between 1 to 9 feet.

Gully Fields and Land Scouring

Approximately 250 acres of land between Main Ditch and blast site 3 (map 11.1) were scoured as the floodway reached the area via the overflowing Main Ditch and flowed into the opening of the Big Oak Tree levee. The fine particles (sand, silt, and clay) were carried into the Mississippi River through the frontline opening (map 11.1), and fine gravels were left behind and covered the soil surface. Significant land scouring took place with many hundreds of acres of land losing inches of topsoil. Deep gully fields were found 5 miles from the Birds Point levee fuse plug, which released floodwater. The water covered the 5-mile-wide floodway to a 3.6-foot depth and traveled 5 miles before ponding in front of the 6- to 8-foot-high O'Bryan Ridge. The floodwater flowed over the 0.6-mile-wide natural levee ridge and then dropped down to lower bottomland Sharkey soil in a few existing natural drainageways. As the water flowed into the drainageways or ditches, it started to cut into the natural levee Dubbs and Dundee soils, and in one case, the water carved a channel 12 feet deep, 150 feet wide, and 0.6 mile long during the first two weeks in May of 2011. After the floodwater cut this huge gully all the way back to the wooded lower bottomland Alligator soil, the trees on the bottomland fell into the 12-foot-deep gully [3]. The gully side slopes were 12 feet high and vertical. These partially gullied fields dried out quickly and were planted to soybeans in 2011 (figure 11.8), but farmers had to stay out of all the gullies since many still had a few feet of water in the bottom months after the flooding. The force of dropping floodwater (hydraulic jumping) created these large gullies and removed any road bed or bridge in its path. The O'Bryan Ridge main gully fields east of Route 77 (map 11.1) were still not reclaimed as of fall of 2012. A smaller gully field was filled in and regraded on the south end of O'Bryan Ridge near the frontline levee in the spring of 2012. The two severed

FIGURE 11.5 Drowned wheat collected significant sediment and protected against soil erosion.

FIGURE 11.6 Organic and clay coating on plants after flooding and floodwater drainage can often be tilled into the soil.

Settlement and Land Use Changes in the New Madrid Floodway

The year 2011 was one of extreme weather, with 14 events in the United States causing losses in excess of one billion dollars each [1]. The southeast Missouri region adjacent to the lower Mississippi River well illustrates the local impacts on agriculture and human settlements when early snowmelt and record rainfall over the Ohio River valley and the lower Mississippi River valley result in saturated soils, extreme flooding, and damage to crop production and community infrastructure [2, 3]. The May of 2011 deliberate breaching of the levees in the New Madrid Floodway, Missouri, was a planned adaptation response to exceptional flooding conditions with the goal to reduce excess river pressure and prevent levee failures along the Ohio and Mississippi rivers [4]. Climate scientists observe that 2011 was not unique, finding that the last decade was likely the warmest globally for at least a millennium, triggering a period of precipitation and heat wave extremes [1]. As long-term weather patterns become more variable and unpredictable, there is much that can be learned from reevaluating past adaptation strategies and exploring new alternatives.

Land Use Changes

The southeast Missouri region adjacent to the lower Mississippi River, aka the Bootheel, was originally alluvial lowlands covered with swamps and old-growth forests historically exposed to seasonal flooding. On September 30, 1850, Congress authorized a survey of the lower Mississippi River from Cairo, Illinois, to the Gulf of Mexico to address navigation and flooding problems. The survey was the precursor to the construction of levees, dredging, and channelization, which eventually transformed the Bootheel into rich farmland [5, 6]. Following World War I, many landowners and sharecroppers switched to cotton when wheat and corn prices crashed [6]. In the early 1920s, as landowners doubled their cotton acreages, a large number of experienced cotton growers, including more than 15,000 sharecroppers from Tennessee, Louisiana, Arkansas, and Mississippi, were recruited to the floodway region. These sharecropping families typically contracted for a plot of land and in return received housing and half of the crop from the owner [6].

The US Army Corps of Engineers (USACE) had completed by 1926 the construction of levees on both sides of the Mississippi River between New Orleans, Louisiana, and Cairo, Illinois, including the Bootheel area of Missouri. After a winter and spring of heavy rains, on April 16, 1927, the Mississippi River at flood stage eroded the levee near Dorena, Missouri, resulting in a 1,200-foot hole [5]. The levee breach near Dorena (map 12.1), located 30 miles south of Cairo, Illinois, flooded more than 175,000 acres. In all, the 1927 lower Mississippi River flooding displaced more than 31,000 people in the Bootheel, mostly poor tenant farmers who suffered irreparable economic disaster and had to find new communities to resettle [6].

The Birds Point–New Madrid Floodway was built as a federal response to the Great Flood of 1927 [6] that flooded millions of acres of rural and urban areas and led to the loss of thousands of lives. The creation of the floodway between 1928 and 1932 was to reduce future flooding pressures on cities with or without levees along the Ohio and Mississippi rivers [2]. All of the land selected for the floodway was river bottomland in Missouri, with most in agricultural use and only a few small towns, villages, conservation areas, and scattered farmsteads. The USACE obtained easements with the help of eminent domain from all floodway landowners giving the USACE the right to pass floodwater into and through the Birds Point–New Madrid Floodway once the frontline levee was topped and to temporarily store floodwater in the basin and then release it back to the Mississippi River through a 1,500-foot gap between the frontline and setback levees.

The entire floodway was flooded once in 1937 after it was created. The USACE operational plan called for 11 miles of the upper fuse plug area to be deliberately degraded by 3 to 5 feet to correspond to a flood stage reading of 55 feet on the Cairo gage [4]. A judge's ruling delayed the USACE from acting on this plan. As a result, the Cairo gage continued to rise, placing extreme pressure on the Cairo levee system. Crews were sent to the Birds Point levee to dig trenches through the levee and induce failure, but that effort failed in 1937 [4, 7]. Finally, TNT was used for the first time to breach the Birds Point levee in four places; subsequently, the Mississippi River widened these breaches, quickly dropping the water level on the Cairo flood gage. Despite the redesign of the levee system to reduce water pressure on Ohio and Mississippi river cities [2] and prevent another disaster, homes and fields of several thousand sharecroppers in the Bootheel region were flooded [6]. These floods and the creation of the floodway resulted in many homeowners and sharecroppers moving from the villages, farmsteads, and sharecropper camps. Some landowners left behind only repaired barns, sheds, and other structures necessary to support farming their land, while others repaired or replaced their homes and continued to live and work in the floodway after 1937.

The flood of 1937 had significant consequences for the unionized sharecroppers near the end of the Great Depression. Historically they grew cotton on the land; however, crop rotations were changing from cotton to soybean, corn, and wheat (see figures 5.6 and 5.7), and machines were starting to replace hand labor previously used to harvest cotton. The flood of 1937 drove the sharecroppers from the land that they had worked and from their sharecropper camps. Many of these sharecropper families set up camps on the side of the road, mostly on US 60 between Cairo and Charleston. The plight of these sharecroppers was captured in northern newspaper articles, and local farm worker union leaders were invited to give talks about the camp living conditions on the edge of the Missouri highways (figure 12.1) [6]. The sharecropper situation became a national embarrassment for the Roosevelt administration, and local leaders were invited to the White House to talk directly with President Roosevelt. The administration eventually responded with a program called the New Deal.

As the years passed with no federal use of the floodway, thanks in part to the creation of the Kentucky and Barkley reservoirs and increased storage capacity for Ohio River basin floodwaters, some landowners probably assumed there would never be another flood and moved forward in building and creating urban expansion plans on the bottomlands [8]. The operation of the floodway and additional modifications were authorized by the 1965 Flood Control Act and included use of fuse plugs and opening with TNT. Once a weather forecast was issued

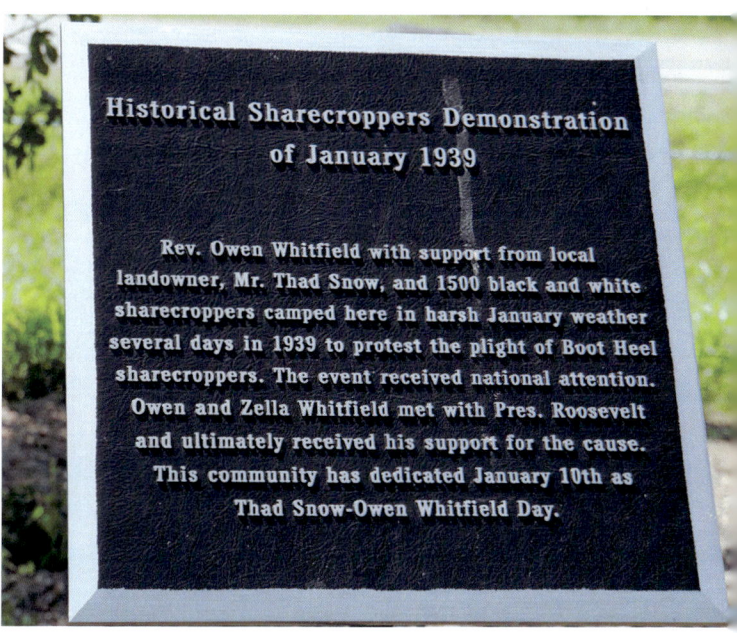

FIGURE 12.1 The flood of 1937 left sharecroppers without land and resources. This site on US 60 between Cairo, Illinois, and Charleston, Missouri, is a reminder of the sharecroppers' loss of livelihoods.

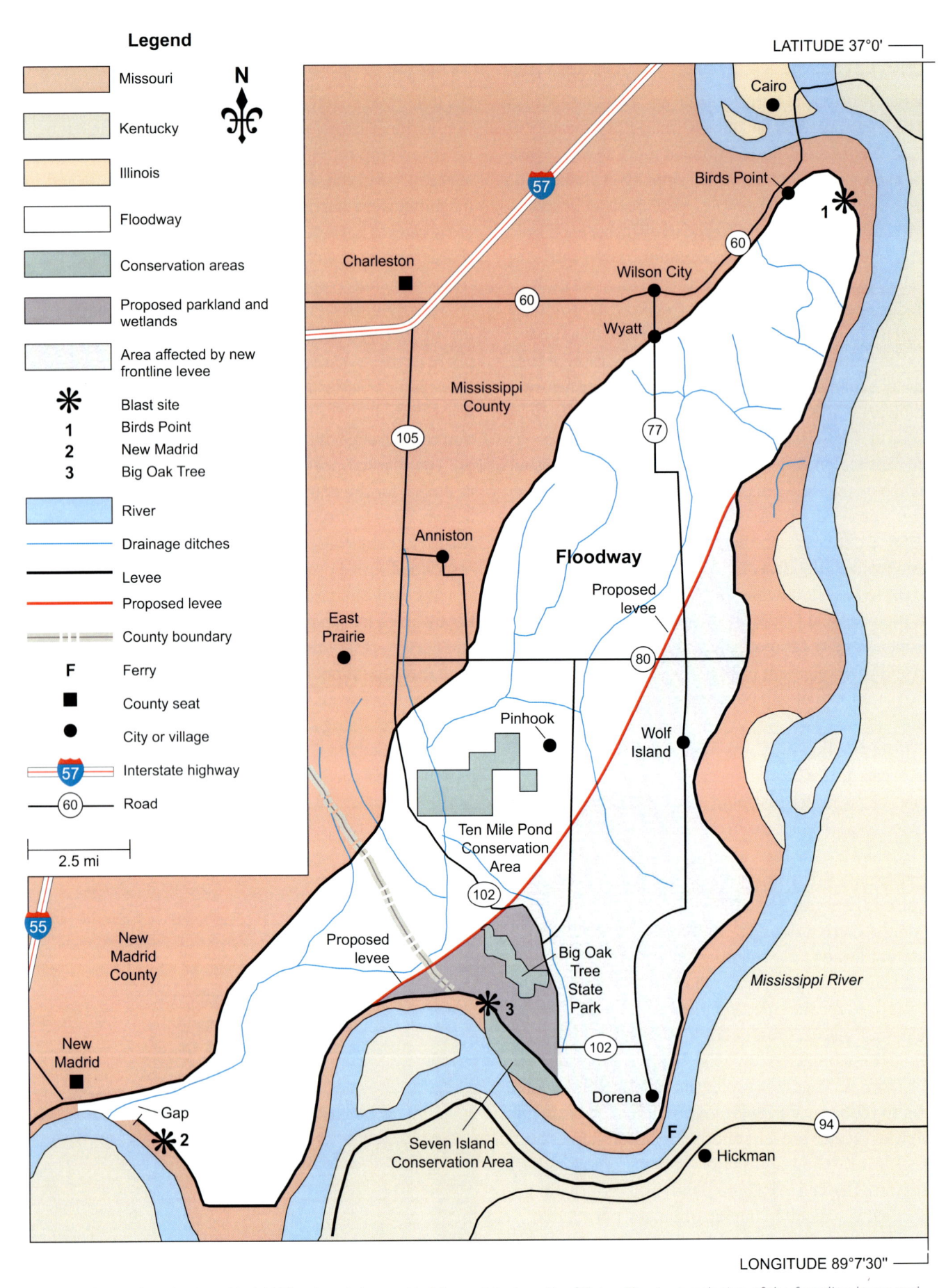

MAP 12.1 The Birds Point–New Madrid Floodway is located in Missouri, just south of Cairo, Illinois. A redesign of the frontline levee and proposed expansion of wetlands and parkland are shown on the map.

FIGURE 12.2 One of the 80 homes in the floodway damaged by the 6 to 10 feet of floodwater in 2011.

FIGURE 12.3 Farm structures including a hog confinement facility were damaged by floodwaters and wind when the floodway was opened in 2011.

predicting a peak flow of greater than 60 feet, the USACE could activate (load TNT into the Birds Point fuse plug) and then open the floodway when the river peak actually reached 60 feet on the Cairo flood gage.

The May 2, 2011, induced breach at Birds Point levee, the first explosion site, and the flooding of the New Madrid Floodway resulted in evacuation of homes and businesses but no loss of life. However, there was severe damage to most of the 80 homes (figure 12.2), a few churches, and 120 barns (figure 12.3) and shed structures that were impacted by the force of flowing water and the 3 to 6 feet of temporarily stored water [9]. Water pressure caused loss of the lower half of walls, damaged most wooden floors, and in over 100 cases completely destroyed structures. Only eight home sites were raised with soil and sand by approximately 10 feet (figure 12.4) by the end of 2014. The local roads and bridges sustained considerable damage from the floodwaters. The Mississippi County transportation department requested $75 million from the USACE to repair the roads and bridges. That request was denied in July of 2012. However, another request for funds was submitted in 2012 to repair the roads damaged by the heavy trucks used to haul soil materials from west of Wyatt to patch the Birds Point levee fuse plug and fill in the crater lakes.

The Village of Pinhook, Missouri

Much of the Missouri Bootheel region, including lands around the village of Pinhook in the center of the floodway (map 12.1 and figure 12.5), was settled by sharecroppers in the 1940s seeking to earn enough money from farming and off-farm work to purchase their own piece of land. In 1943, a plot of land west of Pinhook Ridge freshly cleared of trees by local timber companies was sold at a very low price to five sharecroppers [9].

Platted in the 1960s by Mississippi County, the homes on Pinhook Ridge were on a natural levee about 3.3 feet higher than the lowlands, high enough to avoid local flooding. As the land and property on Pinhook Ridge became available, officers of the Christian Liberty District Association of Southeast Missouri purchased the land and homes. The lots were later sold to families, and additional homes were built and platted. In 2011 the community was a combination of second and third generation families of the original five sharecroppers from Tennessee. The families moved in after the Christian Liberty District Association of Southeast Missouri bought the land on Pinhook Ridge.

Although the floodway was not put into use between 1937 and 2011, there was frequent annual local flooding within the New Madrid Floodway. Families who settled in Pinhook on the lower elevation lands west of Pinhook Ridge experienced annual local flooding events. However, when low-lying areas flooded, residents used tractors and wagons to drive through the flooded areas to leave Pinhook Ridge or to get children to the school bus to East Prairie on the other side of the setback levee. At the time of the 2011 flood and floodway opening, the community had 30 residents living on the slightly higher Pinhook Ridge.

Pinhook is protected by the frontline levee. Localized flooding occurs when water from the Mississippi River backs up into Main Ditch, which flows through a 1,500-foot gap between the front and setback levees northeast of New Madrid and through the nearby drainage ditch near Ten Mile Pond Conservation Area (map 12.1). When the floodway setback levee was built between 1928 and 1932, a 1,500-foot gap was left as a drainage outlet or gap for runoff water from more than 112,500 acres of lands within the floodway [7]. In 1954,

FIGURE 12.4 One of the new homes was built on 10 feet of soil materials to prevent future flood damage.

the USACE gained approval to fill the gap and prevent the backwater flooding that threatened Pinhook almost every year [9, 10]. The proposed $107 million project to close the gap sat pending for decades, with its funding as a point of contention. At that time, Pinhook residents, with a median income of about $15,000 a year, were asked to pay 35% of the cost. In 1993, the local share was reduced to 5%, but another decade passed with no action. A leader of the Village of Pinhook testified before Congress about the living conditions, and in 2005, the work to fill in the 1,500-foot levee gap started. In 2006, the work stopped when the Environmental Defense Fund and National Wildlife Federation sued over its impact on wetlands. A federal judge ordered an injunction to stop work on the levee and ordered the restoration of the area to its previous condition. Nearly $7 million had been spent to fill the levee gap, and $10 million was spent undoing the work [7]. By 2007, the levee gap and main drainage system were restored to 2004 conditions with no change in the local flooding situation.

In 2011 and 2012, the residents of Pinhook searched for land and money to relocate together outside the floodway. Assessment of the Village of Pinhook by the Federal Emergency Management Association (FEMA) determined nearly every home's damage to be greater than 50% of its value. Consequently, the Pinhook residents would have to elevate their homes on stilts if they wanted to legally rebuild there. FEMA offered up to $30,000 to each homeowner for these repairs. Most Pinhook residents, many of them senior citizens, did not accept those conditions. Some homeowners in Pinhook had home insurance but not flood insurance [11, 12]. A FEMA buyout was eventually accepted by owners of 21 properties (figure 12.5) totaling $1.17 million, according to the Bootheel Regional Planning and Economic Development Commission. On November 9, 2011, the community of Pinhook finalized plans for a community Development Block Grant from the Department of Housing and Urban Development. That relocation would cost an additional $1.43 million and would involve the residents moving to a new 40-acre plot of land where homes could be built. The plot of land had to be in either Mississippi County or a neighboring county but outside the floodway. There was no guarantee of funding. If Pinhook relocates, the funds will have to come from the federal government and not the State of Missouri according to a local Missouri congressman [11, 12].

Changing Long-Term Weather Patterns

Extreme events such as the 2011 flooding along the Mississippi River illustrate the challenges ahead for agriculture and communities of rural America as public agencies such as the USACE attempt to anticipate risk and manage emergency and evolving natural disasters related to water resources. The USACE's 2010 National Report acknowledged the increasing frequency and severity of extreme weather events and identified climate change as a key variable. While the average annual local temperature in the Bootheel decreased slightly from 1951 to 2010 [13], average temperatures in the upper Great Lakes region are expected to increase by 3.6°F to 7.2°F, with a 25% increase in precipitation by the end of the century [14]. This has continuing implications for the landowners and residents of the Birds Point–New Madrid Floodway and the USACE as they prepare for future flooding events downstream from the tributaries of the Ohio River basin and the upper Mississippi River.

FIGURE 12.5 Homes in the village of Pinhook in November of 2011 were abandoned after they were damaged by 6 to 10 feet of floodwater.

The conditions that affect climate change are highly complex. However, simple physical laws, such as the relationship between temperature and precipitation, can be sources of weather extremes. Warming leads to more evaporation with the potential of two very different outcomes, depending on conditions: (1) surface drying with potential to increase the intensity and duration of drought or (2) enhanced precipitation as the air holds more moisture and increased atmospheric moisture provides increased latent energy to drive storms [1]. The Bootheel experienced both extremes with the flood of 2011 and extreme drought in 2012. Although particular events cannot be directly attributed to worldwide warming trends, the odds of certain types of weather extremes increase in a warming climate [1]. If rainfall extremes continue, exploration and investment in public and private adaptation strategies will be important if agricultural landscapes and rural livelihoods are to be effectively protected.

Revisiting Current Levee Locations and Land Uses

The floodway acts as a temporary spillway, taking pressure off the levee system under extreme flood conditions, but sacrifices agricultural production in the year of opening and displaces residents to protect loss of life [2]. If the opening of the floodway were to occur more frequently, the public and private costs of reclamation and restoration of agricultural lands and rural communities would also increase. Although returning the entire floodway to its original alluvial floodplain is likely to not be socially desirable or politically feasible, redesign of the floodway could reduce taxpayer and private costs associated with more frequent flooding events. Landowners in the floodway, levee and drainage districts, and the USACE need to continue to work together to explore realignment options that reduce risks and costs. Two potential redesigns include shifting the frontline levee and/or expanding wetland areas.

The floodway is 5 miles or less wide at the top and bottom, with a 10-mile-wide bulge that follows the Mississippi River. A realignment of the frontline levee away from the Mississippi River (map 12.1) and to the west would create a more uniform 5-mile width and enlarge the natural floodplain area available to temporarily store floodwaters during high water events. This reengineering of the river floodplain-levee relationship would provide a hydrologic and biogeochemical buffer to adaptively respond to the large uncertainties in river management [15] and could delay or reduce the frequency in which the floodway is opened and exposes agricultural lands within the floodway to increased soil and crop damage [3, 7]. This new frontline levee location would result in approximately 44,000 acres being removed from the 133,000-acre New Madrid Floodway (map 12.1). The recreated floodplain would no longer be protected from Mississippi River flooding by a USACE levee but could remain in agricultural use similar to many of the unleveed bottomlands of the Ohio River basin.

If the Mississippi River reached flood stage during the growing season, the agricultural lands without levee protection would result in crop loss. Farmers in the area would likely stop growing winter wheat (planted in fall and harvested during the next summer) and shift to corn and soybean or silviculture. Many similar unprotected bottomlands in Missouri, Illinois, and Kentucky remain in agricultural use with limited soil damages from flooding events. However, in the Missouri or west side of the Mississippi River, there would likely be less serious soil damage associated with levee failures (crater lakes, gullies, and sand deposits). These unleveed bottomlands would provide additional water storage capacity during flooding, and the wet soils would help filter pollutants, recharge the water table, capture sediment, and protect adjacent levees from failure. If any parcel or farm in the 44,000-acre area were no longer used for agriculture, it could be eligible for the USDA Natural Resources Conservation Service (NRCS) Wetlands Reserve Program if that program is retained and funded in future farm bills.

A reengineering of the Missouri frontline levee would require hydraulic modeling, including risk and resilience analyses [15] to ascertain the effects on both sides of the Mississippi, including the Hickman levee and adjacent 170,000 acres of bottomlands in Kentucky (map 12.1). If the Hickman levee were to fail because of increased pressure caused by realignment of the Missouri frontline levee, there could be serious land damages in Kentucky bottomlands, including land scouring, crater lakes, gullies, sand deltas, and sediment deposition. The key concept of this redesign is to mimic more closely the natural floodplain capacity to manage highly variable levels of water and decrease pressure on the levee system under extreme flood conditions. USACE modeling would be valuable in the development of a number of engineered scenarios to evaluate the social, economic, and ecological tradeoffs between relocating the frontline levee further from the Mississippi channel and enlarging the floodplain to increase natural storage under flood conditions, thereby delaying the opening of the floodway [15].

A second, more modest redesign would keep the frontline levee in the current location and convert about 2,500 acres of levee-protected but flood-vulnerable acres of low-lying bottomlands from agricultural use to parkland and wetlands (map 12.1). These are the lands located adjacent to blast site 3 (figure 12.6) between the Seven Island Conservation Area and the Big Oak Tree State Park (map 12.1). Part of this area could be purchased by the State of Missouri to expand Big Oak Tree State Park. Landowners in the floodway, the Department of Natural Resources, State of Missouri, levee and drainage districts, and USDA NRCS should explore the potential land use change options for the area surrounding the Big Oak Tree State Park. Other adjacent low-lying bottomlands may be eligible for a USDA NRCS Wetlands Reserve Program permanent easement using two 30-year easements or a restoration cost share agreement if this program is renewed. If purchased with state or federal resources, these wetlands and parklands would provide multiple benefits, retaining current storage capacity at flood stage as well as enhancing local wildlife habitat (map 12.1) with ecological and recreational values. This area historically was late to dry out since it collected the greatest depth of floodwater when the Mississippi River was high and when the floodway was opened, with substantive repair costs following flooding. Returning this 2,500-acre area to its original wetland condition would have an economic impact on current agricultural uses. However, the rebuilding of the ecological infrastructure would also reduce future flood restoration costs in terms of crop land reclamation, damage to roads, and leveed structure repairs. Any future floodway use, and there will be some, would do less damage to the wetland and parkland reserves than would happen with continued agricultural use. This ecosystem is well suited to periodic inundation from heavy rainfall and river flooding.

Both of these options would transform the current land use and be a difficult social and political decision for public agencies and private landowners in the floodway. However, the diverse habitat created by wetlands and parklands could also be an opportunity for the purposeful development of an economic tourism plan to increase the recreational use of the area. The partial return of the Mississippi floodplain to marsh and wetlands offers a compromise, one that attempts to balance the need to protect productive agricultural lands while adapting to changing natural conditions of increased flood pressures on levees and the infrastructure of human settlements.

Changing Conditions and Future Strategies

Changing climate conditions will magnify the risks associated with snowmelt, rainfall, runoff patterns, and flooding [16]. Shifts in global temperatures are likely to have their greatest practical impact via effects on the water cycle, with the amount of water vapor that the atmosphere can hold increasing rapidly with temperature, leading to more extreme rainfall occurrences and flooding [17]. As the odds for certain types of weather extremes increase in a warming climate, farmers, rural residents, and public agencies will need both short- and long-term strategies to address reclamation of agricultural lands and restoration of farmsteads and buildings and to put in place adaptive management plans that anticipate future events.

Landowners and residents of the Bootheel have experienced a number of transformations to their physical, social, and economic landscape over the last 150 years. A former alluvial plain, this region now has some of the most productive soils in Missouri as well as in the upper Midwest, with high corn, soybean, and wheat yields from systematic investments in technologies and modern management practices. The short-term approach is to expect that this region will continue to be a high-profit agricultural region that should be protected at any cost, with a resistance-adaptation strategy of continual levee repair and reclamation of these lands after a floodway opening. This plan is sound when extreme flooding events are few and far between.

Long-term projections of continued global and upstream warming trends suggest higher probabilities of more frequent weather extremes, with both flooding

FIGURE 12.6 Birds build nests in the low, wet areas adjacent to the Big Oak Tree levee blast site where the inflow of floodwater created crater lakes and ponds.

and drought affecting the region. Under these extreme conditions, the resistance-adaptation model is likely to be inadequate, and a more aggressive resilience-adaptation or transformation approach may be needed [18]. Wetlands act like a giant sponge, absorbing rainwater and then slowly releasing it to groundwater and nearby streams [19] mediating both potential drought and flooding. Given the rapid pace of change, there is a need for tradeoffs at multiple scales—from field to farm level to landscape and river basin—in ways that go beyond current short-term land use benefits to create longer term agrobiodiversity with economic and ecological benefits [18].

A comprehensive plan that complements the leveed engineering infrastructure with re-creation of an ecological infrastructure can mitigate flooding and strengthen the effectiveness of well-placed leveed systems [19]. The cost of moving the frontline levee and returning a portion of the floodway to the Mississippi River alluvial floodplain to increase the storage capacity of water during flood stage or the cost of the government purchasing 2,500 acres of the low-lying bottomlands in the floodway will be expensive in the short term but is likely to reduce long-term public and private payouts for property damage and crop losses. These two alternatives and others should be explored as farmers, local leaders, and state and federal agencies evaluate future scenarios, public resources, and political willingness to address the complex interactions among society, land use decisions, and the water cycle.

---

[1] Coumou, D., and S. Rahmstorf. 2012. A decade of weather extremes. Nature Climate Change 2:491-496, doi:10.1038/nclimate1452.

[2] Olson, K.R., and L.W. Morton. 2012. The effects of 2011 Ohio and Mississippi River valley flooding on Cairo, Illinois, area. Journal of Soil and Water Conservation 67(2):42A-46A, doi: 10.2489/jswc.67.2.42A.

[3] Olson, K.R., and L W. Morton. 2012. The impacts of 2011 induced levee breaches on agricultural lands of Mississippi River Valley. Journal of Soil and Water Conservation 67(1):5A-10A, doi:10.2489/jswc.67.1.5A.

[4] Camillo, C.A. 2012. Divine Providence: The 2011 Flood in Mississippi River and Tributaries Project. Vicksburg, MS: Mississippi River Commission.

[5] Barry, J.M. 1997. Rising Tide: The Great Mississippi Flood of 1927 and How It Changed America. New York: Simon and Schuster.

[6] Gellman, E.S., and J. Roll. 2011. The Gospel of the Working Class. Urbana-Champaign, IL: University of Illinois Press.

[7] Olson, K.R., and L W. Morton. 2013. Restoration of 2011 flood damaged Birds Point–New Madrid Floodway. Journal of Soil and Water Conservation 68(1):13A-18A, doi:10.2489/jswc.68.1.13A.

[8] Lowery, B., C. Cox, D. Lemke, P. Nowak, K.R. Olson, and J. Strock. 2009. The 2008 Midwest flooding impact on soil erosion and water quality: Implications for soil erosion control practices. Journal of Soil and Water Conservation 64(6):166A, doi:10.2489/jswc.64.6.166A.

[9] Schick, A. 2011. Farmland behind Birds Point levee recovers faster than expected. Columbia Missourian, November 4, 2011. http://www.columbiamissourian.com/a/142312/farmland-behind-birds-point-levee-recovers-faster-than-expected/.

[10] Schick, A. 2011. Mississippi River town of Pinhook struggles to reclaim its community after levee break. Columbia Missourian, December 22, 2011. http://www.columbiamissourian.com/a/143923/mississippi-river-town-of-pinhook-struggles-to-reclaim-its-community-after-levee-break/.

[11] Moyers, S. 2011. The death of Pinhook; Community will never return after flooding. Southeast Missourian, June 17, 2011. http://www.semissourian.com/story/1737132.html.

[12] Moyers, S. 2011. Buyouts of Pinhook, Morehouse among topics at round-table discussion with Sen. Blunt. Southeast Missourian, September 2, 2011. http://www.semissourian.com/story/1758866.html.

[13] Benning, J., D. Herzmann, C. Ingels, and A. Wilke. 2012. Agriculture and weather variability in the Corn Belt: Bootheel Missouri Bulletin CSCAP0065-2012-MO. Ames, IA: Iowa State University.

[14] USACE (US Army Corps of Engineers). 2010. National Report: Responding to National Water Resources Challenges. Washington, DC: US Army Corps of Engineers Civil Works Directorate. http://www.building-collaboration-for-water.org/Documents/nationalreport_final.pdf.

[15] Park, J., T.P. Seager, P.S.C. Rao, M. Convertino, and I. Linkov. 2012. Integrating risk and resilience approaches to catastrophe management in engineering systems. Risk Analysis doi:10.1111/j.1539-6924.2012.01885.x.

[16] Palmer M.A., D.P. Lettenmaier, N.L. Poff, S.L. Postel, B. Richter, and R. Warner. 2009. Climate change and river ecosystems: Protection and adaptation options. Environmental Management 44(6):1053-68. Epub. PubMed PMID: 19597873.

[17] Hansen, J., M. Sato, and R. Ruedy. 2012. Perception of climate change. Proceedings of the National Academy of Sciences of the United States of America 109(37):E2415-E2423. http://www.pnas.org/content/109/37/E2415.short.

[18] Jackson, L., M. van Noordwijk, J. Bengtsson, W. Foster, L. Lipper, M. Pulleman, M. Said, J. Snaddon, and R. Vodouhe. 2010. Biodiversity and agricultural sustainability: From assessment to adaptive management. Current Opinion in Environmental Sustainability 2:80-87.

[19] Postel, S. 2011. Mississippi floods can be restrained with natural defenses. Water Currents, National Geographic, May 3, 2011. http://newswatch.nationalgeographic.com/2011/05/03/mississippi-floods-can-be-restrained-with-natural-defenses/.

Impact of Levee Breaches, Flooding, and Land Scouring on O'Bryan Ridge Soil Productivity

13

Flooding of agricultural lands after a natural or human-induced levee breach can have large and persistent effects on soils, crop productivity, and water quality, with negative economic, social, and ecological consequences. Many US water management strategies associated with levee-protected agricultural systems are dominated by policies that focus on engineered solutions designed to minimize short-term risk of flooding and breaching while overlooking resilience of the agroecosystem as a whole [1, 2]. A federal damage assessment of the effects of levee breaches and flooding on public and agricultural lands is needed each time a levee fails. Most federal damage assessments only include the levee itself and the adjacent crater lakes, gullies, and sand deltaic deposits but not the remaining flooded areas. Land scouring, sediment deposition in drainage and road ditches, and soil productivity loss are the most severe damages to soils on agricultural lands.

Levee breaches on the Mississippi River in the US interior have occurred since the Great Flood of 1927 [3]. The flood of 2011 [4] on the Mississippi River well illustrates the impacts of flooding and levee breaching on agricultural soil conditions and productivity. This chapter details the effects on a 195-acre field on O'Bryan Ridge (35°51′9″ N, 89°11′3″ W) owned by Levee District Number 3 when the US Army Corps of Engineers (USACE) opened the New Madrid Floodway in Missouri by inducing a breach in the Birds Point fuse plug levee (map 13.1). An unintended conversion of agricultural land to wetlands and ponds occurred within the land scoured bottomlands (figures 13.1 and 13.2) in the O'Bryan Ridge field. These gully lands have since been partially regraded and reshaped in an effort to return the agricultural land to production.

The dramatic changes in land use from levee breaching demonstrate the need for a land scouring and deposition survey, an updated soil survey, and a soil and water conservation plan to reduce further soil loss and gully formation on reclaimed lands. An updated soil survey map with eroded and deposition phases of previously existing soils and new soil series can be used to estimate and compare the crop yields and production levels before levee breaching and after gully field creation, and to guide the periodic reshaping and restoration of the gully fields. Further, gully fields within the New Madrid Floodway are likely to remain vulnerable to the next induced levee breach and subsequent flooding if a revised plan to protect the area is not developed.

Gully Fields on O'Bryan Ridge in the New Madrid Floodway

On May 2, 2011, the Cairo flood gage reached a record 61.7 feet, and for the first time in 74 years, the USACE opened the New Madrid Floodway (map 13.1) using 265 tons of TNT. Approximately, 1.2 miles of the fuse plug were blown-up simultaneously in six places in the frontline levee with one-fourth of the Mississippi River entering the 35-mile-long and 4- to 10-mile-wide New Madrid Floodway. The Mississippi River floodwaters were 13.2 feet above the base of the Birds Point fuse plug levee and adjacent bottomlands when the breach occurred. After the floodway was opened, the floodwaters poured through the levee breach and dropped onto the protected bottomlands creating a crater and then spread out into the 5-mile-wide floodway. As the water from the induced breach flowed south 20 miles, it merged with Mississippi River floodwater that had backed up into the lower third of the floodway through the open gap between the frontline and setback levee at New Madrid. As a result, there was little land scouring in the southern part (New Madrid County) of the floodway.

The floodwater eventually covered 133,000 acres. Five miles south of the newly formed Birds Point levee

FIGURE 13.1 Land scouring of the bottomland located below the gully fields.

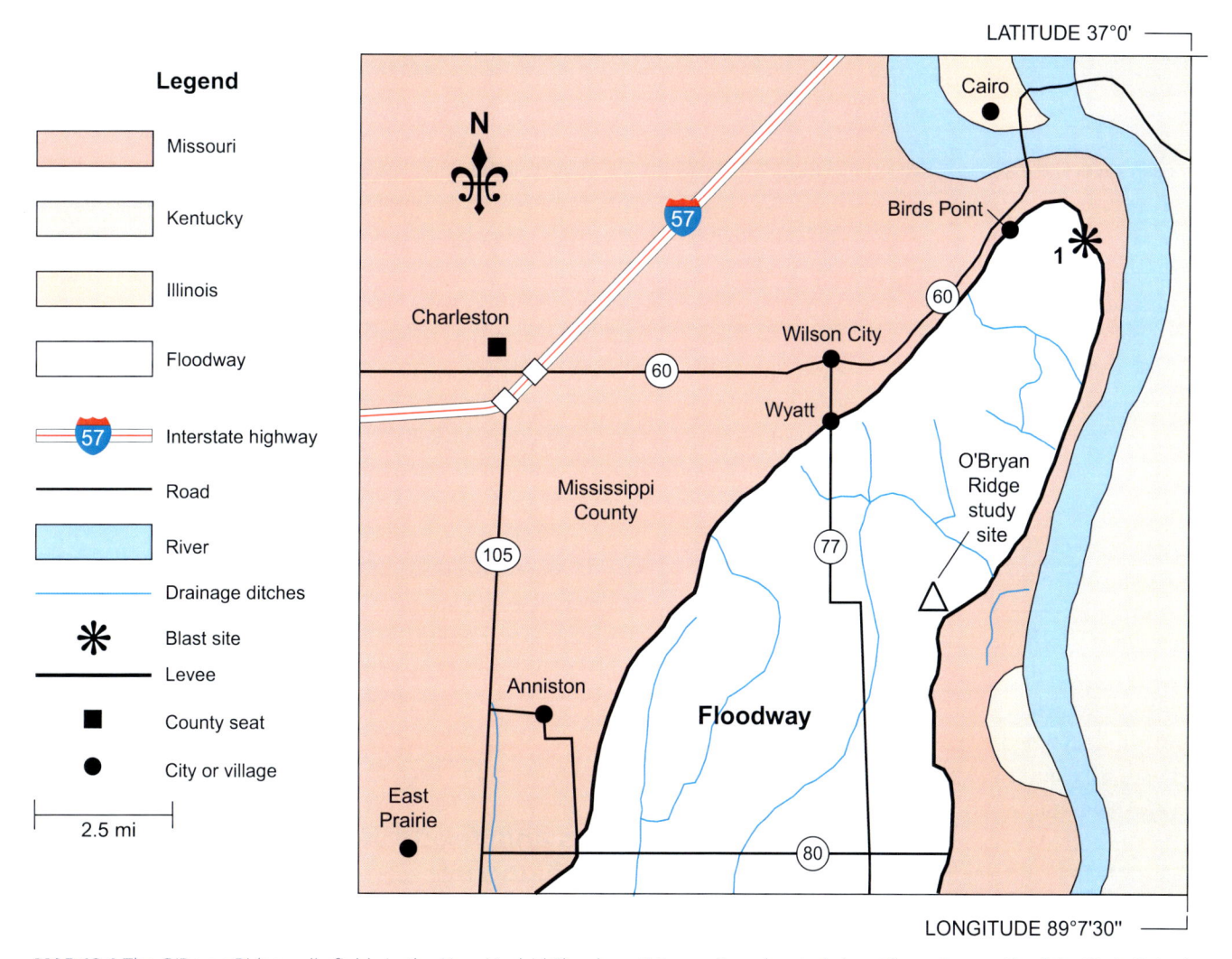

MAP 13.1 The O'Bryan Ridge gully fields in the New Madrid Floodway (Missouri) are located about five miles south of the Birds Point induced breach site.

FIGURE 13.2 May of 2011 aerial view of O'Bryan Ridge gully fields.

breach and craters, the floodwaters ponded in front of O'Bryan Ridge, an old Mississippi River meander bank or natural levee, which was about 6.6 to 8.3 feet higher than the alluvial bottomlands and approximately 6 miles long and 0.6 mile wide. Once the ponded floodwater reached 6.6 to 8.3 feet, the water began to flow rapidly over the soybean field on O'Bryan Ridge. As the floodwater dropped off the ridge, it concentrated in old drainageways and waterways (figures 13.2 and 13.3), cut gullies into the alluvial bottomland (Sharkey soils) on the west side of O'Bryan Ridge due to hydraulic jumping, and created canyon-sized gullies (figure 13.3; see figure 8.8) up to 0.6 mile in length through the entire width of the ridge [5].

Three major gully fields were created on O'Bryan Ridge as a result of the induced breach [6]. Two of the gully fields were reclaimed in 2011 and 2012. The third major gully field, owned by Levee District Number 3 (figures 13.1, 13.2, and 13.3), is examined here to assess the effects of gullies on soil properties, soils, soil productivity, land use change, and agricultural production. This gully field did not qualify as a priority area identified by the USACE as needing immediate restoration and repairs. All the priority areas including levees, crater lakes, and roads were restored by the fall of 2012.

The disastrous consequences of flooding and severe land scouring from the induced breach were probably not anticipated by landowners since the floodway had not been used since 1937 (or in most landowners' lifetimes). The bottom of the trenches eroded more than 3.3 feet below the bottomland surface (figure 13.3) and 12 feet below the surface of the O'Bryan Ridge [6, 7]. A channel was created 0.6 mile from west to east (figure 13.2), which undercut the gravel road (County Road [CR] 310) and extended into the wooded bottomland border area. A series of canyon-sized gullies cut into the O'Bryan Ridge in a dendritic pattern (figure 13.2; see figure 10.5). Some did not cross the entire 0.6-mile ridge. Other gullies cut through the tree line on the south side of the gully field, dissected CR 312 on the south of O'Bryan Ridge, and nearly reached a section of the adjacent frontline levee. The gullies between CR 312 and the frontline levee qualified as a USACE priority area and were reclaimed in the fall of 2011 and spring of 2012.

Crops and Vegetation on O'Bryan Ridge after Flooding

The O'Bryan Ridge field with slightly eroded soils (map 13.2a) had been planted to soybean in 2010. It was returned to soybean production by July of 2011, but 71 acres of the 195-acre field could not be farmed as a result of the deep gullies (figure 13.2 and map 13.2b). Farm equipment had to be kept back from the 12-foot-high vertical edges of the gullies (see figure 10.4), and the land between the gullies could not be cultivated [5]. In June of 2011, the farmers needed access to their fields and the USACE needed access to levees for maintenance, so CR 310, on the east side of the gully field, was reconstructed by partially filling in a 12-foot-deep and 100- to 160-foot-wide gully (figures 13.2 and 13.3). The field was again planted to soybean in 2012, 2013, and 2014. No attempt was made to reclaim the gully fields until the spring of 2013, or two years after floodway use by the USACE.

Ponds and herbaceous wetlands (figure 13.4) spontaneously developed in the gullies by 2013. The ponds that formed in the gully bottoms were 3.3 feet below the first bottomlands and 12 feet below the ridgetops. These deep gullies trapped water as well as whatever nutrients, pollutants, and contaminants the waters carried. The ponds and wetlands were partially filled in 2013 when the gully edges were bulldozed (see figure 8.9) into the gullies to reshape and regrade the sides of the gullies (map 13.3a). Once a soybean crop was planted on the Udifluvents sloping soils (map 13.3a), sheet, rill, and gully erosion occurred, and the sediment was transported by runoff water into the ponds and surrounding wetlands. In 2014, the remaining wetlands were dominated by early successional, herbaceous plant species. Species accounting for the greatest cover at the site included hog peanut (*Amphicarpa bracteata*), a leguminous vine; giant foxtail (*Setaria faberi*), an annual grass; tall boneset (*Eupatorium serotinum*); and beggar's ticks (*Bidens frondosa*). In addition to these early colonizing herbaceous species, several seedlings and saplings of early colonizing wetland trees and shrubs, particularly eastern cottonwood (*Populus deltoides*) and willows (*Salix nigra* and *Salix interior*), were noted. If left unchanged, the gully wetlands would likely follow ecological successional processes and become forested wetlands on the gully slopes, and the shallow bottoms become semipermanently inundated ponds in the deeper gullies.

Assessment of the Resulting Agricultural Productivity Change

Prior to the 2011 floodway use, the entire 195-acre O'Bryan Ridge field was in soybean production with no wetlands or ponds. The soybean crop averaged approximately 44 bushels per acre with a total average of approximately 8,550 bushels of soybean per year. After creation of the gully field and scouring of the bottomlands and ridge, 50 acres of the field became gullies and could

not be farmed, and approximately 30 acres between or adjacent to the deep gullies were not able to be planted. The remaining 115 acres were land scoured, including the bottomland west of the O'Bryan Ridge. Thus, after the gully fields were created, a total of 124 acres were in agricultural production, and 71 acres were wetlands and ponds (map 13.2b). In 2013, 11 acres of sloping madeland, or Udifluvents, were returned to soybean production from re-grading 30 acres of nearly level ridge land and 30 acres of the gully bottoms with ponds (map 13.3a). In the gully bottoms, approximately 40 acres remained in wetlands and ponds. The 60 acres of the Dubbs soil, 14 acres of Sharkey soils, and 60 acres of Udifluvents remained in agriculture—a total of 134 acres. In sum, even after 2013 land regrading, approximately 31% of the land had been unintentionally converted from agricultural lands to wetlands and ponds as a result of floodway use.

The productivity of all the eroded Dubbs, Dundee, and Sharkey soils was lowered. Soybean production was reduced from 45 bushels per acre to 40 bushels per acre on the Dubbs and Dundee soils and from 40 to 30 bushels per acre on the Sharkey soils based on soil properties and erosion phase changes on the remaining 74 acres of nearly level land either on ridge or bottomland scoured area (figure 13.2). The 50 acres of ridgetop that were pushed into the gullies and the 10 acres reclaimed from the gullies yielded about 35 bushels per acre for the 60 acres of Udifluvents (map 13.3a). In 2013 the 60 acres of wetlands and ponds produced 0 bushels per acre of soybean.

In the spring of 2014, deep ditches were dug to drain the ponds and were connected to a culvert under CR 312 at the southwest corner of the O'Bryan Ridge gully field (map 13.1 and map 13.3b). Two bulldozers (see figure

FIGURE 13.3 Gully development in the O'Bryan Ridge soybean field caused trees to fall into channels and undercut County Road 310.

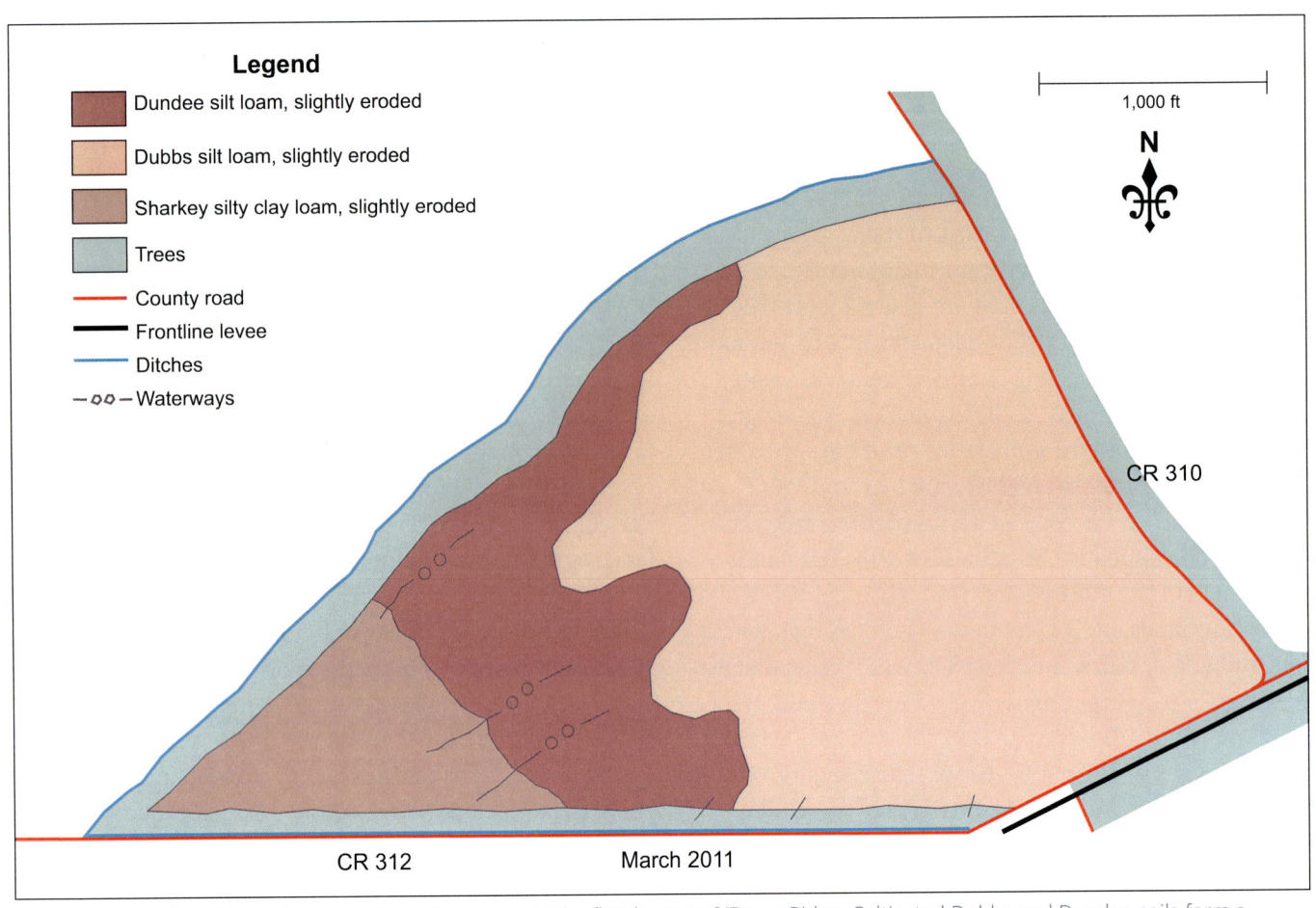

MAP 13.2a March of 2011 soils prior to levee breaching in the floodway on O'Bryan Ridge. Cultivated Dubbs and Dundee soils form a natural levee to the east of Sharkey bottomlands.

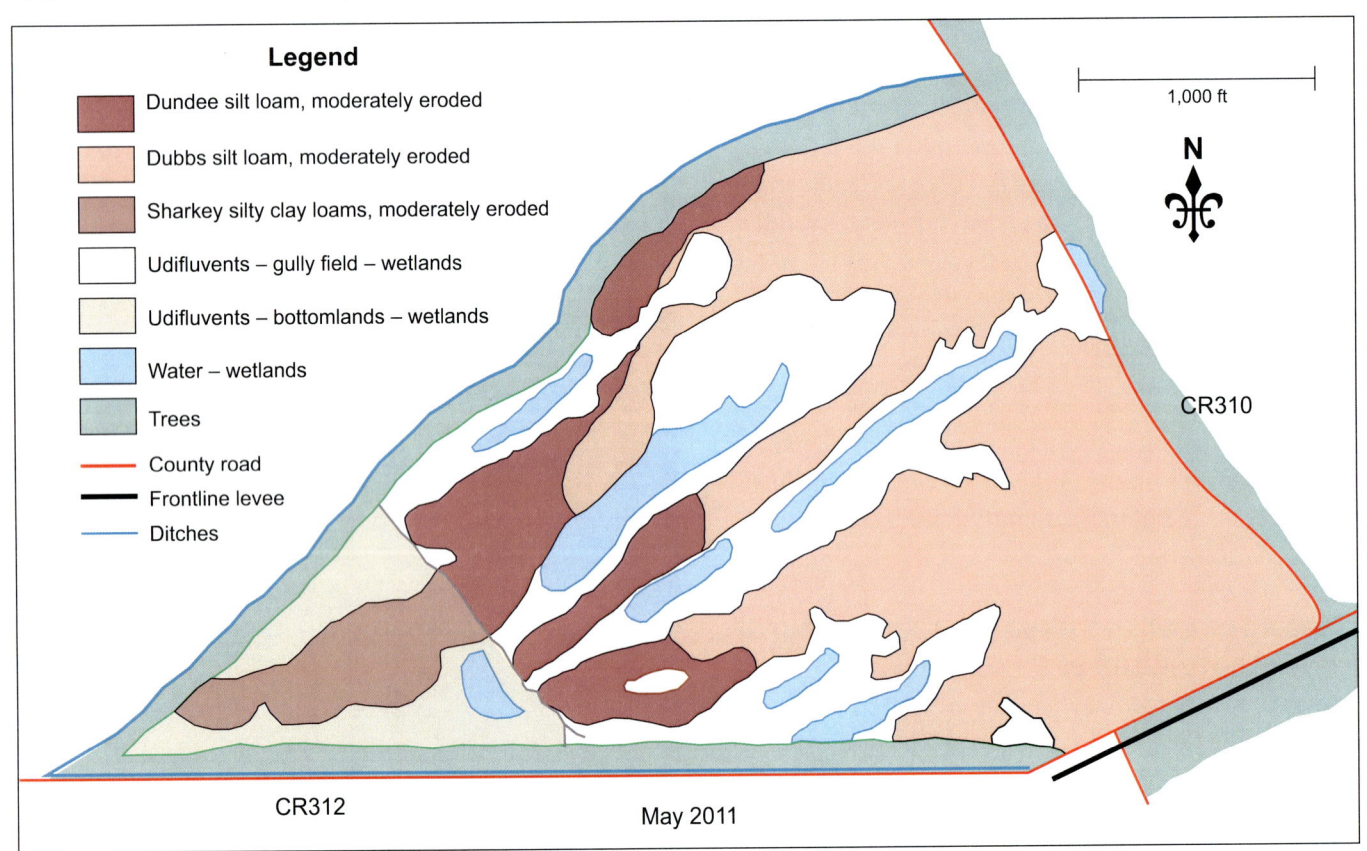

MAP 13.2b May of 2011 soil eroded phases of O'Bryan Ridge show soil loss and degradation with the creation of gully fields during the floodway use and flooding.

8.9) pushed massive amounts of topsoil and subsoil from ridgetops into the gully bottoms, which eliminated most of the wetlands and ponds. This land redistribution continued until there was almost no topsoil and subsoil left on the ridgetops (map 13.3b). This process of land reshaping reduced the slope from steep sloping to gently sloping and reduced the erosion hazard. However, the filled-in gullies are still lower than adjacent ridges and will remain vulnerable to the next use of the floodway. This reclamation changed the land use again, leaving only 25 acres remaining in wetlands and ponds, and increased the agricultural land to 170 acres.

The soybean plant heights in August of 2015 on the Udifluvents soils were much lower than on the adjacent land scoured ridgetops (figure 13.5). Soybean plant biomass and plant height normally correlate well with grain yields. This suggests that four years after the floodway use the soybean yields of the graded soils were most likely significantly lower than the eroded ridgetop soils and the soybean yield loss is likely permanent. Even extensive reclamation efforts in the spring of 2014 (map 13.3b) could only mitigate and restore part of the soil productivity and yield capacity loss. The land use change to wetlands appears to have been temporary, and the land has been returned to agricultural use through grading and reshaping. The 195-acre levee district field, which formerly produced an average of 8,550 bushels of soybeans in 2010, produced an average of 4,820 bushels in 2011 and 2012—a loss of 44% or 3,730 bushels after gully field creation (map 13.2b).

FIGURE 13.4 The deep gullies in O'Bryan Ridge became wetlands and ponds.

After reclamation in spring of 2013 and spring of 2014, the land produced 6,000 bushels, a loss of 30% or 2,565 bushels of soybeans (maps 13.3a and 13.3b). In 2015, Levee District Number 3 planned additional earth moving in an attempt to further mitigate this permanent soil productivity and yield capacity loss. The needed soil materials have to come from offsite as there are only 37 acres of topsoil and subsoil materials left on the tract (map 13.3b).

If the topsoil and subsoil remaining on the Dubbs ridge topsoil are used to fill in the remaining ponds and cover wetlands to restore agricultural use, the 37 acres on the ridgetop will have lower crop yield potential. Depending on depth of removal, the crop yield potential will be lowered by an estimated 5 to 10 bushels per acre

FIGURE 13.5 Soybeans growing on Udifluvents and adjacent land scoured ridgetops.

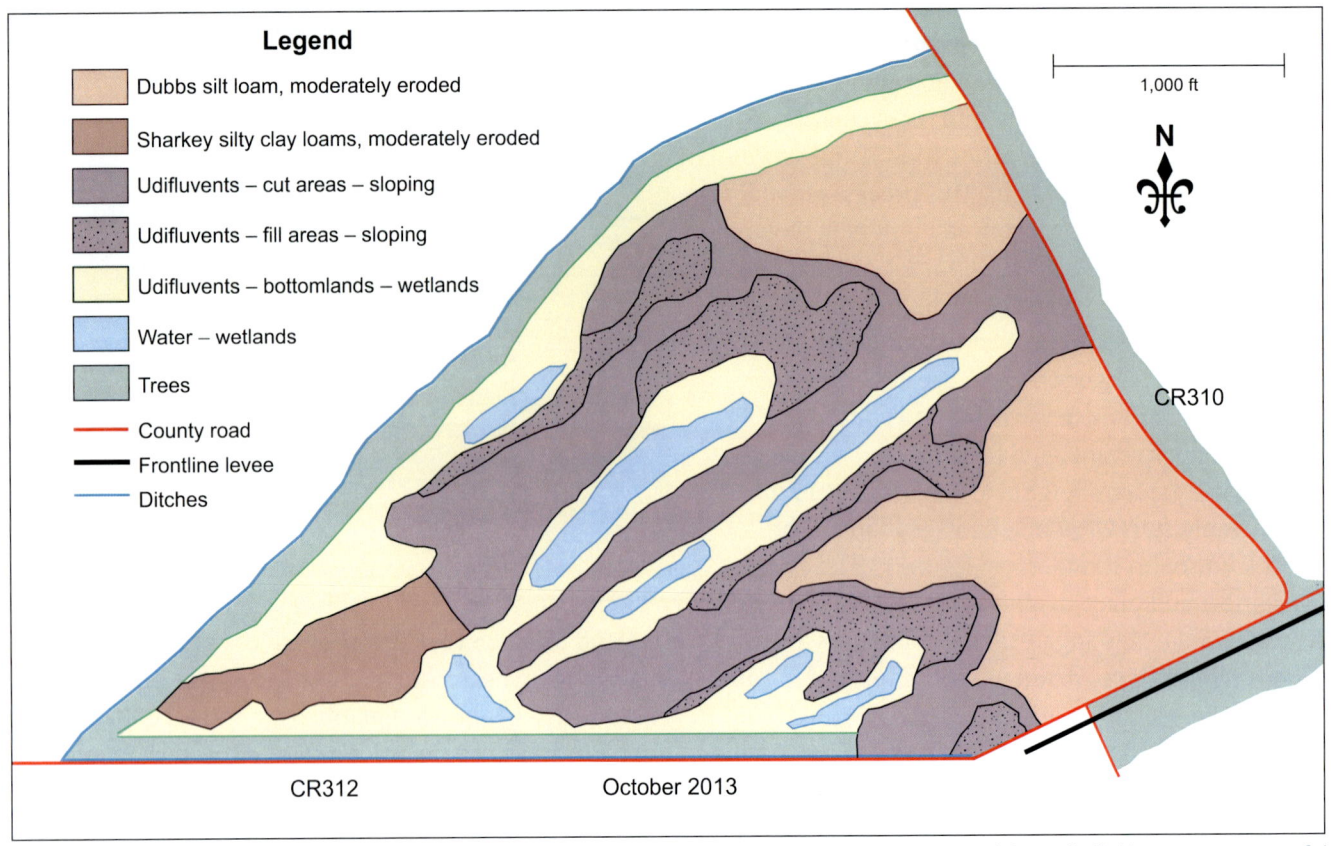

MAP 13.3a October of 2013 moderately eroded soils of O'Bryan Ridge after reshaping and regrading of the gully field to return more of the land to cultivation.

and offset some of the soybean production gains from the conversion of the remaining 25 acres of wetlands and ponds back to agricultural production. An alternative approach would be to transport, at considerable cost, topsoil in from a nonagricultural area to return all 195 acres to agricultural use.

Conservation Recommendations to Improve Productivity

The attempt to reshape the gullies by regrading and filling with soil to increase soybean production had the unintended outcome of creating rills (figure 13.6). The negative impacts of the cutting and filling operations on soil productivity will need to be remediated by restoring natural ecological functions to the affected soils. Organic matter additions to the soil, such as biosolids and composts, decrease the bulk density of soils, increase infiltration rates and porosity, and contribute to better soil structure [8, 9, 10]. The use of cover crops can also be an important management tool for protecting soils from erosion during dormant seasons and for reestablishing key soil quality indicators [11, 12].

The Udifluvents slope had little soil organic carbon and aggregation in the surface layer. These madeland soils are now on 2% to 6% slope with an erosion rate above 30 tons per acre. Many rills were created in 2013 as a result of soybean production on these sloping soils (figure 13.6). A conservation plan is needed for the entire 195-acre area but had not been created as of 2014. The new sloping land created in the spring of 2013 (map 13.3a) is too erosive and no longer suited to continuous soybean production without conservation practices. A terrace system and contour farming with grassed waterways will likely be needed to retain production under continuous soybean. No-till management does not provide a mechanical method to eliminate annual rills, which can quickly turn into new gullies. Alternatively, continuous soybean could be modified in favor of a rotation with corn, forages, and wheat, which could make the soil less vulnerable to erosion. Had the 195-acre field contained grassed waterways and been planted to winter wheat or a cover crop and/or forages when the flooding occurred, the land scouring and gully formation would have been diminished. The middle and lowest parts of the gullies are no longer suitable for row crops and represent a significant land use change to wetlands and ponds.

If one were to have measured to a three-foot depth the soil organic carbon (SOC) stock of the entire 195-acre soybean field on O'Bryan Ridge gully field in 2010 and compared it to the 2015 values, SOC losses would have been significant. An estimated 25% of the SOC in the top three feet was removed (51 of 195 acres) by gully erosion.

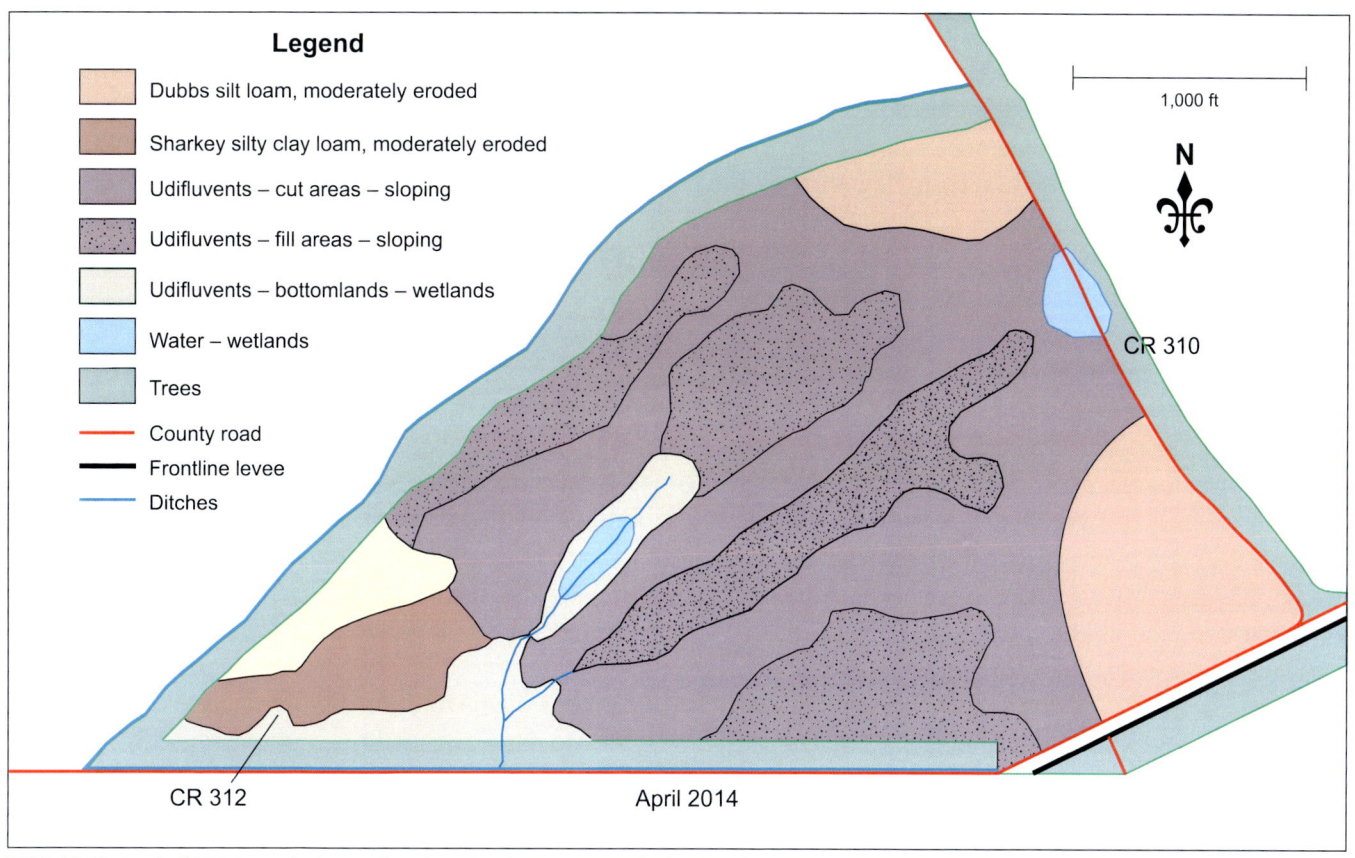

MAP 13.3b April of 2014 ponds drained and wetland areas partially filled with topsoil and subsoil from the Dubbs silty loam to create more cultivated land.

FIGURE 13.6 Rill and gully erosion occurred after planting soybean in 2013 on regraded slopes and filled land.

Further, an additional 25% loss of SOC is likely from the ridgetops due to land scouring, bulldozing of ridgetop topsoils and subsoils into the deep gullies, burying SOC-rich soil below three feet of the soil surface, and as a result of continuous production of soybeans. Without a cover crop or small grain or legume rotation, no significant plant biomass was on O'Bryan Ridge in May of 2011 to help collect SOC-rich sediment and hold the soil on the 195-acre tract in place. Approximately 100% of the SOC-rich sediment (50% of the total SOC stock) was removed from the gully field and delivered to the stream. About 50% to 70% of these eroded and transported carbon stocks were deposited on other alluvial or bottomland soils outside the 195-acre tract, 10% to 30% were retained in the stream, and 20% were lost to the atmosphere.

Road Infrastructure

The main 0.6-mile-long gully remains partially open, and future waters would pond here if the New Madrid Floodway were opened again. Road CR 310 was reconstructed across the largest gully without a culvert (figure 13.2). This road again will block future floodwaters until they flow over the road and with high probability convert the waterways into another gully field. This would dissect CR 310 in many more places. CR 312 was also reconstructed after a gully cut through the road bed (figure 13.2).

Future reclamation efforts after another floodway use would be required and result in even more conversion of agricultural lands to wetlands and ponds. It is recommended that a set of large culverts be placed in the main channel to permit future floodwaters to flow through the partially filled in channel.

Managing Leveed River Bottomlands

The USACE manages more than 14,000 miles of levees protecting Mississippi River and tributary bottomlands in river plains, four floodways, and 12,000 miles of river navigation channel and control structures [1] with goals that include supporting flood risk management activities in communities and restoring aquatic ecosystems [13]. Many of these levees and floodways are financed, built, and maintained cooperatively at the watershed level by local farmers and communities to protect their livelihoods and shared community infrastructures. Extreme flooding events, such as the 1927, 1937, 1993, 2002, 2008, and 2011 floods along the Mississippi River and its tributaries, illustrate the continuing challenges for river communities, industry, and agriculture. These complex issues are related to evolving natural disasters, downstream flooding, and increased water pressure on levee-protected bottomlands [1, 5, 6, 7, 14, 15, 16, 17, 18, 19]. Of particular concern are the vulnerability of low-lying environments that rely on levee protection and the direct impacts of levee breaching on hydrologic patterns, sediment transport and distribution, soil erosion, and land scouring, as well as the indirect impacts on socioeconomic activities, especially agriculture, of flooded areas.

Leveed river bottomlands are designed to protect human populations and various land uses including agriculture from flooding. When a levee fails, the damage caused by floodwaters and contamination of water and land is significant. Water-borne organo-clay sediments often cover plants and soils and fill in road ditches, drainage ditches, and waterways or re-enter water in rivers, streams, and lakes. Usually there are crater lakes created by floodwaters either topping or pouring through the levee breach and substantive gully development [6]. These gullies and land scour areas can extend into the floodplain many miles beyond the breach into fields or along ridges, as seen in the O'Bryan Ridge field study [5].

The case of the O'Bryan Ridge site revealed that 31% of the land use was changed from agricultural use to wetlands and 44% of the agricultural productive capacity was lost as of 2013 due to erosion phase changes and reshaping of the gully sideslopes and soil reconstruction. Additional regrading efforts in 2014 affected the land use and restored some of the soil productivity and yield capacity of the tract to 70%, but there was still a 30% permanent productivity loss. Further, little has been done to prepare the land for the next floodway use. The land scouring and erosional processes remove topsoil, create eroded phases and depositional phases on a soil and sometimes subsoil, and result in less productive soils even if land is reshaped and reclaimed [20]. The effects of sediment deposition and land scouring on soil profiles and productivity need to be determined so agency technical staff, local leadership, and farmers have information to guide decision making in order to effectively return lands to agricultural production and put in place strategies and infrastructure to address future flood events. Findings from the mapping of hydro-geologic patterns, characterization, and measurement of soils and water after being affected by erosion, transport, and sediment deposition as a result of flooding with or without natural and human-induced levee breaches can offer valuable guidance to the restoration of flooded areas and improve decision making, risk analysis, and remedial effectiveness as well as future planning.

The type of vegetation present in the floodplain can have a significant influence on the scouring, transport, and deposition of sediments during a flood event, especially when the floodwater carries a large amount of energy [21]. For example, during the 2011 flood and induced breach of the Birds Point fuse plug levee system, the field closest to the breach contained grassed waterways and a healthy stand of winter wheat, and the soil was mostly protected from scouring (see figure 11.5), whereas an adjacent recently tilled field further from the breach was severely impacted by scouring and loss of topsoil [5]. The wheat residue fields also trapped more sediments than the soybean stubble fields. Feedbacks also exist between natural vegetation and hydrology in floodplains. Floods strongly influence the structure and composition of the vegetation, but vegetation contributes to hydraulic roughness and influences patterns of sediment deposition. As a consequence of these feedbacks, human-caused changes in river hydrology have complex effects on both natural and planted vegetation.

Levees protect public and private lands from the consequences of periodic flooding. However, when they fail naturally or as a result of human induced breaches, the consequences are disastrous and can take different forms. The damages include crop loss; levee damage; crater lakes; gullies; thick sand deltaic deposits; land scouring; irrigation equipment destruction; soil and water degradation; building structure and farmstead damage; filling and blocking of drainage and road ditches; road deterioration; and ecological damage to forests, parklands, and wetlands.

The effects of levee breaches and flooding on soils and soil productivity are seldom determined since updated soil surveys are not routinely made in response to levee breach and flooding. In the case of the O'Bryan Ridge gully field, the damage to soils included the permanent loss of 30% of the agricultural productive capacity as result of land use conversion, land scouring, water erosion, and gully field formation with little deposition of sediments since the rushing floodwaters drained quickly from the 195-acre field except for 21 acres of ponds at the bottom of the 51 acres of deep gullies.

The USACE, the Mississippi River Commission, and the USDA Natural Resource Conservation Service should develop an agreement to immediately update the soil survey maps, conduct a land scouring and deposition survey, and create a soil conservation plan to ensure a rapid federal response after every levee breach and subsequent flooding event. This should be part of the federal government emergency response to a disaster, which provides funds for restoration and repair work, including drainage ditch opening, levee repairs, crater lake filling, gully repairs, restoration of land scoured areas, and sand deposit removal.

---

[1] Morton, L.W., and K.R. Olson. 2014. Addressing soil degradation and flood risk decision making in levee protected agricultural lands under increasingly variable climate conditions. Special issue on Environmental Degradation. Journal of Environmental Protection 5(12):1220-1234.

[2] Park, J.T., P. Seager, S.C. Rao, N. Convertino, and I. Linkov. 2013. Integrating risk and resilience approaches to catastrophe management in engineering systems. Risk Analysis 33(3):356-367.

[3] Barry, J.M. 1997. Rising Tide: The Great Mississippi Flood of 1927 and How It Changed America. New York: Simon and Schuster.

[4] Camillo, C.A. 2012. Divine Providence: The 2011 Flood in Mississippi River and Tributaries Project. Vicksburg, MS: Mississippi River Commission.

[5] Olson, K.R., and L.W. Morton. 2012a. The impacts of 2011 induced levee breaches on agricultural lands of Mississippi River Valley. Journal of Soil and Water Conservation 67(1):5A-10A, doi:10.2489/jswc.67.1.5A.

[6] Londono, A.C., and M.L. Hart. 2013. Landscape response to the intentional use of the Birds Point New Madrid Floodway on May 3, 2011. Journal of Hydrology 489:135-147.

[7] Goodwell, A.E., Z. Zhu, D. Dutla, J.A. Greenberg, P. Kumar, M.H. Garcia, B.L. Rhodes, P.R. Holmes, G. Parker, D.P. Berretta, and R.B. Jacobson. 2014. Assessment of floodplain vulnerability during extreme Mississippi River flood 2011. Environmental Science and Technology 8(5):2619-2625.dx.doi.org/10.1021/es404760t.

[8] Ruehlmann, J., and M. Körschens. 2009. Calculating the effect of soil organic matter concentration on soil bulk density. Soil Science Society of America Journal 73:876-885.

[9] Sloan, J.J., and D. Cawthon. 2003. Mine soil remediation using coal ash and compost mixtures. *In* Chemistry of Trace Elements in Fly Ash, ed. K. Sajwan. New York, NY: Kluwer Academic/Plenum Publishers.

[10] White, C.S., R. Aguilar, and S.R. Loftin. 1997. Application of biosolids to degraded semiarid rangeland: Nine-year responses. Journal of Environmental Quality 26:1663-1671.

[11] Abdollahi, L., and L.J. Munkholm. 2014. Tillage system and cover crop effects on soil quality: I. Chemical, mechanical, and biological properties. Soil Science Society of America Journal 78:262–270.

[12] Stavi, I., R. Lal, S. Jones, and R.C. Reeder. 2012. Implications of cover crops for soil quality and geodiversity in a humid-temperate region in the Midwestern USA. Land Degradation and Development 23:322–330.

[13] National Research Council (NRC). 2012. Corps of Engineers Water Resources Infrastructure: Deterioration, Investment or Divestment. Committee on US Army Corps of Engineers Water Resources Science, Engineering and Planning; Water Science and Technology Board; Division on Earth and Life Studies; National Research Council National Academies Press. http://www.nap.edu/catalog.php?record_id=13508.

[14] Lowery, B., C. Cox, D. Lemke, P. Nowak, K.R. Olson, and J. Strock. 2009. The 2008 Midwest flooding impact on soil erosion and water quality: Implications for soil erosion control practices. Journal of Soil and Water Conservation 64(6):166A, doi:10.2489/jswc.64.6.166A.

[15] Olson, K.R. 2009. Impacts of 2008 flooding on agricultural lands in Illinois, Missouri, and Indiana. Journal of Soil and Water Conservation 64(6):167A-171A, doi:10.2489/jswc.64.6.167A.

[16] Olson, K.R., and L.W. Morton. 2012b. The effects of 2011 Ohio and Mississippi River valley flooding on Cairo, Illinois, area. Journal of Soil and Water Conservation 67(2):42A-46A, doi:10.2489/jswc.67.2.42A.

[17] Olson, K.R., and L.W. Morton. 2013a. Restoration of 2011 flood damaged Birds Point–New Madrid Floodway. Journal of Soil and Water Conservation 68(1):13A-18A, doi:10.2489/jswc.68.1.13A.

[18] Olson, K.R., and L.W. Morton. 2013b. Soil and crop damages as a result of levee breaches on Ohio and Mississippi rivers. Journal of Earth Science and Engineering 3:139-158.

[19] Morton, L.W., and K.R. Olson. 2013. Birds Point–New Madrid Floodway: Redesign, reconstruction and restoration. Journal of Soil and Water Conservation 69(2):35A-40A, doi:10.2489/jswc.68.2.35A.

[20] Olson, K.R., and J.M. Lang. 2000. Optimum crop productivity ratings for Illinois soils. Bulletin 811. Urbana, IL: University of Illinois, College of Agricultural, Consumer, and Environmental Sciences, Office of Research.

[21] Bruneta, R.C., and K.B. Astin. 2008. A comparison of sediment deposition in two adjacent floodplains of the River Adour in southwest France. Journal of Environmental Management 88:651–657.

The 2011 Ohio River Flooding of the Cache River Valley in Southern Illinois

The Ohio River briefly reclaimed much of its ancient floodway through southern Illinois to the Mississippi River in late April and early May of 2011 as heavy rains and early snowmelt over the eastern Ohio River basin raised the Ohio River gage at Cairo, Illinois, to 61.7 feet [1]. The Cache River valley (map 14.1), carved by the ancient Ohio River prior to the last glacial period approximately 14,000 years ago, once again filled with a torrent of water and flooded alluvial bottomlands (see chapter 3). Post Creek Cutoff, a diversionary ditch designed to drain Cache River valley wetlands for agriculture could not drain into the Ohio River as it rose to 21.7 feet above flood stage. As a result, the upper Cache River backed up into Main Ditch (named after the Main Brothers who owned the lumber mill in Karnak, Illinois) and then reversed its flow west into the middle Cache River through the 2002 Karnak breach reaching as far west as the Cache Wetland Center and Route 37. On the west side of the Cache River valley, the rain-swollen Ohio River overwhelmed the upper Mississippi River at the confluence and caused it to back up. The backed-up upper Mississippi water then pushed into the 1950 diversion channel in Alexander County, Illinois, constructed at mile marker 15 on a horseshoe bend in the river. This prevented the western portion of the Cache River from draining local creeks and streams into the Mississippi River.

More than 7.5 inches of local area rain fell from April 30, 2011, to May 2, 2011. The already-saturated soils in Alexander County could not absorb the extra water, and with the diversion channel blocked, local waters had no place to drain and flooded Olive Branch, Horseshoe Lake, and Miller City. Concurrently, the backed-up upper Mississippi River at mile marker 39 threatened the integrity of the Len Small and Fayville levee systems, which protect agricultural lands, Horseshoe Lake, homes, and rural towns (map 14.1) [2]. On the morning of May 2, 2011, the backed-up Mississippi River floodwater caused the Len Small levee to fail. This was just hours before the US Army Corps of Engineers (USACE) deliberately breached the Birds Point levee in Missouri to open the New Madrid Floodway at 10:00 PM on May 2, 2011, and relived the pressure at the Cairo confluence. Sandbagging efforts in Miller City and Olive Branch were able to protect most homes from backed-up local floodwaters but not from the Mississippi River floodwaters that poured through the Len Small levee breach (see chapter 15). The history and geomorphic features of the Cache River basin and the three constructed outlets used to drain the alluvial valley help explain the impacts of the 2011 flood on southern Illinois.

Cache River Valley Alluvial Plain

Formed by the meltwaters of at least four glacial advances and retreats over the last million years, the ancient Ohio River valley is 50 miles in length and 1.5 to 3 miles wide. Today the remains of the ancient Ohio River, which once joined with the Mississippi River northwest of Cairo, Illinois (map 14.1), can be seen in the swamps, sloughs, and shallow lakes of southern Illinois's Cache River valley. During the Woodfordian period (30,000 years BP), the floodwaters from the historic Ohio River watershed drained into eastern Illinois via Bay Creek to the northwest and then west through the Cache River valley (map 14.1) to present-day Alexander County where it converged with the Mississippi River west of the Horseshoe Lake State Fish and Wildlife Area [2]. The middle Cache River valley is 1.3 miles wide as a result of the previous river having been much larger since it carried waters from the ancient Ohio River in addition to the local waters from the upper Cache River valley to the ancient Mississippi River (see chapter 3).

Glacial flooding carved the valley deeply into the bedrock, and then, as the water receded, backfilled the valley with sediments. Deep deposits of gravel and sand on the bedrock floor of the middle and eastern stretches of the valley [2] provide evidence of this glacial flooding. With increasing sediment fill and changes in climate, the Ohio River shifted away from the Cache River into its present course. As a result, the Cache River became a slow-moving stream with extensive isolated, low, swampy areas (figure 14.1) and a water table that ebbed and flowed with seasonal precipitation [3].

1912 to 1915 Post Creek Cutoff

The Cache River basin drains 524,786 acres while meandering 110 miles throughout southern Illinois before emptying into the Mississippi River through a diversion ditch (figure 14.2) southwest of Mound City, Illinois [4]. In 1905, 250,000 acres of the Cache River watershed was considered to be too wet and worthless for farming. The Cache River Drainage District was created in 1911 with the specific purpose of constructing the Post Creek Cutoff to drain the northern region of swamps and sloughs to create agricultural lands [5]. The Post Creek Cutoff (map 14.1 and figure 14.3) was constructed between 1912 and 1915 and rerouted the upper Cache River water into the Ohio River southeast of Karnak, Illinois. By 1916 the 4.8-mile-long cutoff was 30 feet wide and 10 feet deep [6] and diverted 60% of the upper Cache River water due south and into the Ohio River. The Post Creek Cutoff gradient was 12 inches per mile near Grand Chain Bridge, and the diversion has become a canyon in size. The steep gradient and straight channel accelerated flows and started a severe erosion process. The

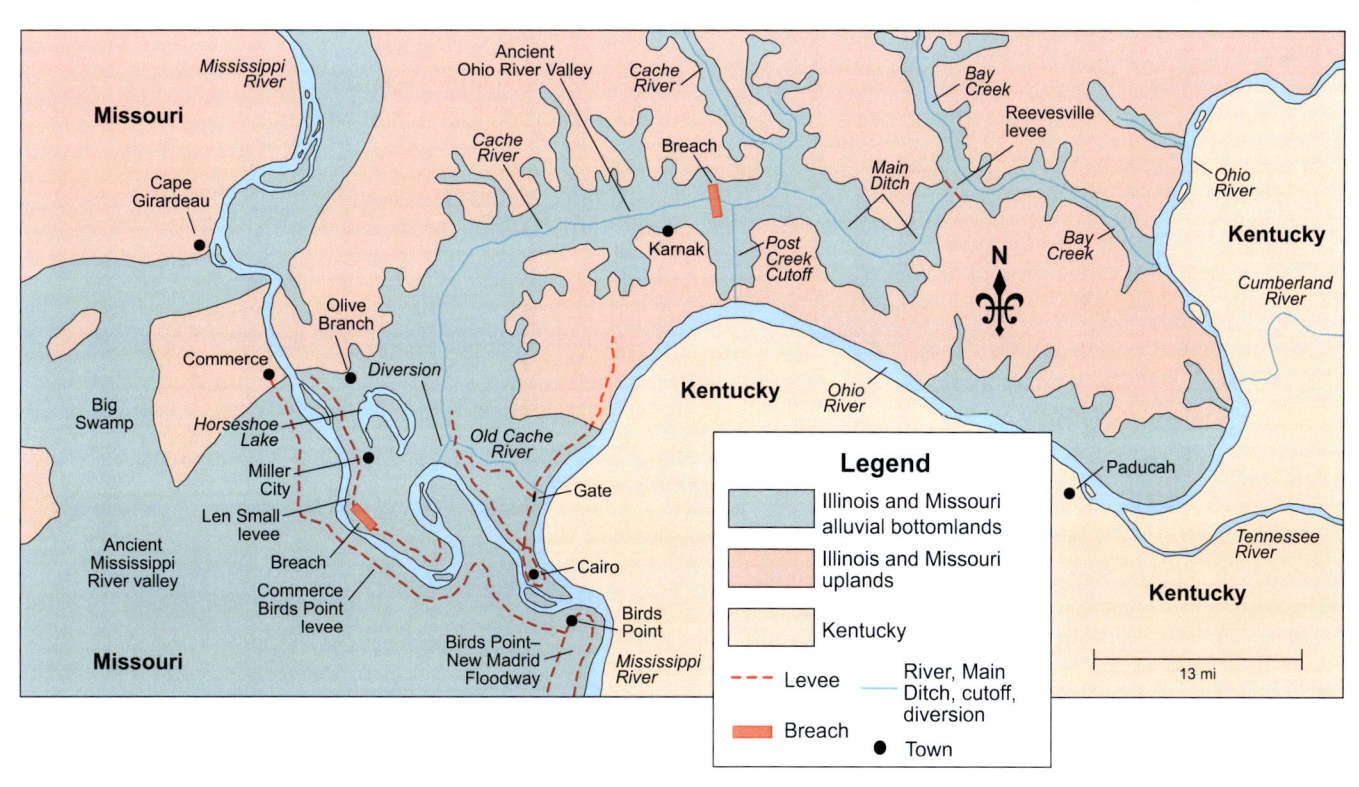

MAP 14.1 Map of the Cache River valley in southern Illinois, including Bay Creek, Post Creek Cutoff, Reevesville levee, and the breached Karnak levee.

FIGURE 14.1 Reestablishing wetland habitats at Grassy Slough near Karnak in the middle Cache River watershed.

FIGURE 14.2 The diversion embankment redirects the middle Cache River water into the Mississippi River and blocks it from draining into the Ohio River via the lower Cache River.

southern part of the channel, an old creek, is now 200 feet wide and 64 feet deep and represents severe gully advance and formation (figure 14.3).

Main Ditch

Following the completion of the Post Creek Cutoff, the Main Ditch (figure 14.4) and other laterals were constructed through the Black Slough area, expanding the agricultural productivity of the region. About the same time period (1912 to 1915), the Reevesville levee (map 14.1) was built by the USACE to prevent Bay Creek from connecting to Main Ditch and the middle Cache River when the Ohio River was high and flooding agricultural lands in the middle Cache River valley. After the 1937 flood overtopped the Reevesville levee, the USACE built a higher structure that has since prevented such flows (figure 14.5) [7].

The Main Brothers Box and Lumber Company used Main Ditch and side ditches to float logs to a sawmill in Karnak where the logs were stored in Heron Pond [3]. Increased water velocities through the Post Creek Cutoff and laterals resulted in headward gully migration, scoured channels, and eroded banks 20 miles up the upper Cache River, the Main Ditch, laterals, and streams. When the swamp levels were higher than the river, the natural levee that separated the swamp and the river degraded and underground piping occurred with gullies threatening to drain Heron Pond as well as other swampy areas.

Over time the Main Ditch (figure 14.4) and Post Creek Cutoff (figure 14.3) deepened and widened, as did many side streams and ditches. These laterals became gullies, some more than one mile long extending into adjacent farm fields, which created field equipment access problems and loss of usable farmland. The Main Ditch and Post Creek Cutoff also lowered the water table and resulted in loss of natural springs. Subsequently, these changes in hydrology have led to significant changes in natural plant and animal communities. The ditches drained water out of Black Slough and reduced the natural flood retention capabilities. Large silt deposits carried by the gullies and the series of laterals ended up at the mouth of the Post Creek Cutoff (figure 14.3) and washed into the current Ohio River. Annual dredging is now required to keep the Ohio River navigation channel deep enough for river traffic.

The approach taken to convert the Cache River valley from forested wetlands to agricultural use was the one used to drain the Big Swamp southwest of Cape Girardeau, Missouri, and west of Commerce, Missouri (see chapters 5 and 6), for agricultural use from 1914 to 1928. Both areas required extensive drainage ditches before the timber could be removed and levees constructed to control the floodwaters. The bottomland alluvial soils are similar in both Big Swamp and Cache River Swamp, but the source of the alluvial sediment is different. Big Swamp was flooded by the ancient Mississippi River (map 14.1), and Cache River Swamp was flooded by the ancient Ohio River. These two swamps joined at the confluence of the two ancient rivers in the area west of Horseshoe Lake and south of Commerce, Missouri, which is 35 river miles north of the current confluence. This occurred before the Mississippi River created a new channel at the Thebes Gap to the north of Commerce, Missouri, and the Ohio River rerouted south joining the Tennessee River at Paducah, Kentucky, after seismic activity and the last glacial advance.

1950 Mississippi River Diversion

In 1950, 11 miles north of the lower Cache River's natural mouth at the Ohio River, a diversion channel and dike diverted the middle section of the Cache River directly into the Mississippi River (map 14.1). The dike on the northeast side of the Mississippi River diversion prevented the middle Cache River water from entering the last 11 miles of the lower Cache River channel. There was also a levee

FIGURE 14.3 The Post Creek Cutoff dug in 1912 has highly eroded banks and sediment loads that are deposited at its mouth as it drains into the Ohio River.

built on the south side of the lower Cache River channel, and the lower Cache River now drains through a gate and into the Ohio River above Cairo.

1952 Karnak Levee and 2002 Breach

At the point where the Post Creek Cutoff diverted the upper Cache River directly south to the Ohio River, the Karnak levee with a floodgate was built in 1952 (figure 14.6). When the upper Cache River levels were high, the floodgate was opened and water flowed west through the middle Cache River valley and channeled into the Mississippi River diversion. Land clearing and drainage efforts increased soil erosion and sediment transport. During the 1970s and 1980s, 12 inches of sediment were deposited near the end of the Post Creek Cutoff. As the Post Creek Cutoff deepened and widened, it resulted in the upper Cache River water flowing directly into the cutoff and bypassing the middle Cache River even when the Karnak gate was open.

In 2002, the 50-year-old Karnak levee gate suffered a catastrophic failure during the spring flood season. The levee breach and gate failure resulted in dropping water levels in the middle Cache River. The effect of this failure is that the gate is now open all the time (figure 14.6) and cannot be closed when the middle Cache River experiences high water. During these high water events, the direction of the water reverses, and it flows east through the Karnak breach and into the Post Creek Cutoff to the Ohio River. When the water level in the middle Cache River drops below the Post Creek Cutoff water level, the flow resumes its westerly travel to outlet at the Mississippi River via the 1950 diversion. As a result of the Karnak breach, which has not been repaired, it is more difficult to control water levels, and the middle section of the Cache River, Main Ditch, and Post Creek Cutoff are subjected to increased risk of flooding. During the flood of 2011, when the Ohio River was 21.7 feet above flood stage, it entered the Post Creek Cutoff. This blocked the outward flow of the upper Cache River, and Main Ditch and Cache River water began to flow west through the breach in the Karnak levee toward the Wetland Center and Route 37.

The middle section of the Cache River valley, between the Karnak levee breach and the Mississippi River diversion and dike, reroutes all its internal water into the Mississippi River. The northern extension of the lower Cache River levee (dike) blocks the middle Cache River from entering the old channel of the lower Cache River. Now, only water in the lower Cache River watershed passes through the old channel and drains through a gate in the Ohio River levee south of Mound City and north of Urbandale. Thus, the current 92-mile-long Cache River (map 14.1) is dissected with the upper section draining into the Ohio River through the Post Creek Cutoff, the middle section draining west through the 1950 diversion into the Mississippi River, and the lower section of the Cache River draining through a gate and into the Ohio River.

The Great Flood of 2011

In April of 2011 the Ohio River approached a record high of 332 feet above sea level on the Cairo gage. It backed up water into Bay Creek at Golconda, Illinois, and pushed into the easternmost reach of the ancient Ohio River valley entrance. The Reevesville levee (figure 14.5), between Bay Creek and Main Ditch held and blocked the Ohio River floodwaters from entering the Main Ditch and flooding the middle Cache River valley from the east (map 14.1). However, 40 miles downstream, as the water rose above flood stage, the Ohio River was able to enter the Post Creek Cutoff and flow north into the middle Cache River valley and Main Ditch, flooding agricultural lands and cypress swamps. Concurrently, water poured into the middle Cache River valley through the previously breached (2002) Karnak levee (figure 14.6) and flowed west into the middle section of the Cache River valley

At the same time, the Ohio River, at 21.7 feet above flood stage, backed up the Mississippi River into the diversion channel, filled the channel, and blocked the drainage of local floodwater. Miller City and Horseshoe Lake were subject to flooding by local rainwater prior to the Len Small levee breach, which occurred on the morning of May 2. At the time of the Len Small levee breach, the Mississippi River was approaching a peak of 332 feet above sea level [2]. The elevation of the water in Horseshoe Lake is normally 322 feet, so there was a potential gradient of 10 feet. However, local flooding had already caused the Horseshoe Lake level to rise, and once the Len Small breach occurred, the Horseshoe Lake area flooded [2]. Eventually, as river levels dropped, these floodwaters drained back into the upper Mississippi River near Route 3 through the 1950 diversion as well as back through the Len Small levee breach.

Approximately 1,000 acres of agricultural lands protected by the lower Cache River levee flooded as a result of the flood gate being closed on the Ohio River levee (map 14.1). This caused a backup in the lower Cache Creek and flooded adjacent forest-covered alluvial soils (such as Bonnie silt loam, wet; Piopolis silty clay loam, wet; and Karnak silty clay, wet) and flooded the slightly higher cultivated soils (such as Petrolia silty clay loam and the Cape and Karnak silt loams) [8]. These cultivated soils drained by the middle of June of 2011 and were planted to soybeans. The floodwaters left a thin silt and clay deposit on the agricultural lands and crop residue when they receded. These coatings included significant amounts of soil organic carbon, microbes, and pathogens [7, 9]. There was little significant soybean damage or yield reduction on lands outside the levees along the Mississippi, Cache, and Ohio rivers since the flooding occurred during the non-growing season. Had winter wheat been planted in the fall of 2010, the wheat crop would have drowned. Illinois farmers are aware of the flooding potential, especially in the winter and early spring, so they do not typically plant winter wheat. Consequently, there was no crop loss in April and May of 2011.

Hydrologic Challenges in the Modern Cache River Basin

The Cache River basin (map 14.1), which once encompassed more than 614,100 acres across six southern Illinois counties, has changed substantively since the ancient Ohio River receded leaving a slow-moving, meandering river; fertile soils and productive farmlands; deep sand and gravel deposits; sloughs and uplands; and one of the most unique and diverse natural habitats in Illinois and the nation. Land use changes, diversion ditches and levees, loss of wetlands and flood-holding capacity, internal channelization of the Cache River and tributaries, and an ever-changing climate have altered the hydrology of the valley, redistributed soil from fields and ditch banks into the river, and transported tons of sediment during flooding events into both the Ohio and Mississippi rivers. The extensive drainage systems throughout the Cache River valley in the spring of 2011 overflowed the internal system of levees and flooded Massac, Pulaski, and Alexander counties. As the 2011 Ohio River floodwater reclaimed much of its ancient floodway, the extent of these hydrologic changes and their social, economic, and environmental impacts have become more apparent.

The Great Flood of 2011 created a need to reevaluate the 1995 Cache River Watershed Resource Plan. Nine resource concerns were previously identified: erosion, open dumping, private property rights, water quality, continuation of government farm conservation programs, Post Creek Cutoff, open flow on the Cache River, dissemination of accurate and timely information throughout the watershed, and the impacts of wildlife on farming and vice versa. Most of these concerns still

FIGURE 14.4 Main Ditch drains water from agricultural lands in the upper Cache River valley.

FIGURE 14.5 The Reevesville levee protects the Main Ditch from Bay Creek floodwaters.

need to be addressed. Since that plan was created, the Karnak levee, no longer managed by the USACE, was breached in 2002 and not repaired, likely due to cost since it is ineligible for federal funding. The Karnak levee is now in the Big Creek Drainage District Number 2. The Cache River Wetlands Joint Partnership, composed of Ducks Unlimited, the Illinois Department of Natural Resources, the USDA Natural Resource Conservation Service, and the US Fish and Wildlife Service, are considering a restoration project that would repair the Karnak levee breach and gate (figure 14.6) as part of its efforts to restore the flow pattern and water table in the middle Cache River valley. Since 2002, two reports on the Karnak levee repair project have been prepared by the Cache River Watershed Resource Planning Committee [6] and made available to the public by the Center for Watershed Science at the Illinois State Water Survey. It is not clear what, if any, impact the 2011 Ohio River flooding of the Post Creek Cutoff, Main Ditch, and the middle Cache River valley east of Route 37 and the Wetlands Center will have on the plans to eventually repair the Karnak levee breach.

Early snowmelt and excessive precipitation resulted in the Ohio River reaching 61.7 feet on the Cairo gage in May of 2011 and created a mighty challenge for residents of southern Illinois, local soil and water conservation districts, state agencies, and the USACE in their attempts to manage the Ohio and Mississippi rivers and the much smaller Cache River to protect towns, farmsteads, agricultural lands, and wetlands of the Cache River floodplain [5]. If the repair and rebuilding of the valley infrastructure is undertaken, there will need to be a significant investment of human and financial resources to reduce the impacts of future catastrophic events.

---

[1] National Oceanic Atmosphere Administration. 2012. Historic crests. Cairo, IL: National Weather Service, Advanced Hydrologic Prediction Service.

FIGURE 14.6 The unrepaired Karnak levee breach allows Cache River floodwaters to flow west toward the Mississippi River diversion.

[2] Olson, K.R., and L.W. Morton. 2013. Impact of the Len Small levee breach on private and public Illinois lands. Journal of Soil and Water Conservation 68(4):89A-95A, doi:10.2489/jswc.68.4.89A.

[3] Cache River Wetlands Center. 2013. Cache River – State Natural Area. Cypress, IL: Illinois Department of Natural Resources. http://dnr.state.il.us/lands/landmgt/parks/r5/cachervr.htm.

[4] Guetersloh, M. 2002. Big Creek Watershed Restoration Plan: A Component of the Cache River Watershed Resource Planning Committee. Prepared for the Cache River Watershed Resource Planning Committee. Springfield, IL: Illinois Department of Natural Resources.

[5] Olson, K.R., and L.W. Morton. 2014. The 2011 Ohio River flooding of the Cache River Valley in southern Illinois. Journal of Soil Water Conservation 69(1):5A-10A, doi:10.2489/jswc.69.1.5A.

[6] Treacy, T. 2011. The Cache: A disconnected river. Sierra Club Illinois website. Resource Plan for Cache River Watershed, Cache River Watershed Resource Planning Committee, December 1995. Cache River Watershed. http://sierraclubillinois.wordpress.com/category/wild-illinois/cache/.

[7] Olson, K.R., and L.W. Morton. 2013. Restoration of 2011 flood-damaged Birds Point–New Madrid Floodway. Journal of Soil and Water Conservation 68(1):13A-18A, doi:10.2489/jswc.68.1.13A.

[8] Parks, W.D., and J.B. Fehrenbacher. 1968. Soil Survey of Pulaski and Alexander counties, Illinois. Washington, DC: USDA Natural Resource Conservation Service.

[9] Olson, K.R., and L.W. Morton. 2012. The impacts of 2011 induced levee breaches on agricultural lands of Mississippi River Valley. Journal of Soil and Water Conservation 67(1):5A-10A, doi:10.2489/jswc.67.1.5A.

Impacts of the 2011 Len Small–Fayville Levee Breach on Private and Public Illinois Lands

15

When the critical Len Small levee failed in 2011, a 2,000-foot breach was created, and fast-moving water scoured farmland, deposited sediment, and produced deep gullies and a crater lake (figure 15.1). The Len Small levee, built by the Farmer Levee and Drainage District on the southern Illinois border near Cairo to protect private and public lands from 20-year floods, is located between mile marker 21 and mile marker 35 (map 15.1). It connects to the Fayville levee that extends to Mississippi River mile marker 39, resulting in a combined length of 18 miles protecting 30,000 to 60,000 acres of farmland and public land, including the Horseshoe Lake State Fish and Wildlife Area. The repair of the breached levee, crater lake, gullies, and sand deltas began in October of 2011 and continued for one year.

Western Alexander County

The Mississippi River is a meandering river with continuously changing paths. Its historic shifting is particularly visible in western Alexander County, Illinois, where a topographical map shows many curves and an oxbow lake, Horseshoe Lake, where the river once flowed south of Thebes (map 15.2) and east of the modern-day Len Small levee. Historically, the region has been a delta, confluence, and bottomlands, and many Illinois lands have been located on both sides of the upper Mississippi River as its channel changed over time (see chapter 3). As a result, the fertile farmland of western Alexander County soils formed in alluvial and lacustrine deposits.

Horseshoe Lake (figure 15.2), a remnant of a large meander of the Mississippi River, is now a state park of 10,200 acres [1]. This oxbow lake, formerly a wide curve in the river, resulted from continuous erosion of its concave banks and soil deposition on the convex banks. As the land between the two concave banks narrowed, it became an isolated body of water cut off from the main river stem through lateral erosion, hydraulic action, and abrasion. With 20 miles of shoreline, the four-foot-deep lake is the northernmost natural range for bald cypress (*Taxodium distichum* L.) and tupelo (*Nyssa* L.) trees (figure 15.2) and has an extensive growth of American lotus (*Nelumbo lutea*), a perennial aquatic plant, and native southern hardwoods that grow well in lowlands and areas subject to seasonal flooding.

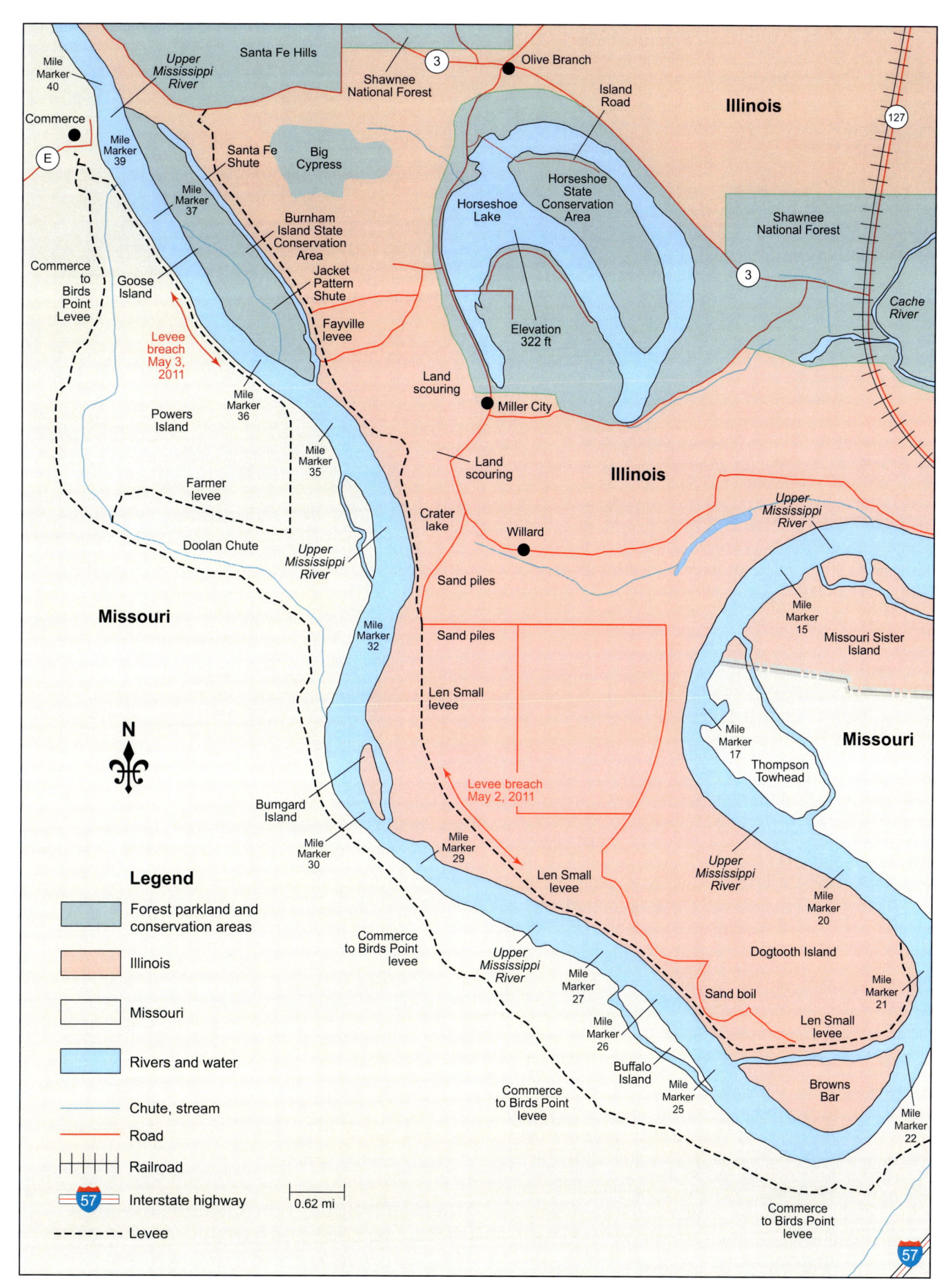

MAP 15.1 Alexander County, Illinois, including the Len Small levee and the northern part of the Commerce to Birds Point levee, Missouri, areas.

115

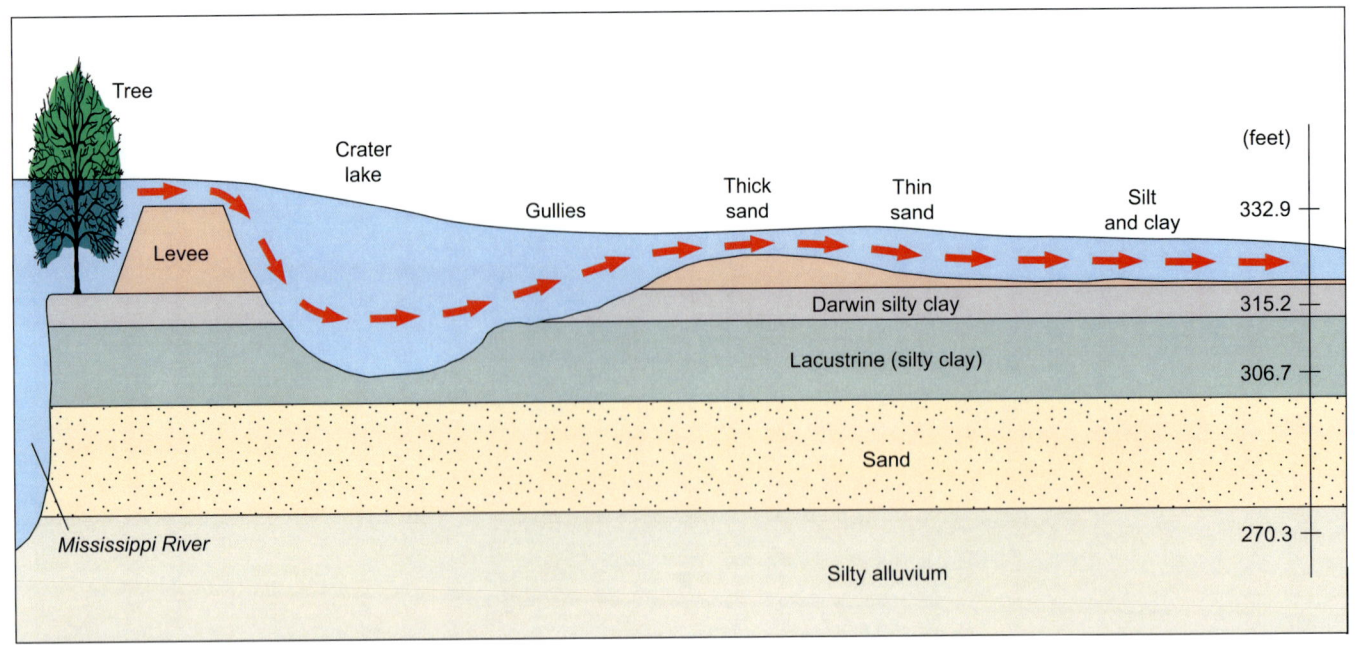

FIGURE 15.1 Diagram of levee topping by the Mississippi River above flood stage, including a crater lake, gullies, and thick sand deposits.

The agricultural lands that surround this oxbow lake are highly productive alluvial soils—mostly Weinbach silt loam, Karnak silty clay, Sciotoville silt loam, and Alvin fine sandy loam. Almost two-thirds of the area (40,000 acres) protected by the Len Small and Fayville levees is privately owned. Corn, soybeans, and wheat are the primary crops, with some rice grown as well.

The Commerce to Birds Point, Cairo, and Western Alexander County Levees

In early May of 2011, the floodwaters at the Ohio River flood gage in Cairo, Illinois, had reached 61.7 feet [2]. The Ohio River, at 21.7 feet above flood stage, was causing a backup in the Mississippi River floodwater north of the confluence at Cairo prior to the USACE opening of the Birds Point–New Madrid Floodway. For more than a month, the Mississippi River backup placed significant pressure on the Len Small–Fayville levees. On the morning of May 2, 2011, approximately 2,000 feet of the Len Small levee breached near mile marker 29 (map 15.1) and flooded agricultural lands.

The flood protection offered by the Len Small–Fayville levees is important to the landowners, homeowners, and farmers in southwestern Alexander County, Illinois. However, the Len Small–Fayville levees are not the mainline levees that control the width and height of the Mississippi River. The controlling mainline levees are the frontline Cairo levee located in Illinois [3] and the Commerce to Birds Point levee in Missouri (figure 15.3). These two frontline levees, by design, are much higher and stronger than the Len Small–Fayville levees. The Len Small–Fayville levees were built by the local levee district and are not part of the Mississippi River and Tributaries Project for which the USACE has responsibility. The Cairo levee has a height of 64 feet, or 334.5 feet above sea level, and levee failure would destroy the city of Cairo. The frontline Commerce to Birds Point levee has a height of 65.5 feet, and its failure would result in the flooding of more than 2.5 million acres of agricultural bottomlands in the Missouri Bootheel and Arkansas on west side of the Mississippi River (map 15.3). The Commerce to Birds Point levee connects to a setback levee on the west side of the Birds Point–New Madrid Floodway, which extends the protection another 33 miles to the south where it joins the frontline levee at New Madrid, Missouri, further extending the protection of the Bootheel bottomlands [3, 4, 5, 6]. The failure of the Hickman (Kentucky) levee on the east side of the Mississippi River would have resulted in the flooding of 170,000 acres of protected bottomlands in Tennessee and Kentucky. The floodwater height and pressure on the Commerce to Birds Point and Birds Point to New Madrid levees has increased over the years during Mississippi River flooding events with the construction of the Len Small–Fayville levees and with a strengthening of the levee near Hickman, Kentucky (map 15.3). This had the effect of narrowing the Mississippi River floodway corridor and removing valuable floodplain storage areas for floodwaters.

The Mississippi River Commission and Its Role in Levee Construction along the Mississippi River and Tributaries

The Mississippi River Commission (MRC) was established by Congress in 1879 to combine the expertise of

FIGURE 15.2 Bald cypress trees and American lotus at Horseshoe Lake conservation area provide wetland habitat for local and migratory birds and a recreational destination for fishing, boating, picnicking, camping, and wildlife observation.

FIGURE 15.3 Routine management of vegetation on the Commerce to Birds Point mainline US Army Corps of Engineers levee protects against woody encroachment and tree roots that can undermine the strength of levee structures.

the USACE and civilian engineers to make the Mississippi River and tributaries a reliable shipping channel and to protect adjacent towns, cities, and agricultural lands from destructive floods [4]. Between 1899 and 1907, the MRC assisted local levee districts in Missouri with construction of a federal levee between Birds Point, Missouri, and Dorena, Illinois. At that time, the MRC jurisdiction was limited to the areas below the confluence of the Ohio and Mississippi rivers [3, 4, 5], which is at the southern tip of Illinois (Fort Defiance State Park). This levee is located approximately where the current frontline levee of the Birds Point–New Madrid Floodway was constructed between 1928 and 1932 after the Birds Point to Dorena frontline levee failed in 1927.

In 1902, the MRC helped Kentucky construct a levee from the Hickman, Kentucky, bluff to Tennessee, where it connected with another levee to extend the levee system five miles to Slough Landings, Tennessee. During this time period, a portion of the natural floodplain near Cape Girardeau was walled off by a local Missouri levee to provide protection of farmland adjacent to the river (map 15.1). These two levees narrowed the river channel during high-water events on the Mississippi River and increased floodwater backup, placing tremendous pressure on the existing systems of levees and floodwalls above and below the Cairo confluence [3, 4, 5].

The Commerce to Birds Point levee (figure 15.3) has long been considered by the MRC and the USACE to be the most critical levee in the Mississippi River valley. The Commerce to Birds Point levee, shown in maps 15.1, 15.2, and 15.3, has had two major threats from past major flooding events (1973 and 1993). During the 1973 flood, a 1,500-foot section of the Commerce to Birds Point levee fell into the Mississippi River. The caving extended to the top of the levee. The USACE Memphis District placed 18,000 tons of riprap stone carried in by barges to prevent additional caving [4]. The Len Small–Fayville levee on the Illinois side of the Mississippi River (map 15.1) and across from the Commerce to Bird Point levee, Missouri, had historically overtopped or failed during larger flooding events, thereby reducing the pressure on the Commerce to Birds Point levee. The local levee and drainage district and owners of the Len Small–Fayville levee strengthened their levee during the 1980s, which increased pressure on the Commerce to Birds Point levee when the river rose above flood stage. As a result, in the 1993 flood event, the Len Small–Fayville levee held, and the Mississippi remained confined as it climbed to within 3 feet of the top of the Commerce to Birds Point levee. Sand boils developed in the Commerce levee and were treated until the underseepage stabilized. In 1995, the USACE Memphis District raised the height of and strengthened the Commerce to Birds Point levee and installed relief wells.

Local and Mississippi River Flooding of Farmland and Towns Located in Western Alexander County

The 2011 flood and record peak on the Ohio River caused the already-flooded Mississippi River near the confluence to back up for many miles to the north and affected all bottomlands in Alexander County, Illinois, that were located on the east side of the upper Mississippi River (map 15.1). Since the gradient on the Mississippi River is between 0.5 to 1 foot per mile, the Mississippi River water rose an additional 18 feet above flood stage further north. This occurred at a time when the Ohio River

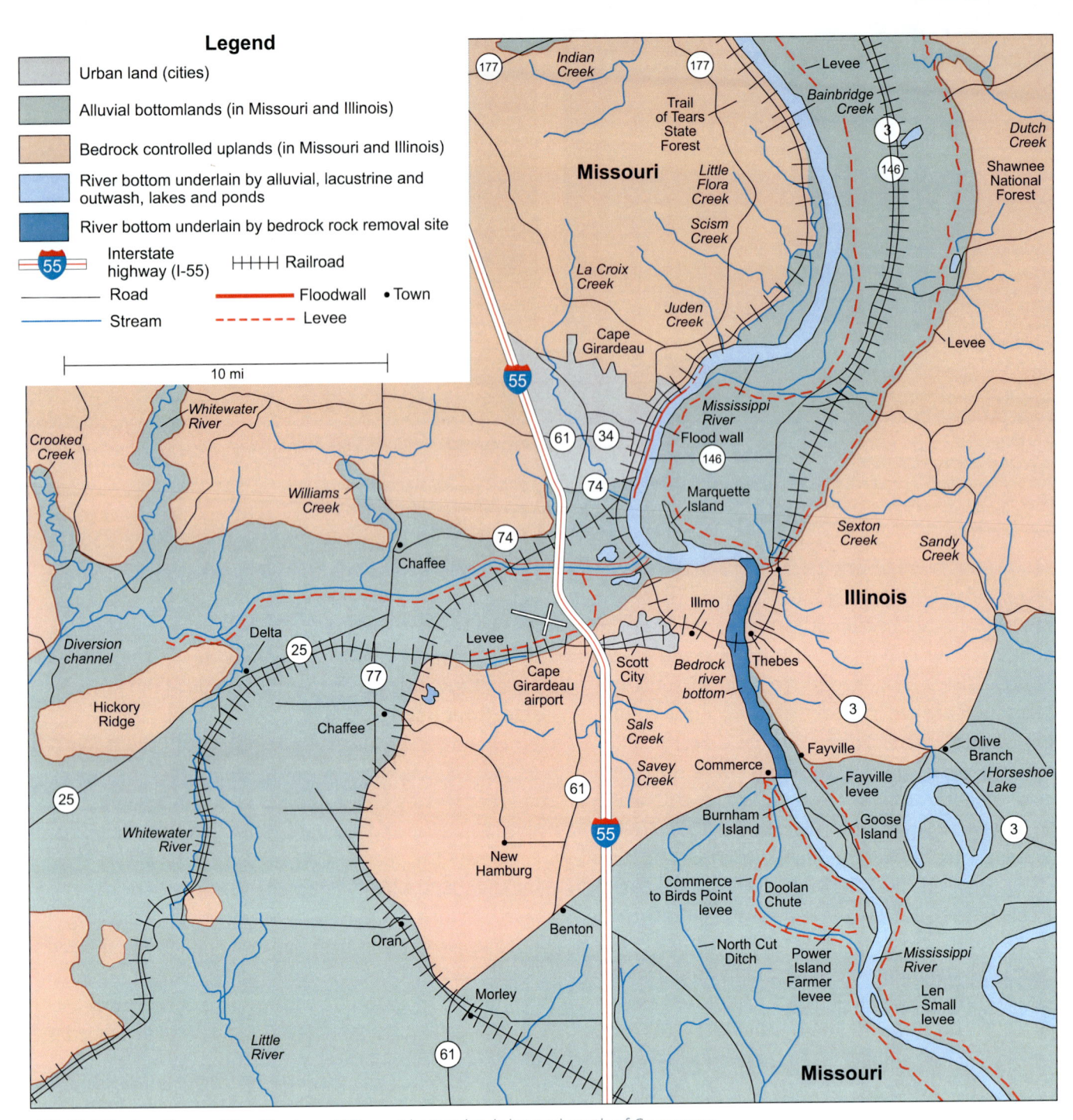

MAP 15.2 Thebes Gap and the Illinois and Missouri bottomlands located south of Commerce.

was 21.7 feet above flood stage and the Mississippi River north of Cape Girardeau, Missouri, was 9.9 feet above flood stage. Cities farther to the north like St. Louis, Missouri, were only subjected to floodwaters 6.6 feet above flood stage as a result of water flowing from the upper Mississippi and Missouri rivers.

The May 2, 2011, topping and breach of the Len Small levee occurred just a few hours before the pressure of record flood levels was relieved with the opening of the Birds Point–New Madrid Floodway. Homeowners in Olive Branch and Miller City are convinced that had the New Madrid Floodway been opened according to the 1986 New Madrid Floodway operational plan they would have been able to save their homes from flood damage. The operation plan called for the floodway to be prepared for operation by the time the river reached 60 feet, which actually occurred on April 30. However, the site preparations had not been ordered by the MRC, so the Birds Point levee fuse plug was not activated and could not be opened on that date. The fuse plug was filled with TNT on May 1 and 2, and New Madrid Floodway was opened about 10:00 PM on May 2. The Len Small levee breach on the Mississippi River occurred on the morning of May 2, 2011, and flooded homes just hours before the

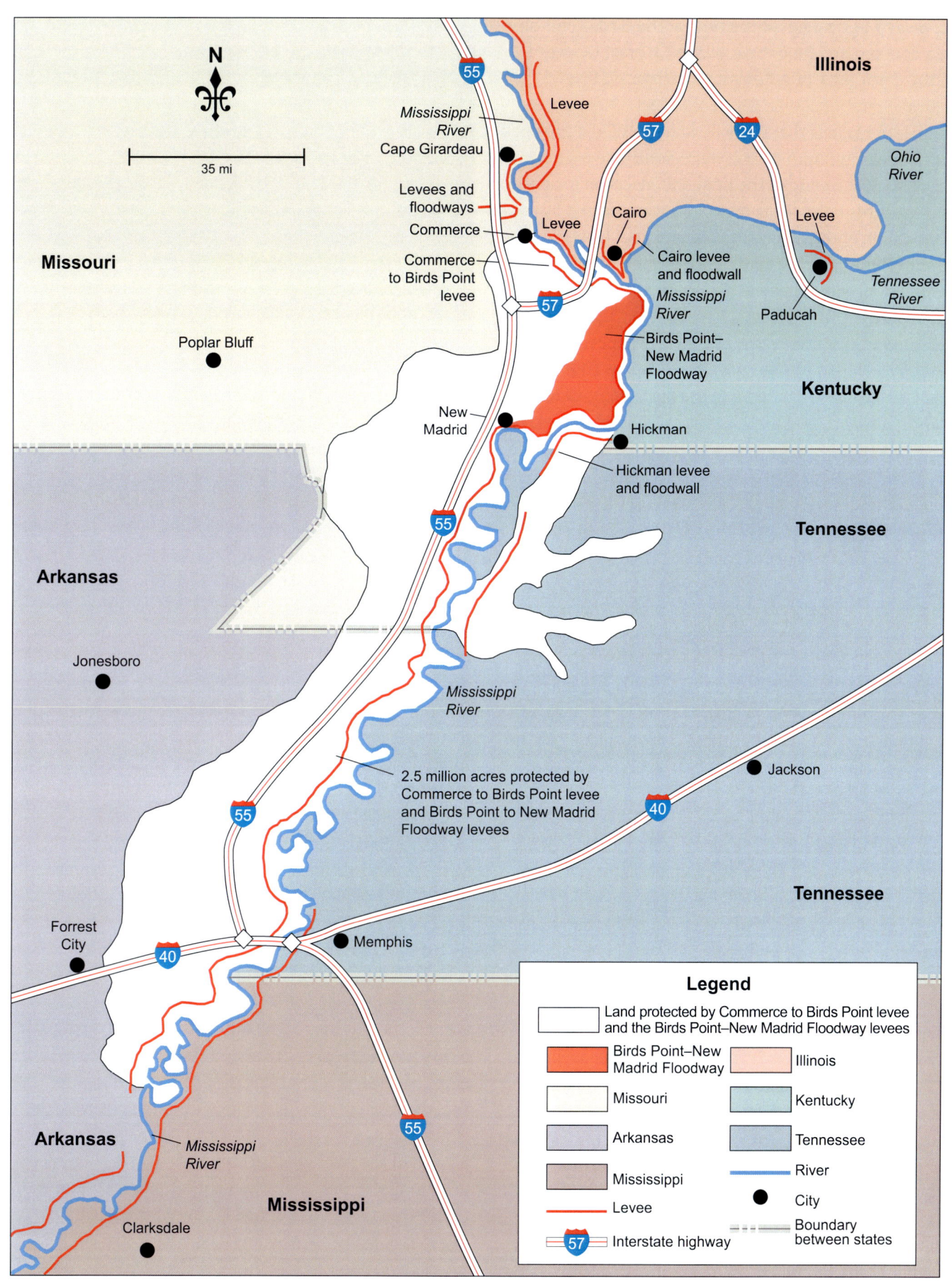

MAP 15.3 Former river bottomlands in Missouri and Arkansas are protected by the Commerce to Birds Point main-line levee, and bottomlands in Tennessee and Kentucky are protected by the Hickman levee.

119

floodway was opened. At the time of the Birds Point fuse plug opening, the Ohio River was at 61.7 feet (a record) on the Cairo gage, or 1.7 feet higher than 1986 New Madrid operational plan depth of 60 feet. There were a number of reasons that the MRC and USACE did not open the floodway on April 30 and waited until the evening of May 2 [4]. The timing of the final MRC decision to activate the Birds Point levee fuse plugs was affected by the reactivation of the mega sand boil in Cairo, heavy local rains in the area of the confluence of the Ohio and Mississippi rivers, and the new peak forecast of 63.5 feet [4]. These on-the-ground events happened on May 1, 2011, the day the Supreme Court refused to accept the lawsuit filed by the Missouri Attorney General in an attempt to block the USACE from opening the Birds Point–New Madrid Floodway to protect Missouri citizens and property (see chapter 10).

Flooding of Alexander County from heavy local rains resulted in some flooding in the towns of Olive Branch and Miller City in late April and on May 1, 2011. This was before the Len Small breach occurred, and there was some damage to private and public lands prior to the breach. Floodwater from the Ohio River backed up the Mississippi River for miles, pushed into the diversion channel at mile marker 15, blocked the flow of the middle Cache River to the west, and prevented the drainage of local floodwater. Sandbagging efforts prevented local flooding of homes, but those efforts failed once the Mississippi River flowed into the Horseshoe area after the Len Small levee breach.

As a result of Cache River valley floodwater flowing west through the Karnak levee breach, local floodwater from heavy rain, Mississippi River water blocking the drainage of the middle Cache River (see map 14.1) through the diversion, and the additional Mississippi River floodwaters pushing through the Len Small breach, 10,000 acres of farmlands lost the winter wheat crop or were not planted in 2011. About half of that land (mostly Weinbach silt loam, Karnak silty clay, Sciotoville silt loam, and Alvin fine sandy loam) [7] had significant soil damages, including land scouring and sediment deposition, or was slow to drain. Crater lakes, land scouring (figures 15.1 and 15.4), gullies, and sand deltas were created when the Len Small levee breached and removed agricultural land from production [5, 8]. Most of the other farmland in Alexander County dried out sufficiently to permit fall planting of wheat in the fall of 2011. All of Alexander County soils dried sufficiently by the spring of 2012 to allow the planting of corn and soybeans. It is not clear how much 2011 farm income replacement came from flood insurance since not all Alexander County, Illinois, farmers had crop insurance. In addition, roads and state facilities were impacted by local floodwaters and the Mississippi River floodwaters that passed through the Len Small breach.

FIGURE 15.4 Land scouring, gullies, and erosion north of the Len Small levee breach resulted in the loss of agricultural productivity in 2011.

Illinois agricultural statistics recorded that 4,500 acres of corn and 6,500 acres of soybeans were harvested in Alexander County in 2011. The area produced 1,570,000 bushels of corn in 2010 but only 710,000 bushels in 2011. The soybean production level was 1,200,000 bushels in 2010 but dropped to 865,000 bushels in 2011 due to flooding, crop, and soil damage. The floodwaters also scoured the agricultural lands in some places and deposited sand at other locations thereby reducing future productivity.

Flooding of Public and Private Bottomlands with and without Levee Protection in Western Alexander County, Illinois

All bottomlands north of the confluence between the Mississippi River and the western Alexander County levees with an elevation of less than 332 feet above sea level were flooded when the Mississippi River backed up from the confluence. Approximately 30,000 to 60,000 acres of public and private alluvial lands, both levee-protected and without levees, were flooded along the east and north sides of the Mississippi River (map 15.1) between mile markers 12 and 39 [9]. The 1957 to 1963 soil maps of the area show alluvial soils consisting of recently deposited sediment that varies widely in texture (from clay to sand) with stratified layers [7]. The natural vegetation of these alluvial bottomlands ranges from recent growth of willows (*Salix* L.) and other plants to stands of cottonwood (*Populus deltoides* L.), sycamore (*Platanus occidentalis* L.), and sweet gum (*Liquidambar styraciflua* L.).

The map (map 15.1) shows the public and private lands of the southwest Alexander County, Illinois, area that were impacted by the flood of 2011. Approximately one-third of the area (20,000 acres) is in public lands, including uplands (the Shawnee National Forest and Santa Fe Hills) and bottomlands (Burnham Island Conservation, Horseshoe Lake State Fish and Wildlife Area, Goose Island, Big Cypress, and the land adjacent to the Len Small–Fayville levee). The unleveed bottomlands and public conservation areas sustained flood damage but were more resilient than the private agricultural and urban lands inside the levees. The Mississippi bottomlands are riparian forests (transition ecosystems between the river and uplands) with fertile, fine-textured clay or loam soils that are enriched by nutrients and sediments deposited during flooding [10]. Bottomlands that experience periodic flooding have hydrophytic plants and hardwood forests that provide valuable habitat for resident and migratory birds. The Illinois Department of Natural Resources has an extensive research program monitoring migratory birds and waterfowl at Horseshoe Lake. Although these alluvial river bottomland species are well adapted to periodic flood cycles that can last several days to a month or more [10], the impact of the 2011 flood duration (2 to 4 weeks) on these wetland habitats and woodlands has not been assessed.

There are a number of towns and villages in western Alexander County, including Olive Branch, Miller City, and Cache. Floodwaters covered roads and railroads and damaged some bridges, homes, and other building structures. In western Alexander County, floodwater destroyed 25 Illinois homes and damaged an additional 175 homes and building structures located on Wakeland silt loam and Bonnie silt loam soils [7] or similar alluvial floodplain soils. The Olive Branch area (map 15.1) was one of the hardest hit according to Illinois Emergency Management Agency.

Agricultural and forest lands on the river side of the Len Small levee are not protected from flooding and store significant amounts of floodwater with minimal damage to the crops such as soybeans, which can be planted later in the spring or early summer after river water levels have dropped. This farmland was under water prior to planting for the entire months of April and May of 2011. After both the Ohio and Mississippi rivers dropped and drained by late June of 2011, these fields were planted to soybeans. Late May and early June is the normal planting time for soybeans in the area, so a small soybean yield reduction was noted.

Repair of Len Small Levee in Western Alexander County

In the fall of 2011, local farmers and members of the Len Small Levee District patched the Len Small levee. They created a sand berm three feet lower than the original levee. They hoped the USACE would cover the levee with a clay cap and restore it at least to the original height. The USACE agreed to do this in August of 2012 after receiving additional funds from Congress. The project was completed in 90 days. Some individual farmers created berms around their farmsteads (figure 15.5) and homes (figure 15.6) to protect their homes and buildings from any future flooding that might occur [9].

In June of 2012, the USACE received $802 million in emergency Mississippi River flood-repair funding for up to 143 high-priority projects to repair levees, fix river channels, and repair other flood-control projects in response to the spring of 2011 flood, which set records from Cairo, Illinois, to the Gulf of Mexico. Both the Birds Point–New Madrid Floodway levee repair and the Cairo area restoration projects were high on the list, with the USACE targeting $46 million to repair the damage to Cairo area, including the Alexander County area flood control systems [3, 4, 5]. Improvements were completed throughout Alexander County, including work on pump stations, drainage systems, and small levees, some of which failed in April of 2011. These projects were funded by the county matching funds with the USACE and a combination of grants from the Delta Regional Authority and the State of Illinois [11]. The creation of a larger drainage system running through northern Alexander and Union counties included large culverts and levees designed to better protect Illinois communities such as East Cape Girardeau, McClure, Gale, and Ware, and help keep water from collecting in low-lying bottomland areas.

FIGURE 15.5 Post-2011 Len Small levee breach, many farmers built small levees around their farmsteads to protect against internal flooding and future breaching.

FIGURE 15.6 This home, less than half a mile inside the Len Small levee, is surrounded by a farmer-built levee to protect against future flooding.

Local Floodwaters and Levee Breaching

In 2011, the Ohio and Mississippi rivers' flooding resulted in the USACE blasting open the Birds Point levee fuse plug as waters reached a critical height on the Cairo gage and in the confluence area. However, this unprecedented flood level at the confluence put tremendous pressure on and under the Mississippi levees to the north in western Alexander County, Illinois. The 48-hour delay in the decision to activate and to blow up the Birds Point fuse plugs and frontline levees had significant adverse consequences for rural Illinois landowners, farmers, and homeowners in Alexander County near the Len Small levee. Local flooding and damage to building structures, crops, and soils initially occurred in late April of 2011 when the Ohio River backed up the Mississippi River, which flowed into the diversion channel and filled the channel, leaving no place for runoff from a heavy local rain to drain. Consequently, the towns of Olive Branch and Miller City were flooded by local water even before the Len Small breach occurred, but homes and building were protected by sandbagging. After the Len Small levee breach, the Mississippi River water topped the sandbags and flooded the homes. Even if the Birds Point–New Madrid levee had been opened two days earlier at a time when the record level floodwaters were 1.7 feet lower, the prolonged record Mississippi and Ohio river floodwater levels and pressure on the Len Small levee, which continued for weeks, may still have resulted in the Len Small levee breach a few days later when the river was lower.

[1] Illinois Department of Natural Resources. 2012. Horseshoe Lake. Springfield, IL: Illinois Department of Natural Resources. http://dnr.state.il.us/Lands/landmgt/parks/R5/HORSHU.HTM.

[2] National Oceanic Atmosphere Administration. 2012. Historic crests. Cairo, IL: National Weather Service, Advanced Hydrologic Prediction Service.

[3] Olson, K.R., and L.W. Morton. 2012. The effects of 2011 Ohio and Mississippi river valley flooding on Cairo, Illinois, area. Journal of Soil and Water Conservation 67(2):42A-46A, doi:10.2489/jswc.67.2.42A.

[4] Camillo, C.A. 2012. Divine Providence: The 2011 Flood in Mississippi River and Tributaries Project. Vicksburg, MS: Mississippi River Commission.

[5] Olson, K.R., and L.W. Morton. 2012. The impacts of 2011 induced levee breaches on agricultural lands of Mississippi River Valley. Journal of Soil and Water Conservation 67(1):5A-10A, doi:10.2489/jswc.67.1.5A.

[6] Olson, K.R., and L.W. Morton. 2013. Restoration of 2011 flood-damaged Birds Point–New Madrid Floodway. Journal of Soil and Water Conservation 68(1):13A-18A, doi:10.2489/jswc.68.1.13A.

[7] Parks, W.D., and J.B. Fehrenbacher. 1968. Soil Survey of Pulaski and Alexander counties, Illinois. Washington, DC: USDA Natural Resource Conservation Service.

[8] Olson, K.R. 2009. Impacts of 2008 flooding on agricultural lands in Illinois, Missouri, and Indiana. Journal of Soil and Water Conservation 64(6):167A-171A, doi:10.2489/jswc.64.6.167A.

[9] Olson, K.R. and L W Morton. 2013. Impacts of 2011 Len Small levee breach on private and public lands. Journal of Soil and Water Conservation 68(4):89A-95A, doi:10.2489/jswc.68.4.89A.

[10] Anderson, J., and E. Samargo. 2007. Bottomland Hardwoods. Morgantown, WV: West Virginia University, Division of Forestry and Natural Resources. http://forestandrange.org/new_wetlands/index.htm.

[11] Koenig, R. 2012. Corps balancing levee repairs on Missouri, Illinois sides of Mississippi. St. Louis Beacon. January 11, 2012. https://www.stlbeacon.org/#!/content/14295/corps_balancing_levee_repairs_on_missouri_illinois_sides_of_mississippi.

The City of Cairo, Illinois, at the Confluence of the Mississippi and Ohio Rivers

The Ohio River began flooding farmland and cities from Pennsylvania to Illinois that were not protected by levees in April of 2011. The US Army Corps of Engineers (USACE) realized as early as March that the torrential rains and heavy snowmelt across the upper Midwest were setting up the Mississippi River basin for an epic flood year [1]. By late April, lakes and reservoirs along the Wabash and Ohio rivers were filled to capacity, and lower sections of cities without levees, such as Metropolis, Illinois, were covered by floodwaters. However, Cairo, Illinois, and many of the cities on the lower Mississippi River are protected by levees and floodwalls (figure 16.1), and residents expected they were safe from flooding. By the end of April of 2011, the floodwaters on the levee and floodwall at Cairo, Illinois (map 16.1), had reached 61 feet and were rising. These floodwaters were starting to put significant pressure on the Cairo floodwall and levees with some seepage and sand boils occurring. The people of Cairo and their city infrastructure would be at great risk if their levee system failed.

On May 2, 2011, the USACE made the decision to blow up Birds Point levee fuse plug on the Mississippi River and flood agricultural lands in the New Madrid Floodway, Missouri, to protect the city of Cairo, Illinois, and prevent its levee and floodwall system (figure 16.2 and map 16.1) from breaching. This was a calculated risk built on a growing body of river science and prior flooding experiences. The decision was a difficult and complex engineering problem with significant social and political trade-offs between loss of human lives and loss of properties in urban and rural areas.

The Ohio and Mississippi Rivers

Cairo has a history of battling two rivers that are its economic lifeblood but also its greatest source of unease and vulnerability. The approximately 10,000 acres between the Ohio and Mississippi rivers where the city is located today were originally an 1818 land trust owned by land speculator investors from New York and Philadelphia [2]. The first levee system around Cairo was completed in 1843 to secure Cairo from the abrasions of the Ohio and Mississippi rivers and make riverfront and interior city lots desirable for purchase. Early engineers noted important differences between the Ohio and Mississippi rivers that had implications for

FIGURE 16.1 The Cairo floodwall is built on the Ohio River side at the bend in the river where an earthen levee would be difficult to maintain. With the river 10 feet above flood stage, a tugboat is visible behind the floodwall.

FIGURE 16.2 An open floodwall gate in Cairo shows city buildings and industries adjacent to the protective wall. The wall has never been breached, and high water crests since 1937 are marked to show the Ohio River height as recorded on the Cairo gage.

construction of levees and offer explanations to Cairo's 2011 vulnerability to flooding and levee crevassing. An 1807 government surveyor wrote, the "Mississippi devours its banks and changes its current from place to place unless restrained..." and its rapid whirling current is loaded with sand and silt [2]. In contrast, the Ohio River had clearer water and slower movement with the origins of its waters from the forested lands of Pennsylvania, Indiana, and Ohio to the northeast as well as the Tennessee River, the largest tributary, from Virginia, West Virginia, Tennessee, and Kentucky to the south. After many years of farming in the Ohio River valley, the sediment load has increased substantively, and the water is now a yellowish brown color, reflecting erosion of timber soils (Alfisols). Since the 1930s, an extensive federal bank stabilization and revetment program has reduced the land scouring and Mississippi River meandering threats to levees in the confluence area.

The Ohio River posed the greatest danger to Cairo in the 1880s. It was observed that the river "claims for itself the right to rise and fall through a perpendicular distance of fifty feet" and filled with heavy early spring rains from the east [2]. The 1867 flood reached Cairo within 24 hours of February 25 and measured 51 feet on the Cairo gage by March. The years 1882, 1883, and 1884 had the highest flood stages ever recorded. The Mississippi River and tributaries north of Cairo often froze over hard and solid with northern spring snowmelts bringing high waters to the Ohio-Mississippi river convergence three to four months (around June 1 or later) after the Ohio peak high waters passed [2].

The historical levees-only strategy of the USACE resulted in construction of levees on both sides of the lower Mississippi River from Cairo, Illinois, to New Orleans, Louisiana [3, 4], as a response to floodwaters from both these rivers and their tributaries. Cairo, located on low-lying alluvial soils at the confluence of the Mississippi and Ohio rivers, did not have a major flood problem until the federal and local agencies extended the levee systems in the vicinity [4]. That levee construction, along with sealing off of the natural diversion through the St. Francis River south and west of Cape Girardeau, Missouri (see chapter 5), represented a source of flood problems at Cairo. In 1843, Cairo was the first city north of New Orleans to build a levee system. The levees-only strategy was modified with the creation of four separate floodways in Missouri (Birds Point–New Madrid Floodway) and Louisiana (West Atchafalaya Floodway, Bonnet Carre Spillway and Floodway, and Morganza Spillway and Floodway). The USACE decision to add floodways was a substantive shift in river management from confinement-only to a dispersion approach intended to divert excess flows during large flood events [4]. In the 1940s Cairo benefited from the USACE building of the Kentucky Dam on the Tennessee River, and in 1960s from the building of the Barkley Dam on the Cumberland River to better control the fast rise of the Ohio River during spring rains (see chapters 19 and 20).

In 2009, Clyde Walton boasted in the foreword to Lansden's 1910 history of Cairo, Illinois, "Today Cairo lives secure behind the mighty levees that protect it from ravage by the two great rivers" [2]. Little did he suspect that the flood of 2011 would challenge the claim that "mighty levees" would protect Cairo. While the reservoirs along the Ohio River tributaries helped lower the 2011 Ohio River high water levels, in the end they were inadequate to prevent downstream flooding and

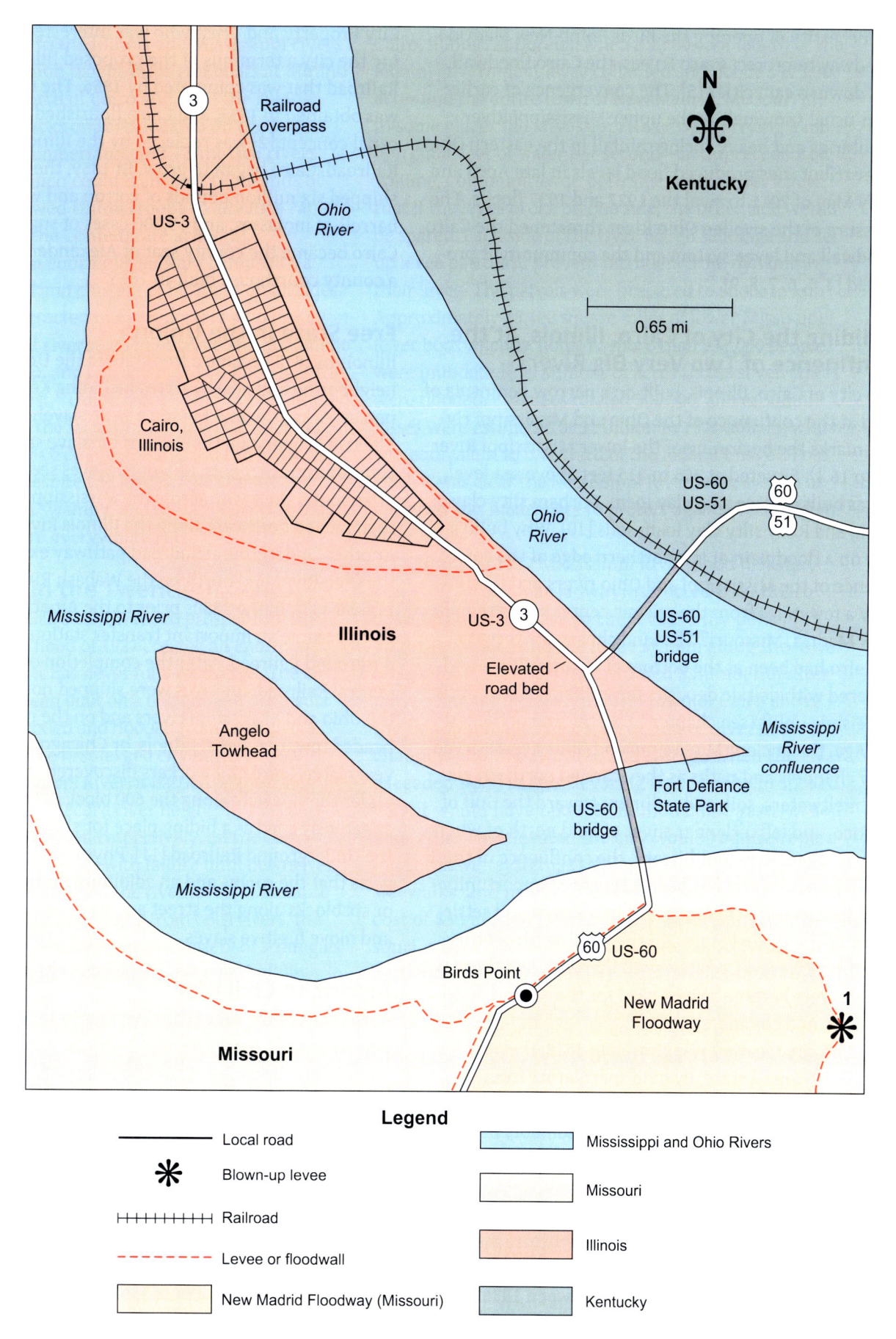

MAP 16.1. Cairo, Illinois, occupies a narrow peninsula created by the confluence of the Ohio and Mississippi rivers. The map shows the proximity of Cairo to Birds Point, Missouri, and the location of the New Madrid Floodway designed to reduce floodwater pressure when the rivers exceed flood stage.

The 2011 Great Flood

The failure of the pre-1927 levees-only policy resulted in the construction of the Birds Point–New Madrid Floodway (map 16.1) in the early 1930s as General Jadwin's solution to managing the flooding at the confluence of the Mississippi and Ohio rivers [3, 4]. The USACE anticipated that a 500-year flood in this area would require the diversion and temporary storage of large quantities of floodwaters. The Missouri bottomlands on the west side of the Mississippi River below the confluence were part of a natural levee system created by the Mississippi when it historically overflowed its banks. A natural levee deposits the heaviest sediment first and thus builds up the land closest to the river. Natural levees extend about 0.6 to 1.2 miles from the riverbank and are adjacent to an extensive landscape of marshes and wetlands. Natural levees along the Mississippi River were fortified and heightened following the draining and ditching of these marshlands to create fertile agricultural lands [6, 15]. The New Madrid Floodway consists of 133,000 acres surrounded by levees and is capable of holding 10 to 12 feet of water or temporarily holding 1,300,000 to 1,596,000 acre-feet of floodwater. When the floodway is opened, river levels quickly drop, and the up- and downstream pressures on levees are substantially reduced.

After the Great Flood of 1927 [3], the USACE also created the current Cairo floodwall and levee system (figure 16.1) on the Ohio and Mississippi rivers. This system was sufficient to handle flood events until the spring of 2011 (figure 16.2) when greater than average rains and snowmelt produced one of the most powerful floods in the river's history and threatened to top and breach the flood control infrastructure [1]. Fort Defiance State Park (figure 16.3), at the tip of Cairo, Illinois, outside of the levee system provided valuable water storage at flood stage between the southernmost point of the Cairo levee and floodwall system and the confluence of the Ohio and Mississippi rivers (map 16.1). The park is the lowest point in the state of Illinois, with an elevation of 279 feet, and was underwater once the Ohio and Mississippi rivers reached flood stage in early April of 2011 and stayed underwater for almost two months. The 500 acres of Illinois farmland with Tice silty clay loam and Gorham silty clay loam soils [10] located west of the park and between the Mississippi River and the levee were also underwater when the Mississippi River was above flood stage.

In 2011 Cairo had a population of 2,900 and about 400 structures covering 1,280 acres. If the levee system failed, it is estimated that a more than 20-foot depth of water would flow in rapidly and cover the city with 24,000 acre-feet of floodwater. The northern boundary of Cairo is raised railroad beds (figure 16.4) that are 20 feet and 40 feet above Illinois Route 3, which passes underneath and through a tunnel. Between the raised railroad beds is a huge metal gate (figure 16.4) that can be lowered to block floodwaters from passing into Cairo from the north if the Cache River levee failed or is topped. During flood events, the entire city of Cairo

FIGURE 16.3 Fort Defiance State Park, the southernmost point of Illinois, was under water most of spring of 2011, with only treetops visible. The park is located outside of the Cairo levee and floodwall system at the confluence of the Ohio (top of picture) and Mississippi (bottom of picture) rivers.

FIGURE 16.4 A walled city. A raised railroad bed serves as a levee for the northern boundary of Cairo, Illinois, and provides an entry to the levee-enclosed city for Illinois Route 3 traffic. Flood gates on this entry could seal the city behind embankments and the floodwall if levees north of the city were to fail.

FIGURE 16.5 The Birds Point levee was the site of the first explosion on May 2, 2011, that opened the New Madrid Floodway and relieved river pressure on the Cairo floodwall. Remnants of the fuse plug levee and the crater lake extend into the adjacent agricultural lands.

can be evacuated and sealed off from the floodwaters. If a levee or floodwall north of Cairo on the Mississippi, Ohio, or Cache rivers failed, it would cause the flooding of Future City, Illinois, and the surrounding 2,000 acres of agricultural lands and cause the Cairo floodgate to be closed.

The breaching of Cairo or Future City area levees or topping by floodwater above the 64-foot maximum protection level would have little effect on peak flow in the lower Mississippi River between Cairo, Illinois, and New Orleans, Louisiana. Cairo and Future City levee or floodwall breaks in 2011 would have resulted in a 15-foot depth of rushing floodwater into Cairo, Future City, and Urbandale, Illinois, with potential loss of life and severe damage to more than 600 buildings. The 1927 Mississippi flood map of overflowed areas [3] clearly shows that Cairo, Future City, and Cache River valley areas were flooded, and it is likely they would have again flooded if the floodwall and levee system had failed in 2011.

Cairo Response to the 2011 Flood

Mayor Judson Childs ordered an evacuation of Cairo, effective Sunday, May 1, 2011. Many of the evacuees were moved to Shawnee Community College as the danger of floodwaters topping the floodwall or breaching escalated. On May 2, 2011, at 10:00 PM, Major General Michael Walsh ordered the Birds Point fuse plug levee (figure 16.5) to be blown up. About 265 tons of dynamite (TNT) were put into 1,100 feet of pipe within the fuse plug levee, and the first sequence of blasts exploded with a force of 3 on the Richter scale. The Birds Point fuse plug blast near Tom Bird Blue Lake, created by a 1937 levee breach, was thought by Missouri farmers to have contributed to the Commerce levee failure in early May of 2011 (see chapter 15). The Birds Point levee fuse plug, two miles to the south of Cairo (map 16.1), was opened in six places by explosions, and the subsequent force of the rushing floodwater from the Mississippi River removed adjacent sections further degrading the levee [6]. The USACE estimated that nearly one-fourth of the entire flow of the Mississippi River entered the New Madrid Floodway at the rate of 396,000 cubic feet per second and reduced the Cairo levee water level from 62 feet and rising to 59 feet and dropping within 48 hours. The city of Cairo, Illinois, was spared from a natural floodwall or levee breach or topping. Most residents returned by May 11, 2011, but were subjected to a nightly curfew. By then, the peak flowed had dropped about 6 feet, and by May 20, 2011, life in Cairo had returned to normal and business activity picked up.

After the deliberate breaching of the Birds Point levee and opening of the New Madrid Floodway, water levels dropped 3.1 feet at Paducah, Kentucky, and 1.9 feet at Cape Girardeau, Missouri. The US 51/US 60

FIGURE 16.6 During the spring of 2011, barges were anchored on the Mississippi River bank next to the flooded Fort Defiance State Park.

bridge over the Ohio River from Illinois to Kentucky (map 16.1) was opened once the Ohio River floodwaters were no longer passing over the US 51/US 60 highway. The Mississippi River was at 46 feet (6 feet above flood stage) on May 20, 2011, and the road and bridge would normally remain closed until the Mississippi River dropped to 43 feet, which occurred by mid-June of 2011. However, the US 60 bridge over Mississippi River, which connects Cairo, Illinois, and Birds Point, Missouri (map 16.1), was closed for repairs during the summer and fall of 2011. The bridge was opened in 2012 but was periodically closed for repairs during the next four years. The local Missouri and Illinois residents and any through traffic continued to travel an additional 17 miles between Birds Point, Missouri, and Cairo, Illinois.

Impact of the Cairo Floodwall and Levee System on Kentucky Bottomlands

The record 62-foot peak on the Ohio River at the Cairo flood gage resulted in the flooding of thousands of acres of unleveed Kentucky bottomlands. The Ohio River, which normally is 0.6 mile wide, was more than 4 miles wide at the confluence. All unprotected bottomlands with an elevation of less than 332 feet above sea level were flooded. These bottomlands are riparian forests—transition ecosystems between the river and uplands [16]. The Kentucky bottomlands provide additional water storage capacity during flooding, and the wet soils help filter pollutants, recharge the water table, and capture sediment before it reaches the main flow of the river [16].

The raised railroad and highway (US 60/US 51) beds on the Kentucky bottomlands, which connect Cairo, Illinois, via elevated highway and railroad bridges across the Ohio River to Wickliffe, Kentucky, were also flooded. The floodwaters covered five miles of US 60/US 51 and the railroad bed for approximately two weeks in May before they were opened to the public and for commercial use. Only a thin silt coating was visible on the road, with little damage to these road and railroad beds. Much of the Ohio River barge traffic was also stopped (figure 16.6), with barges tied up on both sides of the river. Since there are no levees across from Cairo on the Kentucky side of the Ohio River, the bottomlands were not developed for urban uses, and the floodplain was used to temporarily store some of the floodwaters. There were very few building structures on the Kentucky bottomlands and only limited agricultural use. Most of the unprotected Kentucky bottomlands had drained by May having served several multifunctional uses: floodwater storage, nutrient cycling, anaerobic conditions that slow down the decay of vegetative matter (carbon storage), and wetland habitat.

Managing Floodwaters at the Confluence

The purposeful opening of the New Madrid Floodway in May of 2011 reduced the flooding and immense pressure caused by the rain-swollen, fast-moving Ohio and Mississippi rivers on towns like Metropolis, Illinois; Paducah, Kentucky; and Cape Girardeau, Missouri, as well as reduced the pressure on the floodwall (figure 16.1) and levee system at Cairo, Illinois. Unfortunately, the floodway opening was not in time to prevent the Len Small levee from failing (see chapter 15), and the blast may or may not have contributed to the Commerce farmer levee breach immediately after the opening of the Birds Point fuse plug. The peak flow and flood levels at most levees on the lower Mississippi River and along the way to the Gulf of Mexico were lowered by a few feet. None of the lower Mississippi River levees broke, but the historic floodwater levels on the Mississippi River as a result of flooding still required the opening of the Morganza flood gates and the rerouting of floodwater through the Atchafalaya River valley, which is west of Baton Rouge, Louisiana, and a shorter route for floodwaters to enter the Gulf of Mexico. The impact of this floodway opening on wetlands, agricultural land, and urban areas is the subject of other investigations.

Although most of the Mississippi alluvial valley was once covered by bottomland hardwood forests, the hydrology of this landscape has been significantly altered with levees and cleared for urban development and agriculture. There were no levees across from Cairo on the Kentucky side of the Ohio River, and the undeveloped bottomlands and the floodplain experienced minimal land degradation despite temporarily storing a portion of the floodwaters. The Kentucky bottomlands, although not sufficient as a single flood control measure, were an important source of water storage during the 2011 flood and provided valuable ecosystem services. The extensive loss of Mississippi River bottomlands to agriculture and urban development has limited the natural capacity of alluvial bottomlands to manage flooding in main river channels and reduce water pressure under extreme rain and snowmelt events. Without natural wetland buffers, such as the bottomlands, floodwater will push against concrete and levees, increase the height of the river channel, and cause the water to flow faster, intensifying pressures on flood control structures [17, 18]. It appears that the USACE-induced breach of the Birds Point, Missouri, levee and the passing of floodwaters through the New Madrid Floodway dropped the flood level at Cairo by 2.5 feet in 48 hours, which reduced the water pressures on the floodwall and levee system. Upstream, the flood levels dropped more than 3 feet at Paducah, Kentucky, and almost 2 feet at Cape Girardeau, Missouri, within 48 hours. There was no loss of life or property in Cairo, Future City, or Urbandale, Illinois, since the levee and floodwall system held back the record high floodwaters of the Ohio and Mississippi rivers for many weeks.

---

[1] US Army Corps of Engineers. 2011. Great Flood of '11. Our Mississippi. Rock Island, IL: US Army Corps of Engineers. http://www.mvs.usace.army.mil/Our%20Mississippi/ourmississippi_su11_lowres.pdf.

[2] Lansden, J.M (1910) 2009. A History of the City of Cairo, Illinois. Carbondale, IL: Southern Illinois University Press.

[3] Barry, J.M. 1997. Rising Tide: The Great Mississippi Flood of 1927 and How It Changed America. New York: Simon and Schuster.

[4] Camillo, C.A. 2012. Divine Providence: The 2011 Flood in Mississippi River and Tributaries Project. Vicksburg, MS: Mississippi River Commission.

[5] Olson, K.R., and L.W. Morton. 2012. The effects of 2011 Ohio and Mississippi River valley flooding on Cairo, Illinois, area. Journal of Soil and Water Conservation 67(2):42A-46A, doi:10.2489/jswc.67.2.42A.

[6] Olson, K.R., and L.W. Morton. 2012. The impacts of 2011 induced levee breaches on agricultural lands of Mississippi River Valley. Journal of Soil and Water Conservation 67(1):5A-10A, doi:10.2489/jswc.67.1.5A.

[7] Olson, K.R., and L.W. Morton. 2013. Restoration of 2011 flood-damaged Birds Point–New Madrid Floodway. Journal of Soil and Water Conservation 68(1):13A-18A, doi:10.2489/jswc.68.1.13A.

[8] Olson, K.R., and L.W. Morton. 2013. Soil and crop damages as a result of levee breaches on Ohio and Mississippi rivers. Journal of Earth Science and Engineering 3:139-158.

[9] Olson, K.R., and L.W. Morton. 2014. The Ohio River flooding of the Cache River Valley in Southern Illinois. Journal of Soil and Water Conservation 69(1):5A-10A, doi:10.2489/jswc.69.1.5A.

[10] Parks, W.D., and J.B. Fehrenbacher. 1968. Soil Survey of Pulaski and Alexander counties, Illinois. Washington, DC: USDA Natural Resource Conservation Service.

[11] Legends of America Staff. 2015. Cairo Illinois, Death by Racism. Illinois Legends. http://www.legendsofamerica.com/il-cairo.html.

[12] Harrison, L.H. 2009. Civil War in Kentucky. Lexington KY: The University Press of Kentucky.

[13] Welky, D. 2011. The Thousand-Year Flood: The Ohio-Mississippi Disaster of 1937. Chicago: University of Chicago Press.

[14] Castro, J.E. 2009. The Great Ohio River Flood of 1937. Charleston, SC: Arcadia Publishing.

[15] Olson, K.R., M. Reed, and L.W. Morton. 2011. Multifunctional Mississippi River leveed bottomlands and settling basins: Sny Island Levee Drainage District. Journal of Soil and Water Conservation 66(4):104A-110A, doi:10.2489/jswc.66.4.104A.

[16] Anderson, J., and E. Samargo. 2007. Bottomland Hardwoods. Morgantown, WV: West Virginia University, Division of Forestry and Natural Resources. http://forestandrange.org/new_wetlands/index.htm.

[17] Morton, L.W., and K.R. Olson. 2015. Sinkholes and sand boils during 2011 record flooding in Cairo, Illinois. Journal of Soil and Water Conservation 70(3):49A-54A, doi:10.2489/jswc.70.3.49A.

[18] Olson, K.R., and L.W. Morton. 2015. Slurry trenches and relief wells installed to strength Ohio and Mississippi river levee systems. Journal of Soil and Water Conservation 70(4):77A-81A, doi:10.2489/jswc.70.4.77A.

Managing River Pressure from the 2011 Record Flood on Ohio and Mississippi River Levees at Cairo

When the Ohio River began flooding farmland and cities in Ohio, Indiana, Kentucky, and Illinois that were not protected by levees in April of 2011, Cairo, Illinois, and many of the other levee-protected cities on the Ohio and Mississippi rivers did not flood; however, their levees and floodwalls were in danger of failing. The extra weight of the river at flood stage pushed water underneath levees and floodwalls, increased the potential for sand boils, and undermined the strength of the levees and their capacity to hold back floodwater [1, 2]. By early May of 2011, the Ohio River gage at Cairo, Illinois, had reached 61.7 feet [3], and floodwaters were starting to put significant pressure on both Ohio and Mississippi river levee and floodwall systems in the Cairo area (map 17.1).

Officials had to initially decide how to protect the town of Cairo, Illinois, and whether to flood 133,000 acres of Missouri farmland and small communities in the Birds Point–New Madrid Floodway to decrease river pressure on confluence levees. Cairo was evacuated on May 1, 2011, after sand boils and sinkholes appeared inside the floodwall (map 17.1). Commercial Avenue, built on "made land," had a number of sinkholes open up (figure 17.1) where a sanitary sewer line flowed to the sewage treatment plant and parallel to the floodwall. Post-2011, short-term emergency measures to manage the sand boils were replaced with a series of relief wells and slurry trenches to strengthen the weakened levee system.

Sand Boils and Sinkholes in Cairo

Sand boils, including a mega sand boil, occurred near the water treatment plant on 40th Street revealing a weakening levee and floodwall system. Soils mapped as "made land" [4], such as those on Commercial Avenue, are soils high in clay, silt, or sand that were hauled in as fill materials to build up the low bottomlands of Cairo. The soil at the site of the mega sand boil was Darwin silty clay loam [4], which contains more that 40% clay in the upper 6.6 feet. The presence of sand boils on April 29, 2011, especially a mega boil, strongly indicated that the levee and floodwall could fail. Combined with the heavy local rains on May 1 and 2, the Cairo gage forecast was pushed to a peak of 63.5 feet, and the river placed additional pressure on the levees and floodwall [1]. These events were finally enough to convince the president of the Mississippi River Commission (MRC), Major General Michael Walsh of the US Army Corps of Engineers (USACE), that the Birds Point–New Madrid Floodway had to be opened for the first time in 74 years.

When the rivers are at high water, Cairo is like an empty basin sunk to its brim and even a tiny opening at the bottom can force a stream of water to shoot up through the porous earth or sand inside the basin (figure 17.2) [5]. In late April, small sand boils began appearing in Cairo. The first mega sand boil of the 2011 flood was found by the USACE beside a piezometer (a gauge which measures pressure of the groundwater) near 40th Street just south of the raised railroad bed and west of the water treatment plant (map 17.1) [1]. A sand boil (figure 17.2) occurs when extreme water pressure against a levee leads to high hydraulic gradient conditions resulting in excessive seepage and piping of sediment exiting on the inside ground surface in a churning or boiling action (see chapter 8) [6]. Uncontrolled seepage is a major cause of levee failure, creating instability when high water pressure and saturation cause the earth materials to lose strength. A certain amount of underseepage is an expected occurrence and not a major problem as it is incorporated into levee designs. Without some underseepage, the pressure on the levees would be too great; however, piping of

FIGURE 17.1 The Commercial Avenue sinkholes in Cairo were near or above the sanitary sewer lines.

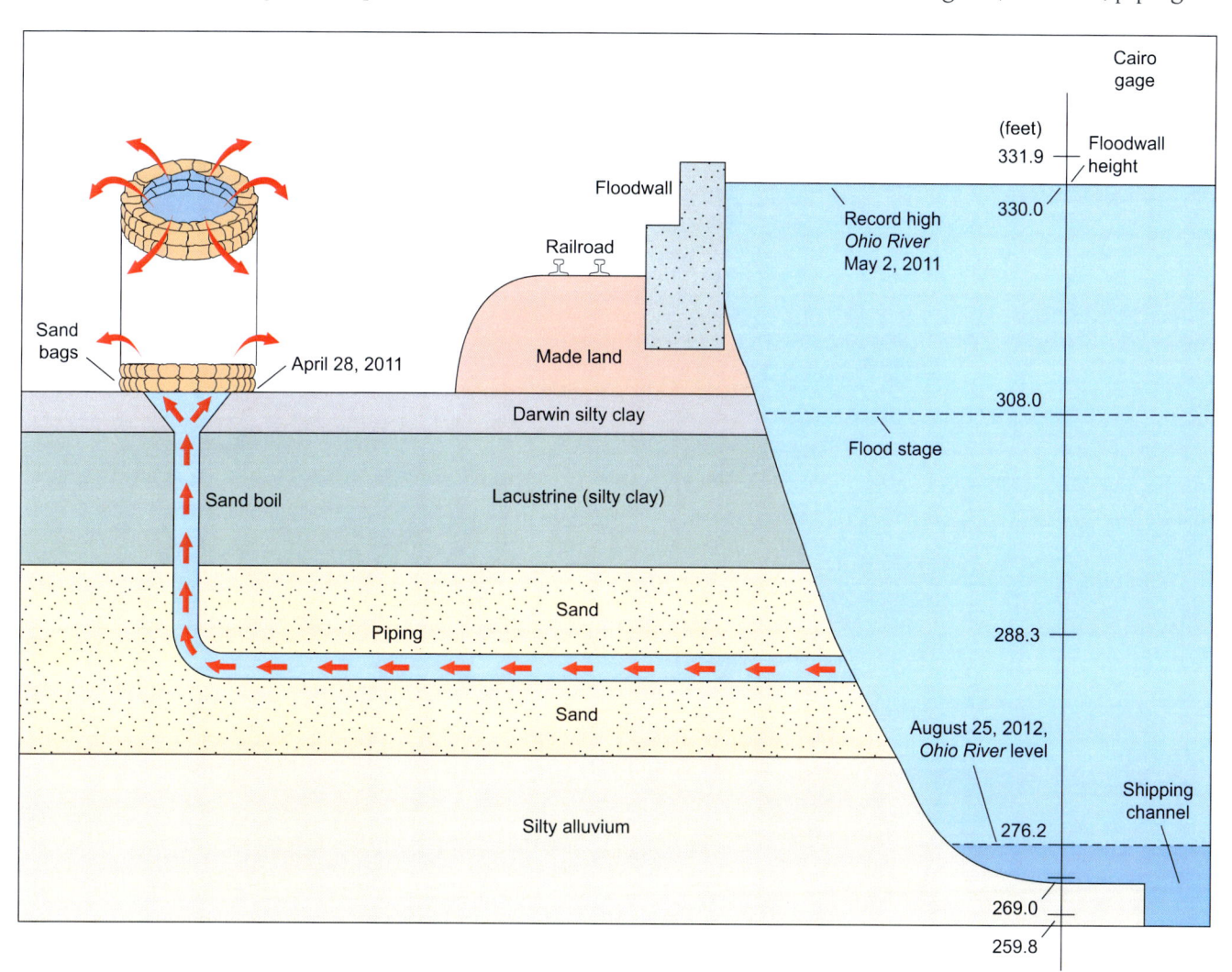

FIGURE 17.2 Diagram of a sand boil at Cairo on April 28, 2011. The elevation of the highest flood level (332 feet) on May 2, 2011, and the lowest level from the drought of August 25, 2012, are noted on Cairo gage.

sediment in the underseepage is a real problem [1]. A sandbag dike is often used as a temporary measure to increase the depth of water over the boil and decrease the hydraulic gradient across the seepage path, thereby reducing the potential for erosion of earth materials along the path, a process called piping [5]. If a boil is piping sediment, the USACE becomes quite concerned and starts ringing the boil with sandbags to apply a counter pressure through the rings or water berms. The engineers are careful not to put too much pressure on the boil or it could serve to completely stop the under-

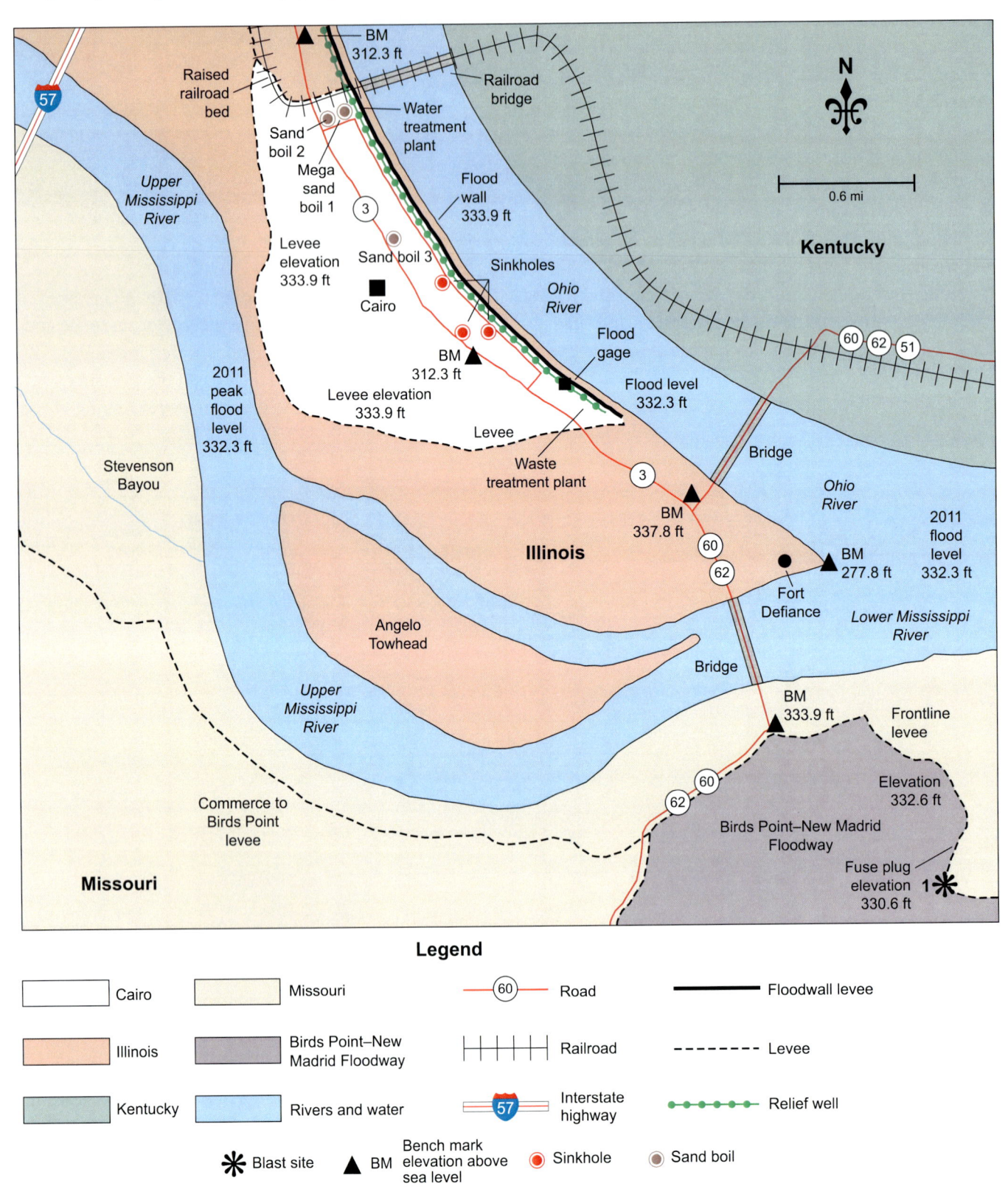

MAP 17.1 Sinkholes and sand boils in April of 2012 occurred just inside the floodwall in Cairo, Illinois, at the confluence of the Ohio and Mississippi rivers. Relief wells were built post-2012 drought to reduce river pressure on the floodwall during future high water events.

seepage or flow of water. Piping is what undermines the levees or floodwalls [1].

The sand boil occurred on Darwin silty clay loam soil [4], and on the evening of April 28, 2011, this sand boil started to grow. In a few hours it enlarged dramatically from a few inches to 2 feet in diameter. The USACE quickly realized that this high-energy sand boil had the potential to cause the Ohio River floodwall to fail at a time when 2,900 people were still living in Cairo. The USACE decided that the traditional treatment of building a sandbag ring around the sand boil to keep water on the boil as a counter weight was insufficient and would not work this time. Instead they constructed a ring berm around the sand boil using a nearby stockpile of fly ash cinders (figure 17.3) owned by the Bunge Corporation [1]. A bank of emergency lights was brought in as well as a bulldozer, backhoes, loader, excavators, dump trucks, and a crew of 40 to contain the large sand boil. The crew constructed a 50-foot ring berm to a height of 6.6 feet. The sand boil continued to pipe sediment at an alarming rate. Since the city of Cairo still had many people who were not yet evacuated, the danger of the city being flooded with loss of life was a real concern. On May 2, 2011, after 48 hours of local rain, the fly ash cinder pile was turning to mush and had to be covered with a tarp [1]. The fly ash cinder pile was then raised to 13.2 feet (figure 17.3). This mega boil was among the most serious in the history of USACE use of earthen levees and floodwalls.

Previously, on April 29, another smaller sand boil was also found near the NAPA auto parts store to the east of Route 3, which was between 40th Street and the raised railroad bed (map 17.1). This sand boil occurred on Cairo silty clay soil [4] and was treated with the traditional 5-foot-high sandbag ring. The Illinois National Guardsmen filled the sandbags in the parking lot of the NAPA store and then moved them through 2 feet of water to the sand boil using flat-bottomed boats [1]. Later an access road had to be built so the National Guard could haul in rock to reinforce the sandbag ring. This effort finally stabilized the second sand boil. A third sand boil on Tice silty clay loam soils [4] was later discovered off of 27th Street approximately 500 feet from the Ohio River floodwall (map 17.1). It was smaller than the other two sand boils and was treated with a 4-foot high ring of sandbags and filled with water.

A number of sinkholes also developed in late April and early May of 2011 in downtown Cairo on Commercial Avenue and 15th Street (figure 17.1). A sinkhole is a hole in the ground surface created by the collapse of overlying material as supporting material below the

FIGURE 17.3 The raised railroad bed is visible in the background of the 40th Street mega sand boil in Cairo near the water treatment plant. The mega sand boil was filled in and covered with a 13.2-foot-high pile of fly ash cinders.

FIGURE 17.4 More than 64 feet of the Ohio River floodwall in August of 2012 were exposed when the Ohio River was at 12 feet during the drought. This was 50 feet lower than the May 2, 2011, high water mark.

surface is eroded or piped by water [5]. Sinkholes have steep sides from the soil shearing as it collapses into the underlying void.

The Cairo sand boils, mega boil, and sinkholes represent uncontrolled seepage conduits that formed under the levee system and threatened the city. Following the 2011 flood event, Cairo and the USACE invested considerable resources to repair the sand boils and to reduce future risk of uncontrolled seepage underneath the floodwall, which undermines the levee system (figures 17.2 and 17.4). Repairing such boils can be complicated, and possible approaches include relief wells, seepage berms, and cutoff walls that involve digging a big trench and adding clay grout (bentonite) to cut off the leak [6]. In October of 2012, the USACE began obtaining the rights to install 31 relief wells adjacent to the Cairo floodwall (map 17.2) with construction completed by the fall of 2013.

Weakened Levee Systems after the 2011 Flood

The city of Cairo and farmland to the north, which is planted to rice, corn, soybeans, and other agricultural crops, are protected from the Ohio and Mississippi rivers at flood stage by an extensive system of levees, a diversion channel, and flood gates. This levee system also protects three towns and the Cairo airport [7]. Failure of any one of these levees can cause severe flooding of Urbandale, Klondike, and Future City, Illinois, and the surrounding 2,000 acres of agricultural lands, and result in the closure of the Cairo floodgate between raised railroad beds over Route 3 [7]. Over 3,000 acres of these levee-protected lands are silty soils (Tice silty clay loam, Riley silty clay loam, Darwin silty clay loam, and Cape and Karnak silty clay loam) at a slightly higher elevation than the city of Cairo. A levee breach could damage a hundred or more buildings in the Future City, Klondike, and Urbandale areas and require the local population to be evacuated. If flooded, this area has the potential capacity to store up to a 10- to 15-foot depth of floodwater, a total of 30,000 to 45,000 acre-feet [7]. In 2011, this amount of temporary floodwater storage would have done little to drop record flooding levels on the Ohio and Mississippi rivers (probably less than 10 inches for several hours) and have had even less effect on the lower Mississippi River between Cairo, Illinois, and New Orleans, Louisiana. The breaching of Ohio, Mississippi, or Cache levees and flooding of the area north of Cairo would also have blocked the evacuation route (Route 3) used by Cairo citizens on May 1, 2011.

The New Madrid Floodway south of the Ohio and Mississippi river confluence at Cairo was deliberately breached on May 2, 2011, and successfully reduced levee pressure. However, the levee system in Cairo and to the north of Cairo; the levee system south of Hickman, Kentucky; and the Ohio River levee near Mound City, Illinois, were considerably weakened and in need of repair and strengthening to withstand future river pressures [2, 7]. After the 2011 flood event, slurry trenches and relief wells were installed to prevent future levee failure from concentrated seepage under levees and sand boils.

Relief Well System

When a levee is built in an alluvial valley, seepage often occurs when the river stage is higher than the adjacent land. The difference in gradient increases water pressure on the underlying sandy soils. Seepage enters beneath the levee through the riverbed, riverside borrow pits, or a weak spot in the river side top stratum of clay soils. When this concentrated seepage becomes an underground channel, it erodes the underlying silts and fine sands, saturates the landside slope of the levee, and reduces the levee stability [8]. Unchecked, this piping can lead to a rupture on the land side of the levee and creation of a sand boil. High uplift gradients and concentrated seepage usually occur along the landside toe of the levee, weak spots in the topsoil stratum, or where old channels or excavations are susceptible to erosion [8]. Water flow through root holes, shrinkage cracks, and burrowing animal holes can also form channels that increase localized piping and potential levee slope sloughing.

There are a number of methods for managing seepage so that levee failures are less likely to occur. These strategies include relief wells, landside seepage berms, impervious riverside blankets, cutoffs, drainage blankets, slurry trenches, and sublevees [8]. Relief wells (figure 17.5), well suited to locations with stratified deposits of sand and gravel, can prevent piping and reduce hydrostatic pressures on the landward side of levees [8]. Relief well engineering is well established with 2,480 installed along 292 miles of Mississippi River mainline and tributary levees during the 1950s. Subsequent studies of relief well systems have found they perform successfully. Assessments after the major 1993 flood event in St. Louis, revealed no levees with relief wells failed due to sand boils or piping. Current postflood best management practices include field observations and record of heavy seepage, piping, and sand boils larger than 3 to 4

FIGURE 17.5 A series of relief wells pump iron-rich, yellowish-red groundwater into a drainage ditch to relieve substratum pressure when the river is higher than the land inside the levee. The water is collected and pumped over the levee.

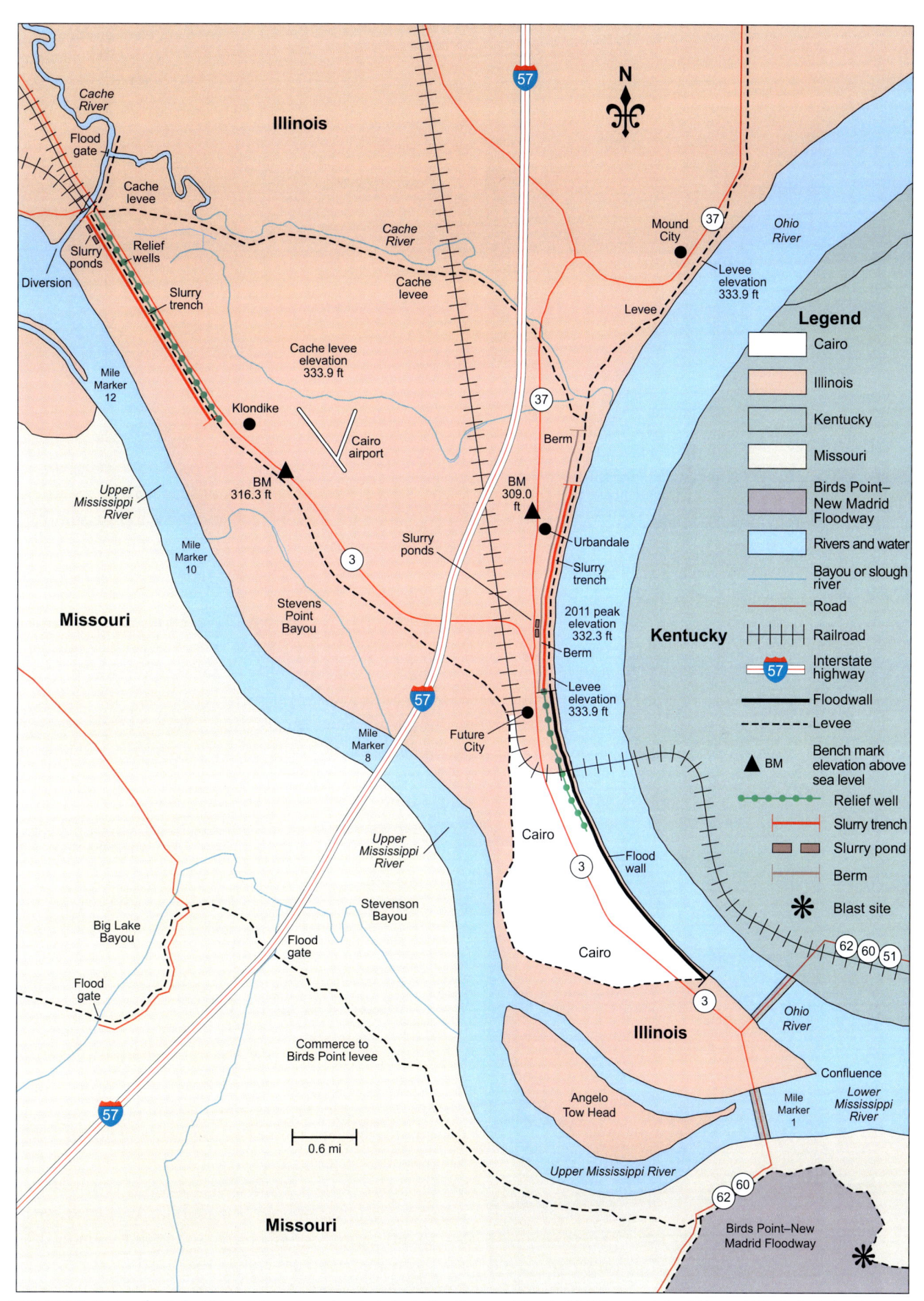

MAP 17.2 Systems of relief wells and slurry trenches installed after the great flood of 2011 protect the Ohio and Mississippi river levees north of Cairo, Illinois, and south of the Cache River against river underseepage and failure.

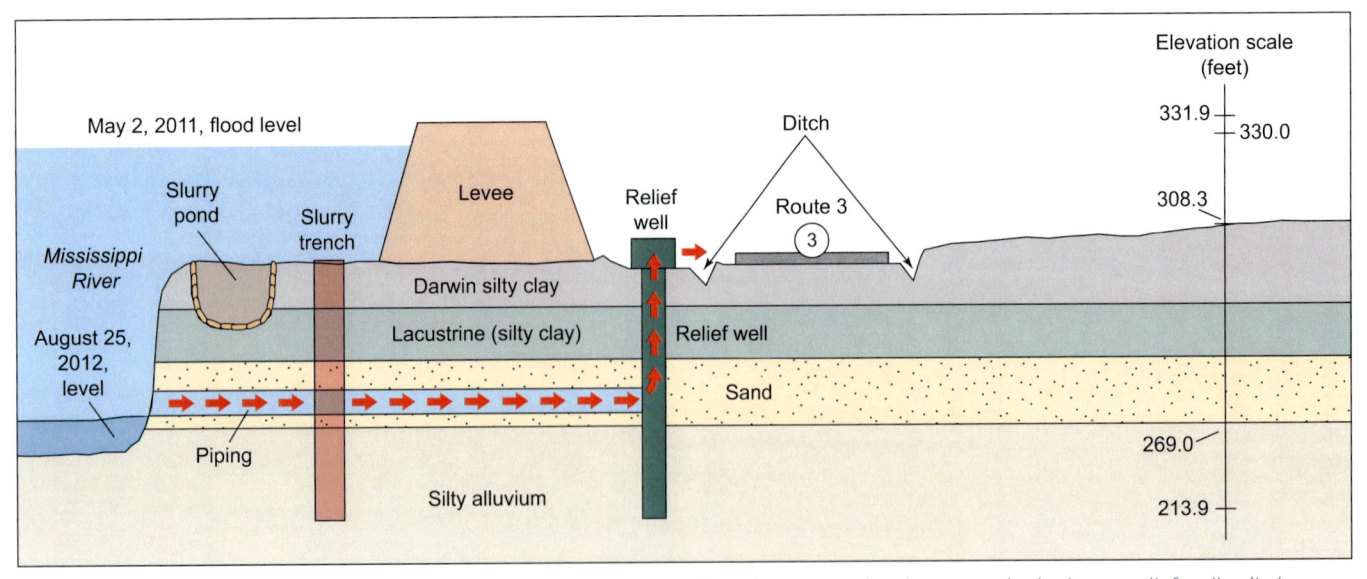

FIGURE 17.6 This diagram illustrates the landscape relationship of the Mississippi River, the slurry trench, the levee, relief wells, ditches, and Route 3 north of Cairo.

inches; as well as evaluation of the adequacy of existing relief wells and the need for new or additional wells [8].

Following the flood of 2011, the USACE allocated $6 million to address the Mississippi River levee seepage in the levee north of Cairo, with work starting in October of 2011. A total of 28 relief wells, which drain into a Route 3 road ditch (map 17.2), were constructed northeast of the Cairo airport and south of the lower Cache River levee. The relief wells (figure 17.5) were built to reduce damaging uplift pressure from excessive seepage through pervious materials that were insufficient to block water flow under the levee [5]. Relief wells, such as the one shown adjacent to Ohio River levee (figure 17.6) control the water pressure and are designed to help prevent piping of sediments, which is the real threat to levees and floodwalls [1, 2]. These wells control the direction and quantity of seepage and together form a collection system to discharge water back into the Mississippi River.

Slurry Trenches

Slurry trenches, typically 2 to 4 feet wide and between 35 to 95 feet deep, are used to create an impervious barrier that cuts off the seepage flow and potential piping of sediment by water [9]. They consist of bentonite or drilling mud clays that are usually formed from volcanic mineral (montmorillonite) clays and are available in powdered form, which form a thick, sticky viscous fluid when added to water [6]. The hydration of the clay particles causes them to bond and swell, forming a gel that can be poured into the slurry trench. Bentonite bags were brought in and mixed in slurry ponds (figure 17.7) for injection into 7,200 feet of slurry trenches (figure 17.8). These trenches reinforce the system of relief wells and restrict the underground flow of water under the levee. The work on the Mississippi River slurry trenches was completed in November of 2012.

The USACE received $802 million in emergency Mississippi River flood-repair funding in June of 2012 for up to 143 high-priority projects to address levee repairs, reconstruct river channels, and repair other flood-control projects in response to the spring of 2011 flood, which set records from Cairo, Illinois, to the Gulf of Mexico. Both the New Madrid Floodway repair and the Cairo restoration project were high on the list, with the USACE targeting $46 million to repair the damage to the Cairo area flood-control system north of the confluence of the Ohio and Mississippi rivers.

FIGURE 17.7 Bentonite bags are dumped into these slurry ponds and mixed with water to create slurry for the trench.

FIGURE 17.8 This slurry trench between the levee and Mississippi River creates an impervious barrier that reduces river pressure and seepage.

Approximately $20 million in additional funding was used to complete the restoration of the Mississippi River levee north of Cairo with repairs including the finishing of the 7,200 feet slurry trench and relief wells (map 17.1 and figure 17.8) started in the fall of 2011. The USACE has responsibility for managing river flows and navigation during flood and drought conditions. Thus another $26 million was used to not only increase protection provided by the Cairo floodwall on Ohio River side by putting in 31 relief wells (figure 17.5) adjacent to the Cairo floodwall (map 17.1) and to help prevent erosion of the riverbank along the Ohio River side of Cairo [10], but also to dredge the Ohio River and Mississippi River channels north of Cairo. The Ohio River water level dropped 7.7 feet when a drought occurred from July of 2012 to February of 2013. The Ohio and Mississippi river channels were dredged in 2012 to maintain a 300-foot-wide and 9-foot-deep shipping channel during this extended dry period [11]. Without a shipping channel of sufficient depth, barges have to be lightened or held back until seasonal rains when the Ohio and Mississippi river flows are restored to more normal levels.

A 5,400-foot slurry trench and two berms were added to the Ohio River levee system between Mound City and Cairo north of the Ohio River floodwall (map 17.2) east of Route 37 and south of the Cache River levee. These projects were designed to prevent seepage during future major flooding, and all work was completed by November of 2013. Another project flattened the slope on the Mississippi River levee and widened its crown. Improvements including work on pump stations, drainage systems, and small levees (some of which failed in April of 2011) were completed in the Cache River valley area north of Cairo. These projects were funded by the county matching funds with the USACE and a combinations of grants from the Delta Regional Authority and the State of Illinois [6]. North of Cairo and south of the Cache River levee, the smaller towns such as Urbandale, Future City, and Klondike are now better protected from major flooding in the future.

Investments in Future Levee Infrastructure Protection

Floodwaters from the Ohio River in April of 2011 were more than 21 feet above flood stage and four miles wide as they approached the confluence. As a result, there was tremendous pressure on and under the Ohio River floodwall and the upper Mississippi River levee in the Cairo area. Sand boils, a mega boil, sinkholes, and seepage occurred, and the entire floodwall and levee system was very close to failure. Failure would have resulted in 22 feet of water covering the city of Cairo at a time when 2,900 people living in Cairo had yet to be evacuated. When the record flood peak reached 61.7 feet, the USACE engineers blasted open the Birds Point fuse plug levee taking pressure off the Cairo system. No flood damage of building structures occurred in Cairo since the floodwall and levee system held. Had it failed, all 400 homes and commercial buildings would have been severely damaged [7].

Two years later the Cairo levee system had been strengthened with considerable investment in repairing infrastructure damaged by sand boils and its integrity restored with the addition of relief wells along the Ohio River side of the Cairo levee. Sinkholes in the city of Cairo roads were repaired by the fall of 2013. The US 60 bridge over the Mississippi River (figure 17.9), which connects Cairo, Illinois, and Birds Point, Missouri, was reopened at the end of 2012 but through 2015 continued to be periodically closed for repairs. The USACE spent $26 million to restore the Cairo area floodwall, levees, streets, and shipping channel. The physical, economic, and social reconstruction of Cairo, Illinois, following the evacuation of the entire city on May 1, 2011, challenged local residents, community leaders, and the USACE to strengthen levees, repair roads and bridges, and improve other basic infrastructure. While the Cairo floodwall and levee system was able to withstand the record-breaking flood of 2011 [3], it took more than two years for sand boils and sinkholes to be repaired and strengthened.

The 2011 flood levels put tremendous pressure on and under the Ohio and Mississippi river levee systems in the area north of Cairo and south of the Cache River levee. The river channel and the systems of levees along these rivers need constant monitoring for structural integrity, continuous routine and remedial repair, and post-disaster assessment to be prepared for future extreme water levels [12]. Relief wells, slurry trenches, and other structures are important strategies for managing the river when it exceeds flood stage. Monitoring of levee underseepage control systems, including observations and careful record keeping during high-river stages, is a critical best management practice. When a river stage is predicted to reach 8 to 10 feet on a levee, piezometers should be read once or twice a week, at the river crest, and until the crest of the flood has passed [8]. After flooding has receded and river levels have dropped, seepage field surveys and data must be assembled to evaluate the capacity of existing slurry trenches and relief wells to handle future 50-, 100-, and 500-year events. Post–high water, routine investments are necessary to assure the relief well system continues to protect the levee. Routine maintenance includes cleaning and pumping every 5 to 8 years; efficiency tests for pumping equipment; and checks on values, gaskets, and piezometer lines [8].

---

[1] Camillo, C.A. 2012. Divine Providence: The 2011 Flood in Mississippi River and Tributaries Project. Vicksburg, MS: Mississippi River Commission.

[2] Morton, L.W., and K.R. Olson. 2015. Sinkholes and sand boils during 2011 record flooding in Cairo, Illinois. Journal of Soil Water Conservation 70(3):49A-54A, doi:10.2489/jswc.70.3.49A.

[3] National Oceanic Atmosphere Administration. 2012. Historic crests. Cairo, IL: National Weather Service, Advanced Hydrologic Prediction Service.

[4] Parks, W.D., and J. B. Fehrenbacher. 1968. Soil Survey of Pulaski and Alexander counties, Illinois. Washington DC: USDA Natural Resource Conservation Service,

[5] Veesaert, C.J. 1990 Inspection of Embankment Dams. Session X in Embankment Dams. Lansing, MI: Bureau of Reclamation. http://www.michigan.gov/documents/deq/deq-p2ca-embankmentdaminspection_281088_7.pdf.

[6] Koenig, R. 2012. Corps balancing levee repairs on Missouri, Illinois sides of Mississippi. St. Louis Beacon. July 30, 2012.

[7] Olson, K.R., and L.W. Morton. 2012. The effects of 2011 Ohio and Mississippi River valley flooding on Cairo, Illinois, area. Journal of Soil and Water Conservation 67(2):42A-46A, doi:10.2489/jswc.67.2.42A.

[8] Mansur, C.I., G. Postol, and J.R. Salley. 2000. Performance of relief well systems along Mississippi River levees. Journal of Geotechnical and Geoenvironmental Engineering 126:727-738.

[9] Nemati, K.M. 2007. Temporary Structures: Slurry Trench/Diaphragm Walls CM420. Seattle, WA: University of Washington, Department of Construction Management.

[10] Ragan, E. 2012. Cairo expects to fare better in next flood. Southeast Missourian. July 29, 2012.

[11] Olson, K.R., and L.W. Morton. 2014. Dredging of the fractured bedrock-lined Mississippi River channel at Thebes, Illinois. Journal of Soil Water Conservation 69(2):31A-35A, doi:10.2489/jswc.69.2.31A.

[12] Olson, K.R., and L.W. Morton. 2015. Slurry trenches and relief wells installed to strength Ohio and Mississippi river levee systems. Journal of Soil and Water Conservation 70(4):77A-81A, doi:10.2489/jswc.70.4.77A.

FIGURE 17.9 After the spring of 2011 floodwaters receded, this soybean field located between the upper Mississippi River and the Cairo, Illinois, levee was planted and produced a crop. In the background, the US 60 bridge over the Mississippi River connects Illinois and Missouri.

Navigation and Flooding on the Ohio River

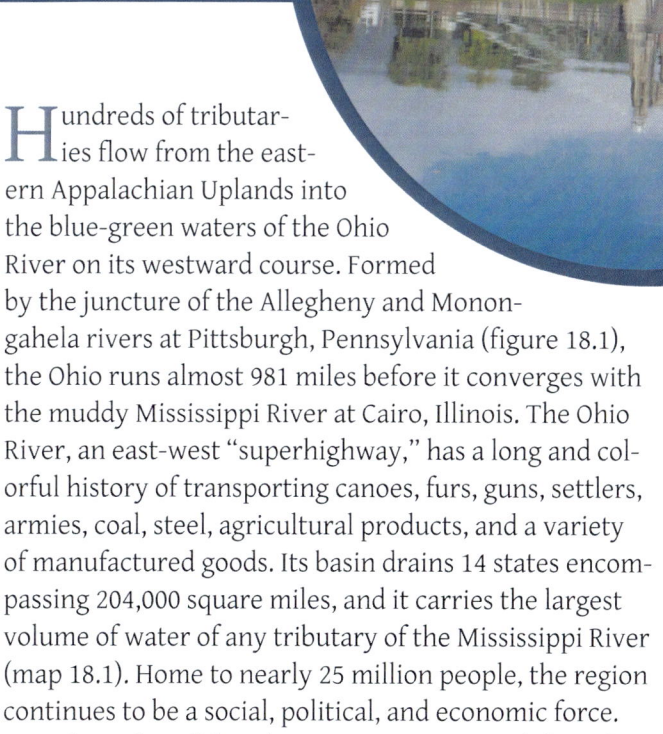

Hundreds of tributaries flow from the eastern Appalachian Uplands into the blue-green waters of the Ohio River on its westward course. Formed by the juncture of the Allegheny and Monongahela rivers at Pittsburgh, Pennsylvania (figure 18.1), the Ohio runs almost 981 miles before it converges with the muddy Mississippi River at Cairo, Illinois. The Ohio River, an east-west "superhighway," has a long and colorful history of transporting canoes, furs, guns, settlers, armies, coal, steel, agricultural products, and a variety of manufactured goods. Its basin drains 14 states encompassing 204,000 square miles, and it carries the largest volume of water of any tributary of the Mississippi River (map 18.1). Home to nearly 25 million people, the region continues to be a social, political, and economic force.

The pulse of the Ohio River—its seasonal drought and flood cycles—is a product of topography, weather, and climate. Historically, the river waters dropped to a depth of 12 inches in an exceptionally dry summer or experienced overbank flooding and a 60-foot crest when spring thaw coincided with moisture-laden storms that dropped their heavy loads over the Ohio River valley. As in the presettlement days of the Mississippi River, heavy winter snows, fast spring melts, and heavy precipitation did little damage to the Ohio River preindustrial valley. However, as steamboats replaced flatboats and keelboats navigating the river and populations grew, the drag on economic potential became an incentive to manage the river levels. Coal barges and steamboats were often grounded in port cities waiting for a high water surge before they could navigate south [1]. In 1875 Congress allocated funds for small dams within the riverbed to ensure a 6-foot channel [2].

While drought can slow navigation and seasonal rains are welcomed to provide high water for boat traffic, extreme and prolonged precipitation across the region poses a substantial threat to river cities and their populations and industries. A series of flood events in the Ohio River valley in the 1800s and early 1900s revealed the increased vulnerability of the region with the growth in river settlements and industrialization [2]. The flood of 1936 to 1937, a natural disaster of unprecedented proportions, drove a million people from their homes, claimed nearly 400 lives, and set a record $500 million in damages [1]. The Flood Control Acts of 1936 and 1938 were the beginning of a

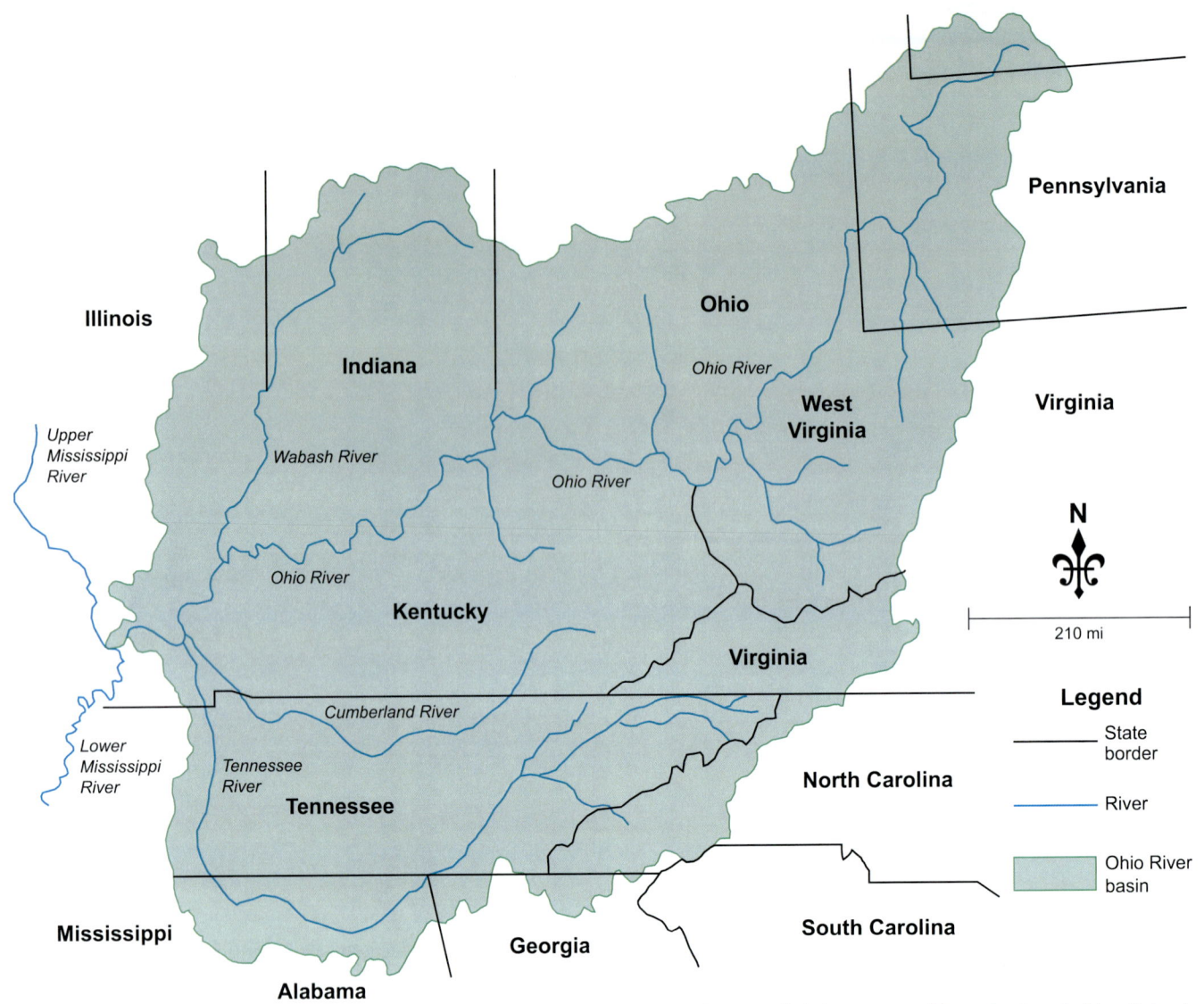

MAP 18.1 The Ohio River watershed includes land in 14 eastern states and their tributaries draining westward from the Appalachian Uplands.

national flood control policy that coordinated efforts to construct floodwalls, levees, and upstream storage reservoirs to protect the economic security of the Ohio and Mississippi river basins. Navigation continues to be a high priority, and the network of locks and low dams on the Ohio River were modernized and enlarged in the 1960s and 1970s to support increased trade and navigation safety (map 18.2). The US Army Corps of Engineers (USACE) currently monitors and maintains the lock and dam infrastructure along the Ohio to provide a nine-foot navigation channel. The new Olmsted Lock and Dam, (figure 18.2) when completed in 2020, will replace the two oldest wicket dams, Lock and Dam 52 near Brookport, Illinois (figure 18.3), and Lock and Dam 53 (figure 18.4) located approximately 11 miles north of the Ohio-Mississippi confluence at Cairo, Illinois.

Growth of the Ohio River as an Inland Waterway

The Ohio River is the main tributary stream of the Mississippi River. Today it drains lands west of the eastern continental divide and runs southwest along the borders of six states, from Pittsburgh, Pennsylvania, to the confluence with Mississippi River at Cairo, Illinois (map 18.1). It was formed between 2.5 and 3 million years ago when glacial ice dammed portions of north-flowing rivers, and smaller than the current Ohio River, it once flowed through the Cache River basin of Illinois (see chapter 14) [3]. At that time, the ancient Tennessee River was not a tributary of the Ohio River but was a main channel flowing into the Mississippi River where the lower Ohio River flows today (see chapters 2 and 3).

The convergence of the Ohio River with the Mississippi River at Cairo, Illinois, made the Ohio a natural transportation route for westward exploration and expansion up the Missouri and upper Mississippi rivers and downstream to the Gulf of Mexico. The Ohio is a fast-flowing river with a rocky, gravel bottom. The water levels change drastically with the seasons as upland snowmelt and spring rains rush off steeply sloping hillsides into tributaries that quickly fill the main stem Ohio to flood stage. Summer and fall have historically received much less rainfall, reducing the volume of water the Ohio carries and slowing navigation as water levels drop. When Meriwether Lewis headed down the Ohio River from Pittsburgh on August 31, 1803, the water depth was very low. Lewis claimed to be able to see pike, bass, catfish, and sturgeon swimming when his keelboat was grounded by ripples and shoals [4]. Farmers often stood on the riverbank waiting to rent out their team of horses to boat owners who needed help being pulled over the wide but shallow and rocky stream.

Like other travelers on the Ohio River, Lewis encountered the Falls of the Ohio River (figure 18.5) near Louisville, Kentucky. It was here in October of 1803 that Lewis, carrying instructions from President Thomas Jefferson to explore the newly acquired Louisiana western lands purchased from France and find the Northwest Passage to the Pacific Ocean, met William Clark at Clarksville in Indiana Territory. The Falls of the Ohio dropped 24 feet over a two-mile-long series of 387 million-year-old limestone ledges formed in the Devonian coral period. Only at high water could boats travel over the falls and continue downstream. Captains with paying customers who did not want to wait for high water would portage their cargo and boats around the falls [2]. Steamboats unloaded passengers and freight on one end of the falls and carried them over land to the other end of the falls to board another boat. Louisville became a key staging location for river travelers and cargo to be transferred to up- and downstream boats. With growth in river traffic, it soon became apparent that some type of canal and lock system was needed if steamboats and other river vessels were to avoid delays and travel unimpeded though the falls.

Intersection of Three Transportation Corridors

The transfer of the Louisiana Purchase (see map 1.3) to the United States in 1803 spurred thousands of pioneers to head west to settle the Northwest Territory. The US government in the 1820s turned the path west into the first federal highway, the National Road. The National Road ran 620 miles from Cumberland, Maryland, on the Potomac River across the Allegheny Mountains and southwest Pennsylvania to Wheeling, Virginia (West Virginia in 1863) on the Ohio River. Feeder toll roads from Cumberland to Baltimore, Maryland, connected the new territories to the third largest city in the new nation and the Atlantic Ocean. At its peak, approximately 200,000 people per year traveled west using the National Road. The Baltimore and Ohio (B&O) Railroad was built along a similar route, reaching Wheeling in 1852. The city of Wheeling on the Ohio River became

FIGURE 18.1 The confluence of the Allegheny (left) and Monongahela (right) rivers in downtown Pittsburgh, Pennsylvania, forms the beginning of the Ohio River, a major tributary of the Mississippi River.

FIGURE 18.2 The Olmsted Lock and Dam at river mile 964.4, the last lock and dam on the Ohio River, is scheduled to open in 2020.

FIGURE 18.3 Lock and Dam 52 is 1.5 miles downstream of Brookport, Illinois, at river mile 939 below Pittsburgh, Pennsylvania. The upper pool above the dam extends 20.5 feet to the Smithfield Lock and Dam.

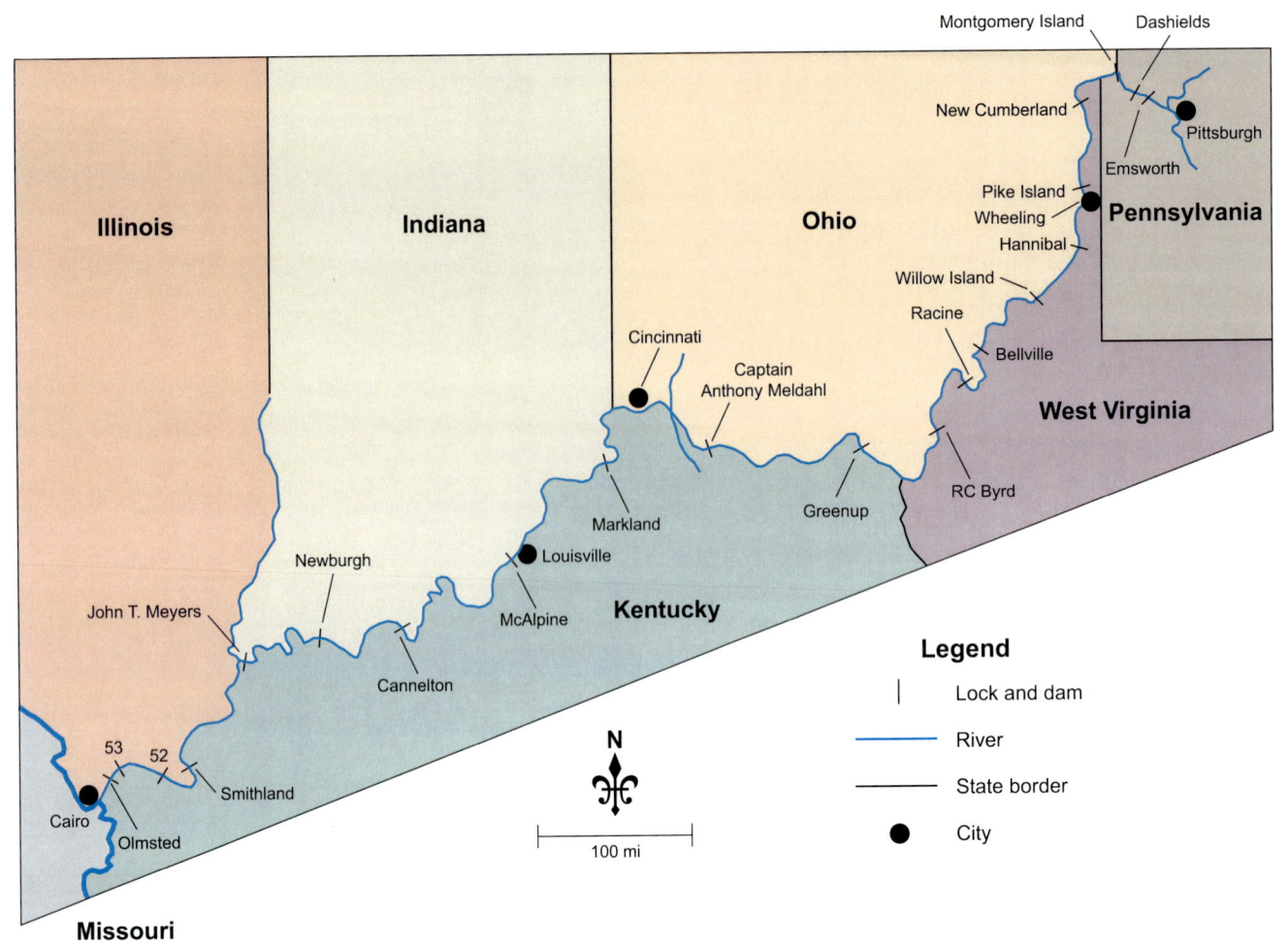

MAP 18.2 There are 20 locks and dams on the Ohio River used to manage river depth to ensure a nine-foot channel for navigation. The location of the new Olmsted Dam, which will replace locks and dams 52 and 53, is shown upstream from Cairo, Illinois.

the gateway to the west where river, overland road, and railroad transportation corridors converged.

The first bridge over the Ohio River was the Wheeling Suspension Bridge constructed in 1849. A 1,010-foot single-span bridge between Wheeling and Wheeling Island, it extended the National Road through Ohio, Indiana, and Illinois. The bridge (figure 18.6), designed after ancient Peruvian rope bridges, had 12 main cables that spanned two stone portals and was anchored in the rock under the city. Each cable contained more than 550 strands of wire. It was a toll bridge with fares as follows: man and a horse, $0.10; six-horse carriage, $0.15; four-horse mail coach, $1.25 per month; hogs and sheep, $0.02 per animal; and western stagecoach, $2,000 per year [5].

The original plan for the National Road was to continue the highway to St. Louis, Missouri, the confluence of the Mississippi and Missouri rivers. This land route west was more direct than following the Ohio River to Cairo, Illinois, and then up the Mississippi River to St. Louis. However, congressional funding faltered, and the National Road ended at Vandalia, Illinois. By 1930, Congress discovered they could not maintain the National Road and turned it over to the states who made it a toll road (a turnpike) since they also lacked funds. The construction of the Wheeling Suspension Bridge (figure 18.6) challenged Pittsburgh, Pennsylvania's claim as gateway to the west and led to a contentious political battle between Wheeling and Pittsburgh. The City of Pittsburgh, Pennsylvania, filed suit against City of Wheeling, Virginia, for building the Wheeling Suspension Bridge, claiming the bridge's height impeded Ohio River traffic to Pittsburgh. The Supreme Court agreed. However, before it was demolished, in 1852 an act of Congress maintained that the bridge was vital to the country and should remain standing. This secured Wheeling's claim as the gateway to the west.

The celebration was short-lived when a violent windstorm caused the cables to snap and took down the suspension bridge in 1854. It was not rebuilt until 1872. The suspension bridge over the Ohio River had brought people and business to Wheeling, making it the second largest city in the Commonwealth of Virginia by the end of the 1850s. In 1863 the State of West Virginia was created from the western third of Virginia. The Federal Aid Highway Act of 1956 led to the formation of the Interstate Highway System and funds for Interstate 70 and

FIGURE 18.4 Lock and Dam 53 at 962 river miles downstream from Pittsburgh has an upper pool above the dam that extends about 23 miles to Lock and Dam 52.

FIGURE 18.5 The McAlpine Dam at Louisville, Kentucky, and Clarksville, Indiana, controls water levels on the Ohio River and allows barges and boats to bypass the falls of Ohio (in the foreground). The dam, an elongated "Z," protects the ancient fossil beds of the falls of Ohio and has "castellations" or waterfalls in the dam to ensure water reaches the wetland next to the dam even in dry months so as to maintain wetland plant and animal species.

the replacement of the Wheeling Suspension Bridge. In 1968 the Fort Henry Bridge was constructed about a mile upstream of the old Wheeling Suspension Bridge on the new Interstate 70 (figure 18.6). Although the suspension bridge on the National Road was scheduled for demolition, local preservation groups and the West Virginia Department of Highways rescued it, making it one of the few historic landmarks in the state. The National Road (US Route 40) now extends from the Atlantic Ocean (New Jersey and/or New York) through Wheeling, West Virginia, to the Pacific Ocean (California).

By the mid-1800s, more than three million people traveled the Ohio River annually. Wheeling became a great manufacturing center and was designated as a US Port of Delivery at the intersection of heavy river commerce, a burgeoning iron mill industry, growing rail traffic on the B&O Railroad, and the National Road. The Wheeling Custom House (circa 1831) housed the US customs service, the US post office, and the federal district courtroom. The Port of Wheeling District controlled the river from Pittsburgh and halfway downstream between Wheeling, Virginia, and Cincinnati, Ohio. The custom house collected duties on foreign goods; licensed ships and vessels used in trade; and conducted federal inspections to validate accuracy of weights, measures, and gauges used in trade.

Improving Navigation on the Ohio River

The Lewis and Clark trip west (1803 to 1806) brought visibility to the Ohio River as the gateway to the western United States. The USACE in 1824 was authorized by Congress to make the river more navigable by removing snags and improving river flow by constructing wing dams or dikes that would concentrate flow into the main river channel. These efforts helped, but navigation on the Ohio was still sporadic and dependent on seasonal rains. The water levels in a dry summer were so shallow in places that horse-drawn wagons could easily cross the river. There were two seasonal rises: one in late October and November and a second rise between February and April [6].

In 1825 construction began on a canal around the falls of Ohio near Louisville, Kentucky. Privately financed by Louisville and Portland Canal and constructed by hand tools with the help of animal-drawn scrappers and carts, it was finished in 1830. The completed canal was two miles long with three locking chambers that created a total lift of 26 feet. Steamboat travel on the Ohio River benefited from the 1830 canal. Navigation became a year-round possibility, even in droughts. River travel was the preferred personal transportation and shipping pathway to the west. East coast and international visitors, including British novelist Charles Dickens, used the Ohio River to get to St. Louis, Missouri, in 1842. The trip required Dickens to travel the 981 miles of the Ohio River from Pittsburgh, Pennsylvania, to Cairo, Illinois, and then turn north at Cairo and the confluence and travel 180 miles up the Mississippi River to St. Louis. Along the way, he wrote the book *American Notes*, which was published in 1842 [7].

The volume of coal transported down the Ohio River from Pittsburgh jumped greatly following the Civil War (1861 to 1865). The size of the tows also grew in length as powerful steam towboats pushed more and more wooden barges of coal. Due to the escalating coal trade, the USACE began studying methods to produce a reliable navigation depth on the Ohio River. After launching an international study to analyze other navigation projects worldwide, they determined that a system of locks and dams would solve the problem.

FIGURE 18.6 A tale of two bridges. The Wheeling Suspension Bridge (foreground) built in 1849 on the National Road (US Route 40) at Wheeling, West Virginia, was the first bridge to cross the Ohio River opening the west to thousands of pioneers. The US Interstate 70 Fort Henry Bridge (green in background) carries four lanes of traffic across the Ohio River and is a major transportation route connecting eastern states to the upper Midwest today.

Ohio Navigation Locks

A lock is an engineered structure used to raise and lower boats between stretches of water that are at different levels on canals or rivers or to bypass rapids or mill weirs using the water for hydroelectricity [8]. It is a chamber with watertight doors or gates at each end that seal off the chamber from the stretch of water between the next upstream or downstream lock. The chamber holds one or more vessels and, when full of water, lifts the boats to the level of the upstream body of water. When empty, the gate is open to the downstream body of water. The lock and dam system is like a flight of stairs going up and down the river using gravity to move the water and maintain a minimum depth for boat traffic. Water drains from an open lock by gravity into a second lock until the water is level. Then barges and boats at the downstream lock can travel upstream to the next lock assured of sufficient water depth for navigation and without expending a great deal of energy against the flow of the river. Once the vessels reach the next lock, they are lifted up in a repeating process. Navigation locks allow towboats, barges, and other vessels to bypass the dams and travel up the river.

Wicket Dams and Locks

Downstream from Pittsburgh, Pennsylvania, Davis Island became the first USACE lock and dam on the Ohio River. Opened for use in 1885, it improved navigation and substantively increased passenger and commercial river traffic. Twenty-five years later, Congress passed the Rivers and Harbors Act, which provided authorization to deepen the navigation depth to 9 feet by constructing a system of lock and dams the entire length of the Ohio River. The project was completed in 1929. The new system had 51 wooden wicket dams and 600-by-110-foot lock chambers. The wicket dams created pools of deeper water to maintain navigation depth in the main channel. The locks enabled vessels to bypass the dams when the river level was low. When the river was high, the wickets could be lowered so vessels did not need to use the locks.

The wicket dam and lock system moved millions of tons of materials throughout the United States during World War II. Diesel-powered towboats replaced steam engines in the 1940s and increased the capacity of tows to pull and push barges that were longer than the 600-foot locks. This led to the hazardous and time-consuming practice of "double locking." One string of barges using one tow now required two turns through the locks. Double locking often delayed other barge and boat traffic waiting to lock through and increased costs to the towing industry. It soon became clear that the lock and dam system needed longer locks. The Ohio River Navigation Modernization Program (map 18.2), begun in the 1950s, enabled the USACE to systematically replace the outdated wicket dams and small locks. A re-engineering of the dams using steel and concrete made them permanently nonnavigable. One 18-foot dam replaced two or three of the wicket dams. Most newly designed dams had two adjoining locks (one 600-by-110-foot lock and one 1,200-by-10-foot lock), which could hold 15 barges and allow them to lock through in one maneuver. The Smithland Locks and Dam (map 18.2) was built with two 1,200-foot chambers. The McAlpine Locks and Dam at Louisville, in 2009, created a second 1,200-foot lock and widened the canal to 500 feet and the lift to 37 feet to accommodate increasing barge traffic.

The Flood of 1937

The locks and dams on the Ohio River serve a navigation purpose only and do not provide flood control. While river navigation came to a standstill during dry seasons, it was the spring rains and melted ice and snow that poured into the mountain gullies and creeks and bloated the river that Ohio Valley residents learned to fear most. During periods of heavy rainfall and snowmelt, the runoff from the nearby hillsides turned the Ohio River into a raging torrent that swept away everything in its path. The growth of cities and industries along the river and the natural flood cycle of the river made the region ripe for natural disaster. Pittsburgh claims to have survived more than 80 floods over a period of 84 years [2]. The worst was 1936 when the river crested at 46 feet, downtown Pittsburgh was under 20 feet of water, and 54 people died. Almost every city along the Ohio has a flood history, and major Ohio River floods are documented in 1862, 1883, 1884, 1901, 1907, 1913, 1936, and 1937. More recently, subsections of the river experienced record flooding in 1945 and 2011 (figure 18.7) [1, 9].

The flood of 1937, called by Welky a thousand-year flood [2], was one of the worst disasters to strike the Ohio River valley. A four-week storm cycle iteratively ravaged the region as two huge weather systems—warm, moist air from the Gulf of Mexico flowing northward and dry polar air from the northwest—collided and dumped rain on frozen ground causing heavy runoff into upstream tributaries of the Ohio River. The first storm began on Christmas of 1936 and continued into the first few days of 1937 when it dropped 7 inches of rain on the Tennessee River basin. A second storm, January 6 through 12, pounded Illinois, Indiana, and Kentucky with intense sleet. The following week 7 to 10 inches of rain fell on the Tennessee, Cumberland, and Wabash valleys and caused rivers in Ohio, Illinois, and Indiana to break 1913 flood records. A fourth storm unleashed the heaviest precipitation with some communities receiving an additional 14 inches of rain.

These sequential storms intensified tributary overflow, stacked the water on top of itself, and created multiple crests on the Ohio River (figure 18.7) [1, 2]. With the Ohio River many feet above flood stage, tributaries could not drain and backed up into surrounding lowlands. By the fourth week of storms, river cities protected by floodwalls realized that even if their floodwall held, the river was likely to come over the wall. Floodwall-protected Portsmouth, Ohio, at the confluence of the Ohio and Scioto rivers, opened their sewers to let the river come in more slowly. Cincinnati, Ohio, docks and low-lying downtown areas flooded and then caught fire ignited by floating gasoline. The fire burned for 12 hours damaging more than 3.5 square miles and property valued at $1.5 million [1]. The only river bridge crossing intact and open to traffic between Steubenville, Ohio, and Cairo, Illinois, was the suspen-

Location	Major Flood Stage Feet	1937 Crest Feet	Record Flood Crest Year	Record Flood Crest Feet	2011 Flood Feet	2011 Flood Date
Pittsburgh, Pennsylvania Confluence Allegheny, Monongahela, Ohio rivers	28.5	35.0	1936	46.4	26.6	March 12
Parkersburg, West Virginia	42.0	55.4	1913	58.9	30.0	March 13
Catlettsburg, Kentucky Confluence Big Sandy, Ohio rivers						
Portsmouth, Ohio Confluence Scioto river	66.0	74.2	1913	67.9	55.4	March 13
Cincinnati, Ohio	65.0	80.0	1884	71.1	55.4	March 13
Ashland, Kentucky	65.5	73.8	1948	65.9	56.1	March 13
Maysville, Kentucky Confluence Limestone Creek	66.0	75.6	1913	68.8	54.8	March 13
Louisville, Kentucky Downstream McAlpine McAlpine upper	73.0 38.0	85.4 52.2	1945 1945	74.4 42.1	62.9 31.1	May 2 April 27
Jeffersonville, Indiana		57.1	1884	47.4		
Evansville, Indiana	52.0	54.0	1913	48.4	46.8	May 5
Shawneetown, Illinois	53.0	65.4	1945	55.6	56.4	May 6
Paducah, Kentucky	52.0	60.8	1913	54.3	55.0	May 5
Cairo, Illinois Confluence Mississippi, Ohio rivers	53.0	59.5	1975	56.5	61.7	May 2

FIGURE 18.7 Major floods and flood crest heights on the Ohio River from 1884 to 2011.

FIGURE 18.8 The Covington to Cincinnati bridge was the only functioning bridge crossing from Steubenville, Ohio, to Cairo, Illinois, during the 1937 flood.

sion bridge (figure 18.8) that connected Cincinnati, Ohio, and Covington, Kentucky [1]. The suspension bridge at Wheeling, West Virginia (figure 18.6), was above the floodwaters, but the west end of the bridge and all of Wheeling Island were flooded.

The 1937 flood destroyed many of the railroads throughout the Ohio Valley and severed telegraph and communications. This disastrous flood only compounded the human and economic toll the region already was experiencing from the Great Depression. Poor and black families were hit the hardest. They lived and worked in the flood-prone lower sections of the river cities; lost what few belongings they had; and experienced discriminatory treatment in rescue operations, disaster aid, and temporary resettlement [2]. Postflood, community leaders, engineers, and the federal government recognized the need to be more proactive in preventing these disasters and began to make flood control investments in levees, floodwalls, and storage reservoirs to better manage river flooding. The Flood Control Act of 1938 redefined the role of government for planning, financing, and operating flood control structures.

Managing the Ohio River Basin for Navigation and Flood Control

For many years river navigation and flood control were considered separate issues competing for local and national resources, with navigation given the highest priority. President Franklin Roosevelt envisioned his New Deal construction project, the Tennessee Valley Authority (TVA), as an opportunity to develop a nationally coordinated watershed-based rivers policy that encompassed planned flood control, hydroelectric power, and improved land management [2]. The TVA did not replace the role of the USACE in managing the Ohio River and the dams and reservoirs on Ohio River tributaries. They work together, with the TVA managing the power infrastructure while the USACE effectively controls water levels to maintain shipping channels and regulates stored floodwaters for slow release. Since the construction of the Kentucky Dam on the Tennessee River in the 1940s and the Barkley Dam in 1960s, the USACE has had the ability to store water and release it during droughts to maintain a minimum depth of water above the 9-foot shipping channel. The water release also increases the flow in the lower Mississippi River. During the 2012 drought, the USACE was able to add 3.3 feet to the existing flow, which reduced the need for dredging.

One strategic reach of the Ohio River is the segment that connects the Ohio, Tennessee, Cumberland, and Mississippi rivers. This area has been described as the "hub" of the Ohio and Mississippi rivers waterway system. Barge traffic moving between the Mississippi River system and the Ohio, Tennessee, and Cumberland rivers must pass thorough this stretch. More tonnage passes this point that any other place in America's inland navigation system with locks and dams. In 2011, more than 90 million tons of goods were shipped through this reach of the Ohio River.

Olmsted Lock and Dam

The USACE role in providing a safe navigation infrastructure on the Ohio River is a mission without an end. The 981-mile-long Ohio River had 20 locks and dams in 2013 (map 18.2). The number will be reduced to 19 by 2020 when the Olmsted Lock and Dam (figures 18.2 and 18.9) replaces locks and dams 52 (figure 18.3) and 53 (figure 18.4). These locks and dams on the lower Ohio are the last of the old wicket dams. The Olmsted Lock and Dam is currently under construction (figure 18.9) at river mile 964.4 (37°11'1.7" N, 89°3'48.6" W) and will greatly reduce tow and barge delays and shorten navigation time from four hours to one hour. The dam will consist of five Tainter gates, a 1,400-foot navigable pass with steel wicket gates, and a fixed weir. In the raised position, the wickets will maintain the required navigation depths from the Olmsted project upstream to Smithland Locks and Dam. When river flows are sufficient, the wickets can be lowered to lie flat on the river bottom and allow traffic to navigate over the dam sill without having to pass through the locks. This reduces delays experienced by locking through the system. The lock chambers, completed in 2002 and located along the Illinois bank, are 110 feet wide and 1,200 feet long [6]. The capacity of this structure is projected to be sufficient to meet demands for tow traffic well into the twenty-first century.

The Olmsted Lock and Dam construction was authorized by the US Congress on November 17, 1988, by the passage of the Water Resources Development Act of 1988 (Public Law 100-676). The cost of this project is being equally shared by congressional appropriation and the navigation industry. The industry pays a tax on diesel fuel, which goes to the Inland Waterways Trust Fund. The trust fund then pays 50% of the project cost, estimated to be $1.45 billion. The Olmsted Lock and Dam was started in 1995 and is scheduled to open in 2020 with a final cost estimate of about $2.9 billion.

Gateway to Western Expansion

Daniel Webster in the 1830 opening session of the US Senate proposed a grand vision for westward expansion for the common benefit: "...a road over the Alleghanies, a canal around the falls of Ohio, a canal or railroad from the Atlantic to the western waters" [5]. This roadmap for economic, political, and social development has guided investments in the development of the Ohio River and the tributaries that feed it. Prior to 1830 the river was not navigable year round. It froze solid in winter, often flooded in March, and by August was a trickle that people could walk across. Today the Ohio River is part of a vast inland water system that is managed to serve many functions including navigation and flood management. The river carries 40% of the commercial river traffic in the continental United States. The Ohio River basin reaches northeast into New York and Pennsylvania, west to Illinois, and south through the drainage area of the Tennessee River in Kentucky, Tennessee, Georgia, Alabama, and Mississippi. The movement to basin-wide management has enabled local and federal agencies to proactively manage the river and its tributaries for navigation, flood control, hydropower, ecosystem protection, and water supply. The variable climate and extreme weather events of the past are projected to continue and even increase into the future. There will be even greater need to invest in coordination, scientific management, and a continuous supply of data to assess system-wide conditions and improve real-time decision making. A continuous supply of data entails the development of better maps based on surveys of the river and reservoir sites; increased connectivity among gaging stations; and better data on weather, precipitation rates, evaporation, water quality conditions, land use, and erosion conditions.

The economic and ecological footprint of this inland waterway affects millions of people. Approximately 125 species of fish are found in the Ohio River. The wetlands along the river and its tributaries are important feeding and nesting areas for migratory birds such as wood ducks, great blue heron, and Canada geese. The grasslands and woodlands that drain into the river provide habitat for a wide variety of bird species from the black crowned night heron and kingfishers to osprey. Over 270 different species of birds are found at the Falls of the Ohio alone. Frogs and crayfish are important food sources for wetland birds and animals. Gulls and terns circle the river current at high water and dive for food. Long-legged killdeer and sandpipers skitter across shallow rocks, sandbars, and bank beaches exposed at low water poking for insects, worms, and crustaceans. As we plan for the future of the Ohio River, the ecological footprint of this unique water resource must be better understood, protected, and valued alongside its economic contribution.

FIGURE 18.9 A giant 5,304-ton lift built on the construction site of the Olmsted Lock and Dam is used to carry cement frames to barges on the Ohio River for placement in the river.

---

[1] Castro, J.E. 2009. The Great Ohio River Flood of 1937. Charleston, SC: Arcadia Publishing.

[2] Welky, D. 2011. The Thousand-year Flood: The Ohio-Mississippi Disaster of 1937. Chicago, IL: University of Chicago Press.

[3] Cache River Wetlands Center. 2013. Cache River – State Natural Area. Cypress, IL: Illinois Department of Natural Resources. http://dnr.state.il.us/lands/landmgt/parks/r5/cachervr.htm.

[4] Ambrose, S.E. 1996. Undaunted Courage. New York, NY: Simon and Schuster.

[5] Wheeling Visitors Convention Center. 2015. Historical archives. October 8, 2015. Wheeling, WV: Visitors Convention Center.

[6] US Army Corps of Engineers. 2012. The Upper Mississippi River. Nine-foot channel navigation project. The US Waterway System. Transportation Facts and Information. Navigation and Civil Works. Decision Support Center.

[7] Dickens, C. 1842. American Notes. New York, NY: John W. Lovell Company.

[8] Nelson, S.B. 1983. Water Engineering. In Standard Handbook for Civil Engineers, ed. FS Merritt. New York, NY: McGraw-Hill.

[9] National Oceanic Atmosphere Administration. 2013. Historic crests. Morris, IL: National Weather Service, Advanced Hydrologic Prediction Service.

Managing the Tennessee River Landscape

The network of reservoirs and dams built on tributaries hundreds of miles upstream in the headwaters of the Ohio and Mississippi river basins is one of the most powerful weapons for averting flood disasters and reducing the maximum height of floodwater downstream [1, 2]. The Kentucky Dam on the Tennessee River, 22 miles upstream from the confluence with the Ohio River at Paducah, Kentucky, is one of the largest human-constructed reservoirs built to store excess floodwater and produce electricity. In the late 1800s the most pressing inland river policy issues focused on encouraging industrialization and westward expansion, and ensuring navigation on the Ohio and Mississippi rivers [1]. Although high water was an expected seasonal occurrence, large flood events only became a major concern as they collided with increasing settlement and industry development along river valleys. The disastrous flood of 1937 caused a number of US politicians, engineers, and citizen leaders to reimagine river basins as unique ecological units and to more purposefully manage them as watersheds. A 1908 report by the Inland Waterways Commission called for multipurpose dams, but it took another 30 years and a sequence of major river flood disasters (1913, 1927, 1936, and 1937; figure 19.1) before the federal government, with public support, took leadership in constructing the system of dams and reservoirs we have today in the Ohio River basin (see chapter 18) [1, 2].

During the 1937 flood, many communities along the Ohio River experienced water 15 feet above previous records in their downtowns; loss of lives and property; disruptions to railroad traffic and the regional economy in the eastern half of the United States; and loss of telegraph connections and communication functions throughout much of the Ohio Valley. As December of 1936 came to a close, despite rising water, cities like Cincinnati, Ohio; Shawneetown and Cairo, Illinois; and Jeffersonville, Indiana; were confidently secure behind their floodwalls. However, by January of 1937, the reality that levees and floodwalls were inadequate to manage the Ohio River and its tributaries became apparent too late to avert disaster. Paducah, Kentucky (map 19.1), evacuated more than 27,000 of their 33,000 population [2] and lost most of their downtown. Paducah today is a thriving city that owes much of it character and revival to the painful destruction of its city in the 1937 flood. Its back-to-the-river renaissance now celebrates its river history with new industries, downtown redevelopment, riverfront stores and museums, and a floodwall that documents historical river events [1]. This spirit of recovery for Paducah is made possible by the 13-foot floodwall (figure 19.2) and 1940s construction of the Kentucky Dam (figure 19.3), a 184-mile-long reservoir, which can hold back and store 4,008,000 acre-feet of water from the Tennessee River before releasing it into the Ohio River.

Tennessee River Name, Location, Water Rights, and Border Disputes

The headwaters of the Tennessee River, the largest tributary of the Ohio River, originate in the Appalachian Mountains of the eastern United States. Created by the confluence of the French Broad and Holston rivers (figure 19.4) at Knoxville, Tennessee, the Tennessee River runs approximately 652 miles to Paducah, Kentucky, where it drains into the Ohio River (map 19.1 and figure 19.5). The Cherokee Indians called the river *hogohegge*, meaning "Big River," while Europeans called it "Old French Fork" and "River of the Cherakees," or Cherokee River [3]. By the eighteenth century, it was known as the Tennessee River. Over millions of years, the river carved a path through the Appalachian Escarpment in eastern Tennessee as it flowed southwest toward Chattanooga before crossing into Alabama. Today the river cuts through steep forested hillsides and the grasslands of the central plains as it loops through northern Alabama and intersects three state borders (Mississippi, Alabama, and Tennessee) before turning northward through Tennessee and Kentucky where it merges with the Ohio River at Paducah, Kentucky (map 19.1).

The starting point of the Tennessee River, as well as state land and water rights, has been the subject of much debate. In 1890 a federal law declared the Tennessee River to begin where the French Broad River and the Holston River come together southeast of Knoxville (figure 19.4). In 1796, when Tennessee was admitted to the Union, the border between Georgia and Tennessee was originally defined as the 35th parallel, thereby ensuring that at least part of the river would be located in Georgia (map 19.1). As a result of an erroneous survey conducted in 1818, the actual border line was set one mile south of the 35° latitude, which placed the entire river in Tennessee. Georgia has made many attempts between the 1890s and present to correct the erroneous survey line that resulted in considerable Georgia land and population belonging to Tennessee. Eventually, if state negotiations continue to fail, the border issue could end up in the US Supreme Court.

The Tennessee River was the western boundary for lands open to settlement until 1818 when the Jackson Purchase pushed many Native Americans west

FIGURE 19.1 The floods of 1884, 1913, and 1937 and their crests at Paducah, Kentucky, on the Ohio River are marked on a downtown building. John Crivello, one of the many Paducah Ambassador Club volunteers who are leading the effort to revitalize the city as a port on the Ohio River, shows past flood crests.

FIGURE 19.2 The river side of the floodwall at Paducah, Kentucky, is a park to view the Ohio River at low water and a protection against the river at high water. The floodwall was constructed after the 1937 flood, which destroyed most of downtown Paducah.

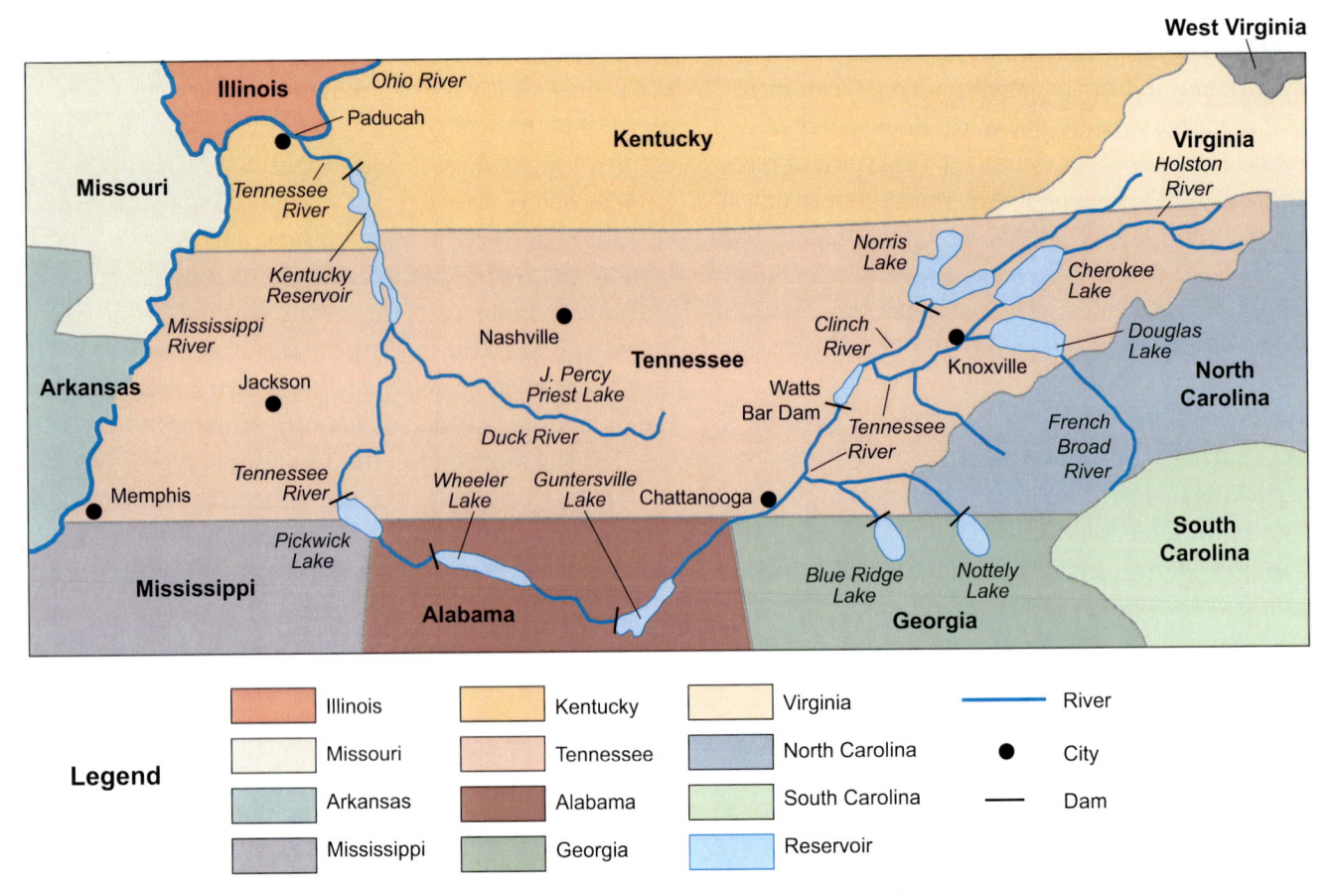

MAP 19.1 The Tennessee River basin drains upland waters from seven states into the Ohio River.

of the Mississippi River [3]. The endorsement of the 1830 Indian Removal Act by President Andrew Jackson forced the Cherokee, Muscogee, Seminole, Chickasaw, and Choctaw nations from their ancestral homelands in the southeastern United States to the designated Indian Territory west of the Mississippi River. The Tennessee River, part of the Trail of Tears, was a key water route traversed by the Indians in a difficult and deadly 2,000-mile forced relocation trip west.

During the Civil War Chattanooga Campaign, November 23 to 25, 1863, Union forces led by General Ulysses S. Grant forced the Confederate troops away from the Tennessee River, the railroads, Lookout Mountain, and Chattanooga (figure 19.6). Chattanooga became the base camp for Major General William Sherman's Atlanta campaign, and the Tennessee River was used as part of the supply line during Sherman's march to Atlanta, Georgia. In August of 1864 General Sherman departed Chattanooga and was temporarily delayed by Confederate troops. However, on September 1, 1864, the Confederate Army surrendered Atlanta.

Paducah, Kentucky, at the Confluence of the Tennessee and Ohio Rivers

Paducah, Kentucky, was built on the southern banks of the Ohio River at the confluence of the Tennessee River. This port city, incorporated in 1830, became a thriving metropolis and transportation hub fueled by steamboat and towboat traffic and an expanding network of railroads hauling agricultural commodities, coal, iron ore, and other raw materials for manufacturing, steel, and finished products to markets. Paducah was protected from river flooding by earthen levees during the 1884, 1913, and 1927 floods. However, these earthen levees were inadequate in 1937 when the Ohio River reached

FIGURE 19.3 The Kentucky Dam at Gilbertsville, Kentucky, was built on the Tennessee River 22 miles upstream from its confluence with the Ohio River.

record levels, breached the levees, and flooded the entire downtown (figure 19.1 and see chapter 18) [2].

Heavy precipitation from the Gulf of Mexico and a polar vortex from the northwest collided over the Ohio Valley from Christmas of 1936 into January of 1937 and dumped 165 billion tons (41 quadrillion gallons) of ice, snow, sleet, and rain over a four week period [1]. In the third week of the storm, 7 to 10 inches of rain fell on the Tennessee, Cumberland, and Wabash valleys breaking 1913 flood records throughout the Ohio Valley. The following week a fourth wave of storms added up to 14 inches of rain, and on January 21, 1937, the Ohio River at Paducah rose 50 feet above flood stage. The swollen Ohio River rushing south to Cairo, Illinois, was a wall of water that blocked the Tennessee River from draining into the Ohio. Without an outlet, the Tennessee backed up and flooded surrounding lowlands.

The Paducah gage on the Ohio River marked the crest at 60.8 feet on February 2. The river water did not drop below flood stage until February 15. For 26 days, the 27,000 residents who had been forced to flee Paducah had to stay with relatives and friends on higher ground in McCracken and other adjacent counties. The American Red Cross and local churches provided shelters for those who had no place to go. A total of 1.5 feet of rain over 16 days and sheets of swiftly moving ice on the rivers created the worst natural disaster in Paducah's history. Congress authorized the US Army Corps of Engineers (USACE) to build the current floodwall (figures 19.2 and 19.7) 3.3 feet above the 1937 record peak in 1938 to prevent future flooding of downtown businesses, industries, and residences.

Tennessee Valley in the 1930s and 1940s

The 1937 flood was a disaster of unparalleled magnitude for Paducah and many other cities in the Ohio Valley and much of the lower Mississippi Valley. Over a million people were displaced; nearly 400 lives were lost; and millions of dollars in damage to industries, businesses, and communities were reported [2]. Although levees and floodwalls are critical infrastructure in protecting cities, industries, and agricultural lands, the flood of 1937 demonstrated the need to control tributary and upstream runoff volume and velocity. Further, it became rather clear that the Ohio and Mississippi river basins were interconnected and needed a unified, comprehensive engineering and land use plan encompassing watershed and basin scales.

However, controlling main stem inland river flooding and tributary back flow were not the only problems facing the central United States in the 1930s. Many

FIGURE 19.4 The Tennessee River begins at the confluence of the French Broad and Holston rivers near Knoxville, Tennessee.

FIGURE 19.5 The Tennessee and Ohio river confluence east of Paducah, Kentucky.

rural areas lacked rural electricity, year-round passable roads, and other public infrastructure. Malaria and yellow fever carried by mosquitos breeding in swamps and wetlands and cholera from contaminated drinking water caused illnesses and deaths. Joblessness and rural poverty were exacerbated by the Great Depression. Eroded hillsides from years of poor farming practices and depressed agricultural markets made it difficult for rural families to produce and sell enough to meet basic food and housing needs. During the 1800s timber and tobacco were main income sources in the Tennessee River valley. However, by 1937 much of the land had been farmed too intensively without good conservation practices resulting in depleted and eroded soil that was unable to produce the agricultural yields of the past. As crop yields decreased so did farm incomes. Concurrently, the timber industry peaked and declined as the best of the hardwood forests had been clear-cut.

Faced with a large portion of the US population unemployed and unable to support themselves, President

FIGURE 19.6 The Tennessee River in Chattanooga as viewed from Lookout Mountain meanders and loops throughout its alluvial valley.

Franklin Roosevelt and Congress created a number of social, environmental, and economic government programs. Legislation in 1933 authorized the Tennessee Valley Authority (TVA), a public-private institution designed to produce low-cost electricity using hydropower and revitalize poor rural regions throughout the south. The largest public power provider in the United States, the TVA served an area covering 80,000 square miles ranging from Tennessee, parts of Alabama, Georgia, Kentucky, and Mississippi to North Carolina and Virginia (map 19.1 and figure 19.8). During World War II, TVA hydropower produced electricity for critical war industries including aluminum plants that built bombs and airplanes. At its peak production in 1942, 12 hydroelectric projects and steam plants were under construction at the same time, and design and construction efforts employed 28,000. Today, the TVA directly produces electricity for 56 industries and federal facilities and 155 local distributors who provide power to over 9 million people.

The first chairman of the TVA, Arthur Morgan, was an engineer with a vision for large-scale, comprehensive planning that integrated dam building projects, public works employment, and the transformation of poor rural communities in the Tennessee Valley [4]. Morgan's social engineering and dam projects on the Miami River (Ohio) after the flood of 1918 offered a prototype for his problem-solving approach to address resource management and community development.

The TVA attempted to integrate social, economic, land use, and natural resource management of the region in the development of electric power production, navigation, flood control, malaria prevention, reforestation, erosion control, and employment opportunities. The TVA continues to operate on the principle of integrated solutions even as issues have changed over time.

Dam construction to harness the region's rivers was the centerpiece of the TVA. The dams controlled floods, improved navigation, generated electricity, and provided jobs. Concurrent to building dams and rural electrification, the TVA undertook rural development by teaching farmers erosion control practices to prevent soil loss and introducing fertilizers that improved crop yields. The TVA also invested in the reestablishment of forests, better forest fire management, and preservation of wildlife habitat. Electricity generated by TVA dams (figure 19.8) modernized rural farms and homes by providing electricity for lights, motors, and other power equipment.

Construction of the Kentucky Dam on the Tennessee River

Upstream dam and reservoir construction is the most straightforward strategy for downstream flood protection. Dams hold back high volume runoff water from upland spring snowmelt and extreme and prolonged periods of precipitation. This allows for the slow release of water downstream, thereby managing river height

FIGURE 19.7 Murals on the Paducah floodwall celebrate its colorful river history.

FIGURE 19.8 Watts Bar Dam on the Tennessee River between Knoxville and Chattanooga, Tennessee, generates hydroelectricity, contributes to downstream flood management, and enables navigation on the Tennessee River via its locking system.

and velocities and limiting potential flood damage. Five dams built in the early 1900s in the Miami Valley (Ohio) by a consortium of public and private interests withstood the 1937 flood in the Ohio River valley and demonstrated the effectiveness of river surveys, topographical maps, a network of gaging stations, and well-sited reservoirs [1]. Lessons learned from the construction and management of these early dams provided the TVA with engineering and survey mapping knowledge and reaffirmed the value of comprehensive planning.

As early as the 1920s and prior to the flood of 1937, area communities along the Tennessee River and their leaders lobbied Congress to have a dam constructed. The Kentucky Dam, a hydroelectric dam on the Tennessee River on the county line between Livingston County and Marshall County in the state of Kentucky, was sited in Gilbertsville, Kentucky, after extensive geologic evaluation. The dam is 22 miles above the mouth of the Tennessee River, which empties into the Ohio River east of Paducah, Kentucky (map 19.2). After absorbing the Tennessee River, the Ohio River flows another 46 miles before meeting the Mississippi River at Cairo, Illinois. The Kentucky Dam (figure 19.3), started in 1938 and completed in 1944, is the lowermost of nine dams designed to hold back the Tennessee River. Above 55 feet (the original river depth), the dam controls the release of water into the Ohio River and helps manage the fast rise of the Ohio and lower Mississippi rivers during spring snowmelt and rains. At 206 feet high and 8,422 feet long, it is the widest dam on the Tennessee River and in the TVA system. The dam impounds Kentucky Lake, which covers 160,000 acres, making it the largest of the TVA reservoirs and the largest artificial lake by area in eastern United States. Kentucky Lake has 2,064 miles of shoreline and the most flood storage capacity (4,008,000 acre-feet) of any lake in the TVA system.

The construction of Kentucky Dam (figure 19.3) and its reservoir required the public purchase of 320,244 acres of land including 48,496 acres of forest, which had to be cleared. Prior to the flooding of the land to make the reservoir, over 2,600 families, 3,390 graves, and 365 miles of roads had to be relocated. Bridges over some roads needed to be rebuilt, and 65 new ones were constructed on the new roads. The Illinois Central Railroad was rerouted to cross the top of the Kentucky Dam. Several small communities were submerged by the new reservoir (Johnville and Springville, Tennessee, and Birmingham, Kentucky), and a dike was constructed around Big Sandy, Tennessee, to protect from reservoir backwater [5]. Kentucky Dam, located in the New Madrid Seismic Zone, was built to withstand major earthquake shocks of 7.0 to 7.9 magnitude. Construction was finished on August 30, 1944, at a cost of nearly $118 million.

Navigation on the Tennessee River

Watts Bar Lock and Dam construction was started in 1939 and completed in January of 1942, three weeks after the Pearl Harbor attack. The lock and dam and shipping channel moved raw materials and manufactured products throughout the region, increasing the value of the Tennessee River as an important inland transportation system during World War II. The reservoir has a dam that is 112 feet high and 2,960 feet across, which creates 39,090 acres of surface water and holds 379,000 acre-feet of water. The winter pool has a sea level elevation of 745 feet and a summer elevation of 740 to 741 feet to ensure navigation. There are five hydroelectric generators (figure 19.8) capable of generating 182 megawatts per day. The reservoir provides 72 miles of navigation on the Tennessee River with 20 miles of slack water on the Cinch River and 12 miles on the Emory River. The Watts Bar Lock is 60 by 360 feet with a lift of 70 feet to the downstream Chickamauga Reservoir. The Watts Bar Nuclear Power Plant (figure 19.9) is located on 1,770 acres along

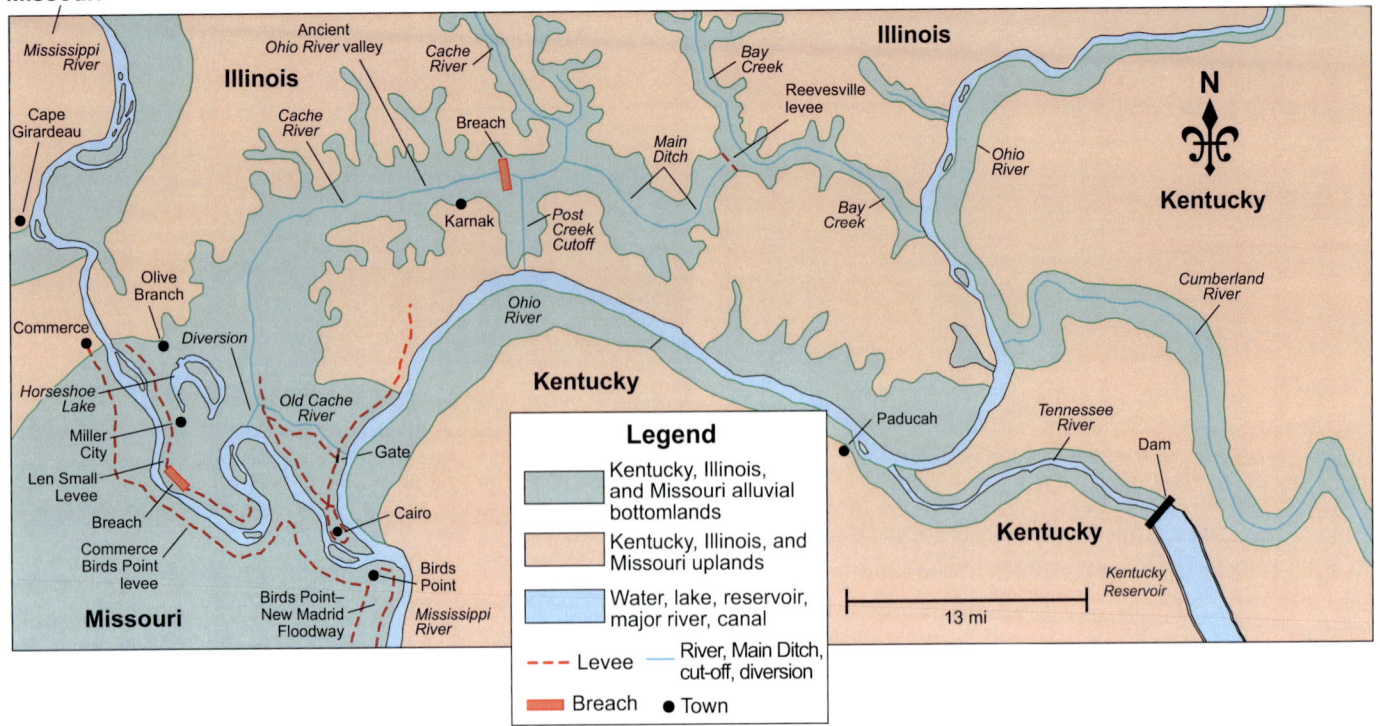

MAP 19.2 The Kentucky Dam and lake on the Tennessee River (lower right), upstream from Paducah, control the release of water from the Tennessee River into the Ohio River. This reservoir can store excess floodwater or ensure sufficient downstream navigation depths during drought periods.

the Tennessee River immediately south of the lock and dam. Unit 1 was constructed between 1976 and 1996. Unit 2 was completed in December of 2015, and is currently undergoing power ascension testing. It is scheduled to begin commercial operations in the summer of 2016. Unit 2 cost $4.7 billion and will be the first US nuclear plant to go online in the twenty-first century.

By the end of World War II, the TVA had created a 652-mile navigation channel the length of the Tennessee River using a system of dams and locks. The locks and dams have a significant impact on the economy of the region. Historically, goods were shipped by rail or truck. The use of barges reduced shipping costs by about $500 million each year depending on volume and year. The railroads reduced shipping costs to stay competitive with barge transport. This lower cost has reduced consumer prices for many products transported long distances. One barge can transport as much tonnage as 60 semitrucks or 15 rail cars. This water transportation system also reduces highway truck traffic, fuel consumption, air pollution, wear and tear on highways, and the number of tires buried in landfills.

The Tennessee River originates at Knoxville, Tennessee, and drops 513 feet before it joins the Ohio River at Paducah, Kentucky. Much like the Ohio River prior to the construction of the lock and dam system, the Tennessee River depth had a great deal of variability. Shoals and rapids made navigation almost impossible depending on drought, seasonal rainfall, and timing of snowmelt.

The Tennessee River has nine main-river locks and dams, which create a continuous series of pools the entire length of the Tennessee River and enable navigation. River navigation is possible from Knoxville all the way up the Ohio River to Pittsburgh, Pennsylvania, or downstream to the Ohio-Mississippi river confluence and south to the Gulf of Mexico. Commercial navigation extends into three major tributaries of the Tennessee: up the Cinch River 61 miles, up the Little Tennessee 29 miles, and up the Hiwassee River 22 miles. Recreational boating is popular on the river, and over 13,000 recreational craft annually lock through the system. In addition to the 9-foot channel commercial vessels use, about 374 miles can be used by recreational boaters but are too shallow for commercial traffic. River ports have become regional centers of social and economic vitality and sources of industrial activity. River transportation of food products for processing has proved to be inexpensive and efficient, reducing the price of groceries in the Southeast and across the United States.

Ecological Impacts of Changes to the Tennessee River Landscape

Dam construction on the Tennessee and Cumberland rivers and their tributaries has changed river flow patterns, temperatures, and sediment transport thereby modifying aquatic and terrestrial habitats [6, 7, 8]. Hydropower demands resulted in water releases timed to meet power production needs often without regard to aquatic ecosystem impacts. As a consequence, water depth and veloci-

FIGURE 19.9 Nuclear cooling towers near Watts Bar and adjacent to the Tennessee River contribute to the power production of the region.

ties in tailwaters (the river below the dam before it flows into another reservoir) during nongeneration periods declined to very low levels. Further, reservoir conditions and the dam structure produced low dissolved oxygen (DO) concentrations with low flows exacerbating the low DO in tailwaters [6]. In 1991, the TVA put in place a Reservoir Releases Improvement Plan to address deteriorating ecological conditions in the Tennessee River watershed. Modifications included the installation of fish ladders; experiments in adding DO; and changes to the minimum flow to tailwaters by modifying timing, velocity, and quantity of periodic water releases. A variety of strategies are now used to increase DO including pumps, turbine venting, air blowers, forced air turbine venting, infuser weirs, and line diffusers [6].

A number of studies have begun to monitor dam modifications and impacts on abiotic and biotic conditions and the abundance and diversity of fish species to better understand the extent to which these modifications mitigate and restore the ecology of the river. Of particular interest are changes in discharge fluctuations into tailwaters associated with peaking hydropower operations. In a recent study, data monitoring below nine dams revealed that yearly mean DO and mean minimum velocity increased following dam modifications. However, flow changes alone had a smaller benefit than the combined effects of modifications that increased flow and DO [6].

The Tennessee and Cumberland river basins—considered a single aquatic ecoregion—are thought to contain one of the greatest diversities of temperate freshwater species in the world, with 231 fish species, mussels, crayfish, and salamanders [7]. However, urbanization, mining, logging, agriculture, river channelization, and dams have accelerated erosion and sedimentation, changed stream flows, and degraded or destroyed aquatic habitat putting many of these species at risk. Research on ecological flows of the Tennessee and Cumberland rivers and their tributaries is examining how changes in climate, land cover, soil properties, and physiography drive stream flow responses [7]. Knight et al. found that basin characteristics explained over half of the variation in streamflow with daily temperature range, geology, and rock depth major factors [7]. One of the most influential factors is regional climate variables—mean monthly precipitation, January precipitation, daily temperature range, and August temperature. The relationship between flow and aquatic ecology is complex and not yet well understood [8], and much more research is needed. The TVA continues to monitor and use research findings as feedback information to adaptively manage not only their lock and dam network but also the land and river uses throughout the Tennessee River basin.

The River and Regional Revitalization

Public funding of the Tennessee Valley Authority power program ended in 1959, and its environmental and economic development activities were phased out in 1999. Electricity sales and other power initiatives currently fully fund all the TVA's activities. The Tennessee River and its series of locks and dams are owned by the US federal government and jointly managed by the TVA and the US-ACE. The US Coast Guard works closely with both agencies to assure reliable and safe navigation for commercial and recreation vessels and enforces maritime laws, marine safety, and investigations of marine accidents [9]. The Coast Guard is also responsible for installation and maintenance of lights, buoys, and shoreline markers along 800 miles of the river's commercial channel.

With the construction of the Kentucky Dam on the Tennessee River, water can be stored and released during droughts to maintain a minimum 6.6-foot depth of water above the 9-foot shipping channel. These water releases also increase the flow in the lower Mississippi River. During the 2012 drought, the USACE was able to add 4 feet to the existing flow, which reduced the need for dredging. During major flooding events between the 1940s and 2011, the New Madrid Floodway in Missouri did not have to be used until 2011. The USACE was able to mitigate the flood impact by not releasing water from upstream reservoirs including the Kentucky and Barkley. This strategy worked until the record flood of 2011 when the Kentucky Reservoir was not able to hold any more water and had to release some into the Ohio River, contributing to the record peak at the confluence of the Ohio and Mississippi rivers. Without the use of the Kentucky Reservoir and later the Barkley Reservoir on the Cumberland River to manage peak flow, the New Madrid Floodway would have had to be opened during major flooding events between the 1940s and 2011, and substantive soil and crop damages would have occurred.

---

[1] Welky, D. 2011. The Thousand-Year Flood: The Ohio-Mississippi Disaster of 1937. Chicago, IL: University of Chicago Press.

[2] Castro, J.E. 2009. The Great Ohio River Flood of 1937. Charleston, SC: Arcadia Publishing.

[3] Land Between the Lakes National Recreation Area. 2014. October 22, 2014. Golden Pond, KY: Visitor Center, Land Between the Lakes National Recreation Area exhibits.

[4] Tennessee Valley Authority (TVA). Bringing the Land to Life. https://www.tva.gov/About-TVA/Our-History/Bringing-the-Land-to-Life.

[5] Tennessee Valley Authority. 1951. The Kentucky Project: A Comprehensive Report on Planning, Design, Construction, and Initial Operations of the Kentucky Project. Technical Report no. 13. Washington, DC: US Government Printing Office.

[6] Bednarek, A.T., and D.D. Hart. 2005. Modifying dam operations to restore rivers: Ecological responses to Tennessee River dam mitigation. Ecological Applications 15(3):997-1008.

[7] Knight, R.D., W.S. Gain, and W.J. Wolfe. 2012. Modelling ecological flow regime: An example from the Tennessee and Cumberland river basins. Ecohydrology 5:613-627.

[8] McManamay, R.A., D.J. Orth, C.A. Dolloff, and D.C. Mathews. 2013. Application of the ELOHA framework to regulated rivers in the upper Tennessee River basin: A case study. Environmental Management 51:1210-1235.

[9] Olson, K.R., and L.M. Morton. 2014. Runaway barges damage Marseilles Lock and Dam during 2013 flood on the Illinois River. Journal of Soil and Water Conservation 69(4):104A-109A, doi:10.2489/jswc.69.4.104A.

Managing the Cumberland River Landscape

Heavy rainfall over the Cumberland River basin during the night of January 1, 1937, averaged 2.5 inches with several stations reporting in excess of 4 inches [1]. It continued to rain, and on January 2 flood warnings on the rising Cumberland River were issued from Burnside, Kentucky, to Nashville, Tennessee. Twenty-five successive days of rain over the Ohio River valley created one of the greatest floods of record in 1937. The 688-mile-long Cumberland River (map 20.1), one of the three largest tributaries of the Ohio, was a major source of flooding as it drained over 18,000 square miles of upstream lands and poured into the Ohio River at Smithland, Kentucky (figure 20.1). A 65.5-foot crest at Clarksville, Tennessee, set a record, exceeding the 1927 flood crest by 5.5 feet. Downstream at Eddyville, Kentucky, the January of 1937 crest marked an all-time record at 76.9 feet.

Twenty-six years later, the Barkley Dam on the Cumberland (map 20.1) was constructed east of the Kentucky Dam on the Tennessee River to reduce Ohio River flooding and control early snowmelt and precipitation flowing from the Appalachian Mountain headwaters of the Cumberland River. Named after Senator Alben W. Barkley, who was instrumental in passing the 1938 Flood Control Act, the Barkley Dam and Reservoir is a key component of the US Army Corps of Engineers (USACE) and Tennessee Valley Authority (TVA) flood control and navigation plan for the Mississippi River basin. A string of locks and dams upriver from the Barkley Dam protect Clarksville, Nashville, and other small towns from Cumberland River internal flooding and enable river transportation from Lake Cumberland, Kentucky, to the Ohio and Mississippi rivers all the way to the Gulf of Mexico. Although the construction of Barkley Lake (as the reservoir is called) displaced 4,400 people from their homes along the Cumberland River, it was part of a rural development strategy for southern Kentucky and north-central Tennessee by TVA to provide power, employment, and economic growth for the region.

Ancient Ohio and Cumberland River Valleys

Millions of years ago the ancient Cumberland River carried snowmelt and precipitation runoff water from the Appalachian Uplands westward and drained into the ancient Ohio River on its way through southern Illinois (map 20.2). The Green and Cumberland rivers flowing

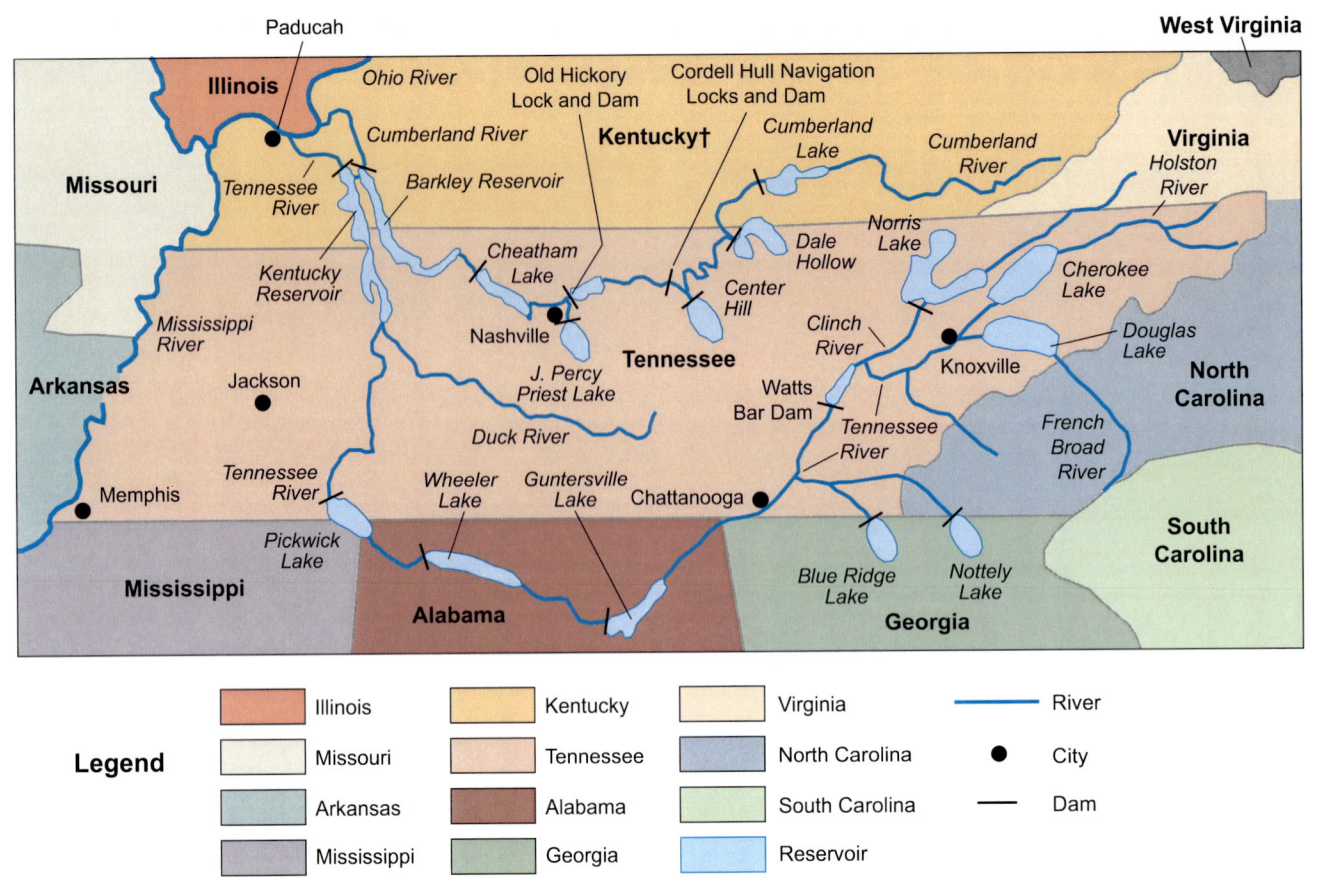

MAP 20.1 The Cumberland River watershed begins in the Appalachian Mountains of southeast Kentucky, draining southern Kentucky and middle Tennessee lands before confluencing with the Ohio River in western Kentucky.

in from the south were major tributaries of the ancient Ohio River while the Wabash, White, and Vermillion rivers were blocked by glacial ice dams draining northward into the ancient Teays River valley (see map 2.5). The ancient Tennessee River (west of the Cumberland River) ran in the modern-day Ohio River channel and confluenced directly with the ancient Mississippi River south of modern-day Cairo. Seismic activity and glacial meltwater cut through the land bridge (see map 2.6), and the ancient Cumberland, Green, Tennessee, and Ohio rivers combined to create the current Ohio River, which was formerly the ancient Tennessee River channel. About 12,000 to 15,000 years ago at the end of the glacial period these combined rivers redirected their flows into the current Ohio River channel leaving the ancient Ohio River valley in southern Illinois (modern-day Cache River and Bay Creek valleys; map 20.2) without a major river and only local drainage.

Cumberland River

Three separate forks (Martin's Fork, Clover Fork, and Poor Fork) flow out of the Appalachian Mountains in southeast Kentucky near the Virginia border to form the headwaters of the Cumberland River near Harlan, Kentucky. Steamboat traffic on the Cumberland River increased substantially in the 1800s as coal fields expanded and Tennessee produce began to be shipped throughout the region. The Cumberland River (map 20.1) was surveyed during this period, and between 1832 and 1838 Congress appropriated $155,000 for improving commercial navigation. With this infusion of money, the USACE could clear the river of snags and build wing dams to deepen the channel.

During the Civil War, Kentucky and Tennessee were critical border states. Both the Union and Confederate armies sought to control river traffic to ensure troops and supplies reached strategic locations. Fort Donelson was built by the Confederates on the Cumberland River 50 miles upstream from its confluence with the Ohio River in order to protect Nashville from Union approaches by river. However, the upper Cumberland was not easy to defend, and the fort was taken in February of 1862 by Union troops. As the Civil War came to an end in 1865, the region began to rebuild and redirect wartime resources into industries that would bring economic and social prosperity. The timber industry boomed as logs from the hardwood forests of the Appalachian Mountains and foothills were cut and transported by river to lumber companies to finish for the growing construction industry. Without locks and dams, lumber rafts (up to 100 feet by 30 feet by 8 feet) had to wait for the spring rise in channel depth to float and pole their cargo downstream to markets.

After surveying the Cumberland, USACE engineers began to identify lock and dam sites and make plans to modernize the river for commercial navigation. Construction on Lock and Dam 1 located upstream from Nashville, Tennessee, began in 1888. A few years later Lock and Dam A at Harpeth Shoals was authorized by Congress. By the early 1900s, eight stone or concrete and timber dams were built above Nashville and six below. By 1924, the Cumberland River's main channel had been raised to a minimum of six feet from the Smithland confluence with the Ohio River upstream to Burnside, Kentucky. New technologies in powering towboats replaced steamboats with gasoline and diesel engines enabling raw and finished materials to be moved more quickly up- and downstream to manufacturing plants and consumers. Simultaneously, railroad construction accelerated and expanded commercial transportation throughout the United States. This growing network of rail and river transport provided complementary and competitive options for moving freight and fueled the industrialization of the nation in the twentieth century.

The USACE selected sites in 1936 for building four reservoirs on the Cumberland and its tributaries: Wolf Creek on the upper Cumberland (1952), Dale Hollow on Obey (1948), Center Hill on Caney Fork (1951), and Stewarts Ferry (1968) on Stones River. The Wolf Creek Dam was designed to manage the upper Cumberland River snowmelt and drainage from the mountains of Kentucky to control flooding and create hydroelectricity. Lake Cumberland, the 100-mile-long reservoir behind the dam, has an average depth of 89 feet and the capacity to hold 6.1 million acre-feet of water, which can be strategically released to manage downriver flooding. The 1,255-mile shoreline and 65.5-thousand-acre lake has become a tourist destination for fishing, houseboats, and other types of recreational boating. Two additional reservoirs were also built, the Cheatham near Ashland City (1959) and the Barkley in western Kentucky (1966), for hydropower and flood control. In total, eight dams controlled the river from Burnside, Kentucky, to the Ohio River by the 1970s.

The Barkley Dam, Reservoir, and Canal

A series of storms after Christmas of 1936 dumped heavy precipitation over the Ohio River valley and its tributaries as warm moist air from the Gulf of Mexico collided with dry, cold arctic air over a 26-day period extending into late January of 1937. The January crests on the Cumberland River ran from Burnside, Kentucky, at 54.3 feet; to Nashville, Tennessee, at 53.8 feet; to Eddyville, Kentucky, at 76.9 feet; making it one of the worst floods the

FIGURE 20.1 The Cumberland River (right) and Ohio River (left) confluence at Smithland, Kentucky.

FIGURE 20.2 A railroad built above the Barkley Lock and Dam on the Cumberland River transports iron ore and manufactured products.

region had ever experienced [1]. Kentucky and Tennessee were not alone. This sequence of storms resulted in one of the most destructive flood disasters ever recorded along the Ohio River and the lower Mississippi River. The Flood Control Act of 1938 led to the construction of scores of upstream dams and reservoirs on tributaries throughout the Ohio River valley, including the Kentucky Dam on the Tennessee River (see chapter 19).

The USACE determined that the small locks and dams up and down the Cumberland River were not sufficient for flood control, so the construction of the Barkley Dam began in 1959 and was completed in 1964. Between 1958 and 1963 about 1,400 families in western Kentucky and Tennessee were moved from their homes along the Cumberland River to make way for a second large dam and reservoir, the Barkley [2]. The Barkley Locks and Dam (figure 20.2) were con-

FIGURE 20.5 Puddingstone (quartz pebbles) lines small streams in the Land Between the Lakes.

production in the nineteenth century. With the construction of the Barkley Dam and Reservoir in the 1960s, almost 20 years after the Kentucky Dam and Lake was built, the land between the two lakes was purchased by the federal government and became known as the Land Between the Lakes. From 1963 to 1968, about 3,000 people were moved from their homes in the area between the lakes to create a national recreation area. The lakes are connected where they run closest together by the Barkley shipping channel (figure 20.4), making this 170,000 acre area the largest inland peninsula in the United States.

The Tennessee and Cumberland river basins are separated by a dividing ridge of limestone known as the Tennessee Divide [2]. Rounded, water-worn "puddingstone" consisting of quartz pebbles cemented in a matrix of iron oxide, calcium carbonate, and clay can be found in creek bottoms throughout the Land Between the Lakes area (figure 20.5). Historical accounts and ecological research in the Land Between the Lakes indicate that upland oak forests and grasslands were more prevalent than they are today. In recent years, oak-grassland restoration demonstration areas have been established. A tourist destination, the Land Between the Lakes offers camping, hunting, fishing, off-highway vehicle riding, horseback riding, and environmental education programs. Featuring "outside play," the national recreation area provides access to over 300 miles of natural shoreline, 200 miles of paved road, and 500 miles of trails.

Managing for Navigation and Flood Control

Management of the Ohio and Mississippi river landscapes continues to be a challenge for the USACE, the TVA, and cities and towns along these rivers and their tributaries. An increasingly variable climate and extreme weather events, such as the flood of 2011 and the drought of 2012, are expected to become the new norm. The Kentucky Dam and the Barkley Dam are important infrastructures in this inland waterway. They can store water during floods and release water during droughts to maintain a minimum 6.6-foot depth of water above the 9-foot Ohio River shipping channel (see chapter 18). The water release also increases the flow in the lower Mississippi River. The reservoirs are one of the first lines of defense for managing excessive upland snowmelt and spring rain runoff. This strategy reduced the need to open the New Madrid Floodway for many years until the record flood of 2011. With the Kentucky and Barkley reservoirs at full capacity, it was necessary to release excess water, which contributed to the record crest at the confluence of the Ohio and Mississippi rivers.

---

[1] Williamson, R.M. 1937. Cumberland River and Tributaries. Monthly Weather Review 65(2):81.

[2] Golden Pond Visitor Center exhibits. 2014. Golden Pond, KY: Land Between the Lakes National Recreation Area.

[3] Camillo, C.A. 2012. Divine Providence: The 2011 Flood in Mississippi River and Tributaries Project. Vicksburg, MS: Mississippi River Commission.

[4] Barry, J.M. 1997. Rising Tide: The Great Mississippi Flood of 1927 and How It Changed America. New York: Simon and Schuster.

[5] Morton, L.W., and K.R. Olson, 2014. Addressing soil degradation and flood risk decision making in levee protected agricultural lands under increasingly variable climate conditions Special issue: Environmental Degradation. Journal of Environmental Policy 5(12):1220-1234.

[6] Olson, K.R., and L W. Morton. 2012. The impacts of 2011 induced levee breaches on agricultural lands of Mississippi River Valley. Journal of Soil and Water Conservation 67(1): 5A-10A, doi:10.2489/jswc.67.1.5A.

[7] Olson, K.R., and L W. Morton. 2012. The effects of 2011 Ohio and Mississippi River valley flooding on Cairo, Illinois, area. Journal of Soil and Water Conservation 67(2):42A-46A, doi:10.2489/jswc.67.2.42A.

Managing the Upper Mississippi River to Improve Commercial Navigation

The majestic Mississippi River has its origin in Lake Itasca in the state of Minnesota, the land of 10,000 lakes. The river falls 1,475 feet from northern Minnesota to the Gulf of Mexico through a pre–Ice Age gorge from Minneapolis south for 300 miles, accounting for nearly two-thirds of the drop. Early post–Ice Age people used the upper Mississippi River extensively for travel to follow their food supply as seasons changed and for trade, transporting furs, pottery, stone-tipped arrows, tools, and their burial mound culture throughout the region. The upper Mississippi River basin drains portions of Minnesota, Wisconsin, Iowa, Illinois, northern Indiana, and northwestern Missouri (map 21.1). The western edge of the upper Mississippi River basin abuts the Missouri River basin (map 21.1), and together these two basins carry headwaters from north-central US lands east of the continental divide into the main stem Mississippi River at St. Louis, Missouri (figure 21.1). Today, the upper Mississippi River is an 860-mile inland navigation system with 29 locks and dams used for commercial and recreational traffic (map 21.2) running from Minneapolis–St. Paul, Minnesota, past St. Louis, Missouri, to Cairo, Illinois, where it joins the Ohio River flowing south to form the lower Mississippi River.

Geologic and Hydrographic History

The beautiful and dramatic landscape of the upper Mississippi River valley likely originated as an ice-marginal stream during what is referred to as the Nebraskan glaciation. Current terminology places this as pre-Illinoian stage. The upper Mississippi River is a portion of the now-extinct Glacial River Warren, which cut deep river channels into the Minnesota sandstone bluffs when it melted and drained the immense glacial Lake Agassiz (see map 2.7) south to join the world's oceans at the Gulf of Mexico. The melting of ice at the end of each glacial epoch (see chapter 2) led to cycles of erosion and sediment deposition. Torrential meltwaters scoured valleys more than 200 feet deep below current river levels at the beginning of each interglacial period. Sand and gravel transported by these fast-moving floods were deposited in the valleys as the interglacial warming advanced, and the volume of river water was greatly reduced after the ice had melted.

165

Similarly, the collapse of ice dams holding back glacial Lake Duluth and glacial Lake Grantsburg carved out the dells of the Saint Croix River. Before the last glacier, the ancient Wisconsin River drained the northern part of Wisconsin. About 18,000 years ago, the Green Bay Lobe of the glacial ice sheet pushed in from the east and butted up against the Baraboo Hills. The ancient Wisconsin River was blocked, and water backed up filling the basin to the north and west creating glacial Lake Wisconsin. This glacial lake existed for a few thousand years with storms and ice scouring sand off the sandstone bluffs. Then, 14,000 years ago, the climate warmed and the glacier retreated. The meltwaters raised the ancient lake level and opened a path around the Baraboo Hills. Eventually, the stream cut through a thin dam or plug in a few days near the Wisconsin Dells. In a catastrophic flood, most of the lake drained out to the south, and flowing floodwater cut new channels through the lake bottom sand, then cut canyons through the weakly cemented sandstone.

A portion of the upper Mississippi region where northeast Iowa and western Wisconsin intersect is known as the Driftless Area (see map 2.3). This area was left unglaciated at the height of the Ice Age. Characterized by sandstone and limestone bluffs, the Driftless Area has a three-to-seven-mile valley trench with a deep meandering riverbed and steep slopes that were not smoothed out or covered over by glacial processes. The Wisconsin glaciation to the north formed lobes that met and were ice blocked where the Mississippi now flows. Geologists posit that the bursting of ice dams explains this region's topography and the modern-day

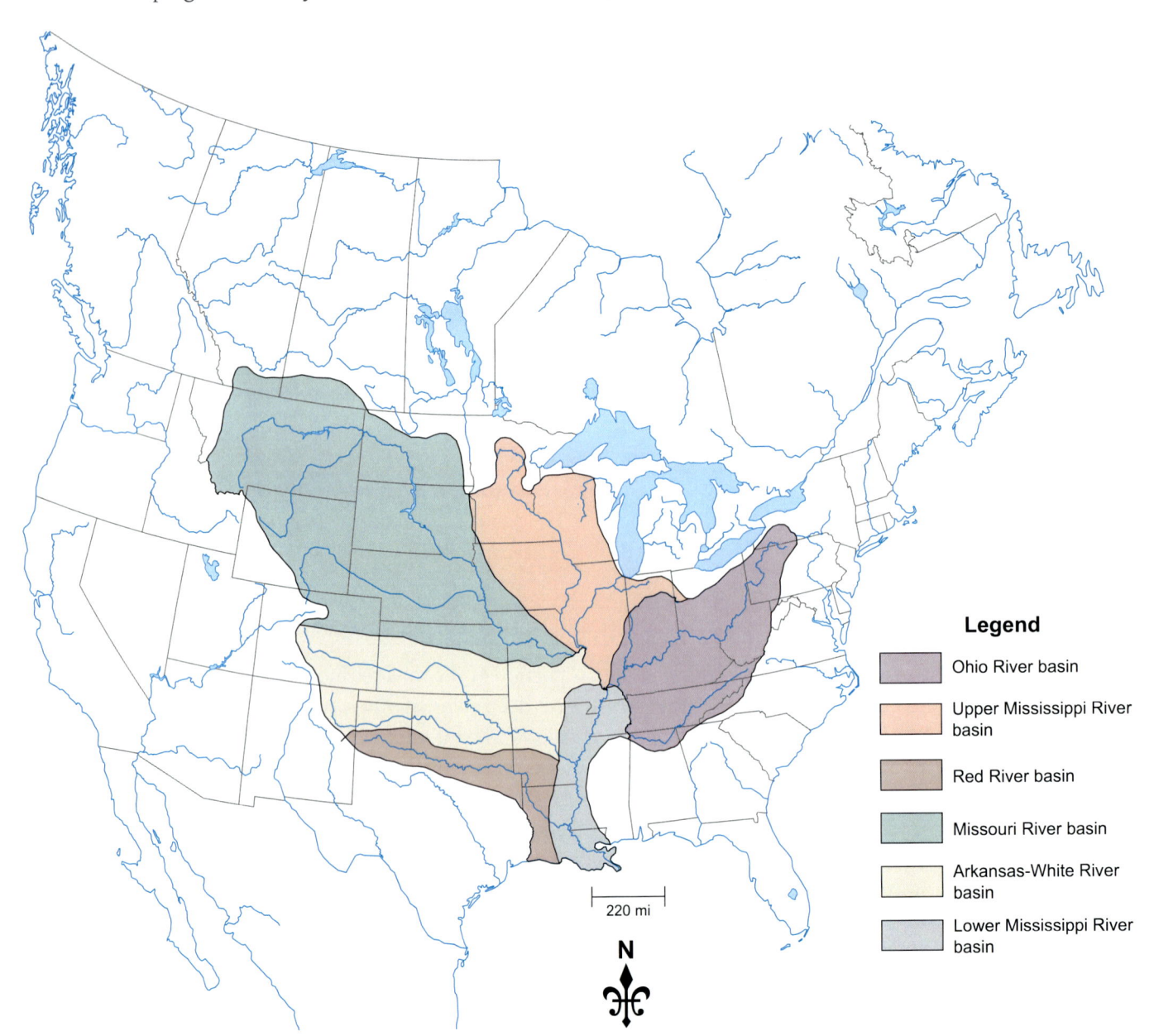

MAP 21.1 Six major subwatersheds make up the Mississippi River basin.

FIGURE 21.1 The confluence of the rapid-moving Missouri River (left) and Mississippi River (right) occurs just north of St. Louis, Missouri.

Mississippi River channel. This is based on the huge amounts of glacial meltwater that were flowing into the Driftless Area during interglacial periods and the absence of any lakebed. The history of glacial Lake Missoula seems to reflect similar geologic processes.

The western boundary of Illinois is the Mississippi River. However, before the glacial periods, the ancient Mississippi River passed much farther to the east. The ancient Mississippi River entered Illinois south of Davenport, Iowa, and flowed east into the valley where the Hennepin Canal is currently located (see map 2.4). Then it joined with the ancient Illinois River and turned south near the current city of Peoria, Illinois, and flowed toward present-day St. Louis, Missouri. The ancient Mississippi River was eventually blocked by the terminal meridian of the Wisconsin glacier about 12,000 to 15,000 years ago. The ancient Mississippi River then shifted to its current position and became the western border of the state of Illinois. If the Mississippi River had not realigned, a little less than half of the current 7.5 million Illinois acres would now belong to Missouri and the balance to Iowa (see chapter 3) [1].

The Mound Builders

Early people of the upper Mississippi were Woodland dwellers (AD 700 to 1300) who hunted, fished, and gathered wild plants. Spring and summer were spent in small nomadic settlements in river valleys and alongside lakes. They followed their food supply and moved to sheltered upland valleys as cold weather approached. Remnants of their culture, including unique burial mounds, are found at a number of places in the Driftless Area along the rivers [2]. Many of these effigy mounds are animal shaped with feet facing downstream of the Mississippi River. Effigy Mound National Monument, just north of Marquette, Iowa, at Harper's Ferry, has over 200 Indian mounds that contain pottery, triangular stone arrow tips, and other personal tools of daily living.

Farther south at Cahokia, Illinois, across from modern-day St. Louis, Missouri, another mound building culture, the Mississippians (AD 900 to 1300), left behind larger mounds including Monks Mound (figure 21.2), which was built on Darwin silty clay soils with low agricultural productivity. This mound seems to have served ceremonial and burial functions [3, 4]. The Cahokia settlement coincides with large-scale intensive maize agriculture and population growth as people from smaller villages moved into larger settlements [5]. In addition to maize, scientists have found evidence of crops such as squash, sunflower, barley, and several other seed-bearing plants. The Mississippians had a very complex food supply ranging from crop cultivation to hunting, fishing, and gathering of wild food plants from AD 800 to 1200. The decline of the prehistoric Cahokian culture and communities is not well understood. Geographer and anthropologist William I. Woods suggests that global cooling from AD 1200 to 1400 may have affected the climate and made crop production on the Wakeland silt loam soils (on bottomlands, streams, and alluvial fans) no longer viable and resulted in site abandonment. There is some speculation that Mississippian culture moved upriver from the south and supplanted

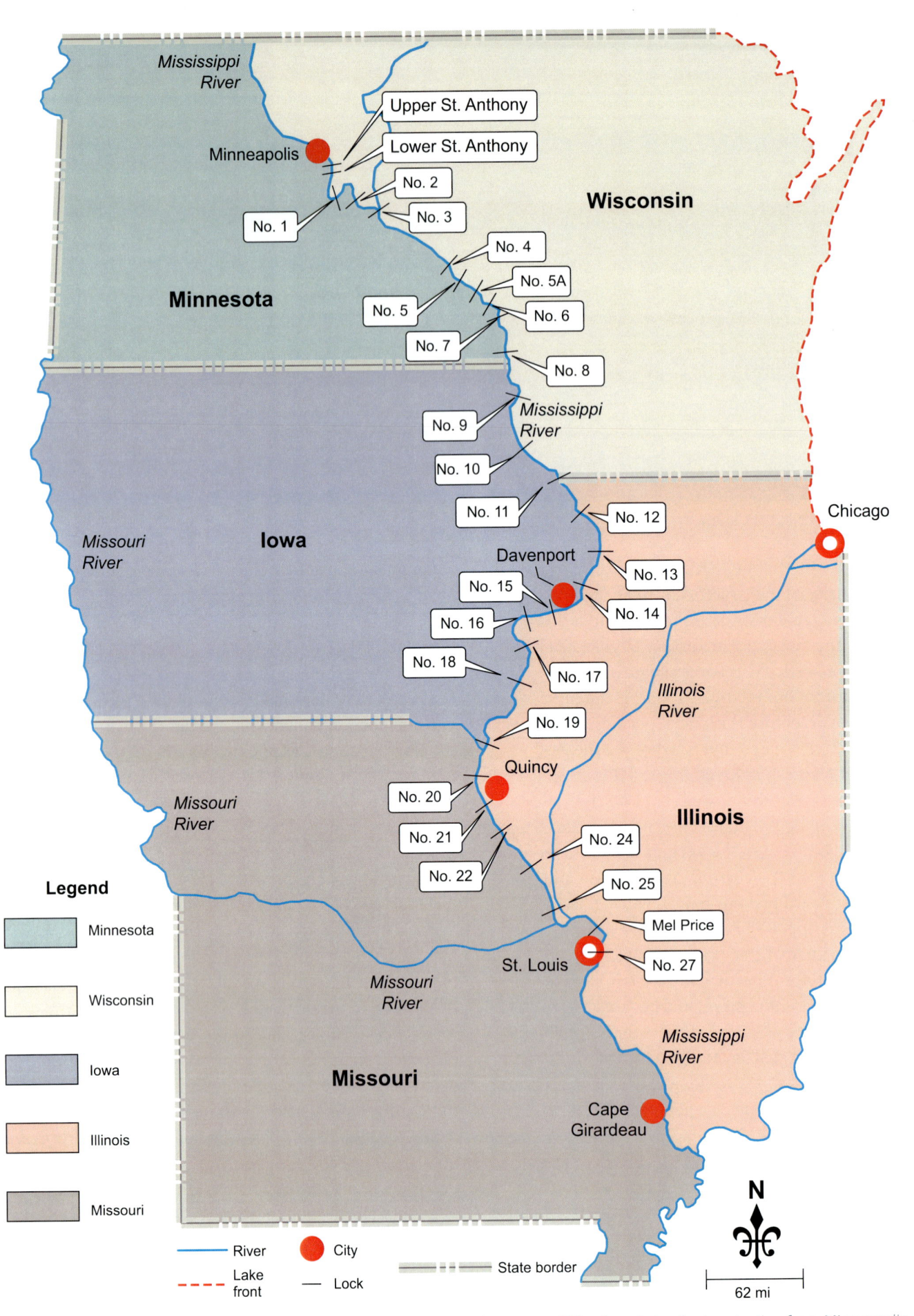

MAP 21.2 There are 29 locks and dams on the upper Mississippi River that ensure 850 miles of nine-foot navigation from Minneapolis, Minnesota, to Cairo, Illinois.

the older Woodland lifestyle (AD 800 to 1000) with the increase of maize cultivation replacing the hunting and gathering of resources from streams and forests during a time of global warming.

The Falls of St. Anthony

French explorers Hennepin, Marquette, and Joliet opened the upper Mississippi River (1670 to 1730) to European in-migration. Glowing reports of the abundant water and other natural resources in the region led fur traders, early settlers, and military surveyors to follow the river north in keelboats. The first steamboat, *Virginia*, reached St. Paul, Minnesota, in 1823 and initiated the golden age of steamboat travel and trade on the Mississippi River. St. Paul was settled in the 1840s and soon became the northernmost destination of steamboat navigation. Twelve miles upstream, the Falls of St. Anthony provided the waterpower for sawmills that cut the pines of northern Minnesota forests into boards. For a brief six years Minneapolis had the largest sawmill center in the United States. However, by 1910, nearly all the mills had closed as the timber supply of northern forests waned [6]. The agriculture of Minnesota's western prairies soon supplanted the economic importance of sawmills with flour mills making Minneapolis the flour milling capital of the nation (1880 to 1930). Water was channeled from the Mississippi River into underground raceways to drop into turbine pits. The force of its fall rotated turbines that drove the milling machinery. The Pillsbury A Mill, completed in 1881, had the largest direct drive waterpower system ever constructed, with two Victor turbines each generating 1,200 horsepower. At its peak, the A Mill produced more than 17,000 barrels of flour per day [6].

With the invention of the electric light, St. Anthony Falls was soon harnessed for electricity to power street lights and street cars, and allowed industries to grow even when they were not located on the banks of the river. A dam and hydroelectric station were constructed between 1887 and 1898 below the Falls of St. Anthony to capture the gravitational force of the falling water. Hydroelectricity became a leading industry in Minneapolis and one of the new technologies replacing direct drive waterpower for milling in the early twentieth century. Northern States Power Company still delivers hydroelectricity to 9,000 homes in the city of Minneapolis today.

Natural erosion over many centuries moved the falls upstream to their present location. Lumbering and milling activities increased the pace of erosion to about four feet a year as logs crashed into the limestone falls and excavation broke off limestone ledges exposing the soft, easily erodible sandstone underneath [6]. A disastrous tunnel project almost destroyed the falls in the 1860s and threatened the economic viability of waterpower-dependent Minneapolis. The US Army Corps of Engineers (USACE) built a concrete dike under the river and a wooden apron over the ledge to protect the face of the falls. This apron is now a concrete spillway (figure 21.3). Congressional authorization in the 1930s to improve navigation on the upper Mississippi led to the construction of an upper lock that bypasses the Falls of St. Anthony and enables navigation on the Mississippi River above Minneapolis.

Upper Mississippi River Navigation

The dawn of the golden age of steamboat travel and trade on the Mississippi River brought politicians and industrialists to the realization that this unique inland waterway had huge economic and social value. The Mississippi River below the confluence of the Mississippi and Missouri rivers at St. Louis (figure 21.1) had few impediments and was deep enough for steamboats and other river vessels to navigate. However, river traffic above the confluence was dangerous and difficult with snags, rapids, sandbars, and other obstructions in the water and along the banks. In the 1830s the USACE began removing snags and sandbars, and dynamited several rapids to improve navigation. In 1878 the government began to construct canals and locks to deepen the upper Mississippi River main channel to a uniform 4.5 feet. Canals with navigation locks were built to bypass rapids near Keokuk, Iowa, and between Rock Island and Moline, Illinois. Prior

FIGURE 21.2 Monks Mound, a Mississippian mound located at Cahokia, Illinois, across from St. Louis, Missouri, is part of the 2,200-acre Cahokia Mounds State Historic Site, which contains about 80 mounds.

to these efforts, it was almost impossible for riverboats to pass through this area because the rocky riverbed was so shallow. The Mississippi River became fully navigable from New Orleans, Louisiana, to St. Paul, Minnesota, in 1907 when the Moline Lock was opened and boats could easily bypass Rock Island Rapids.

As settlers continued to move west to farm and entrepreneurial men and women built cities and new industries, the river became an important transportation highway for moving raw materials, food, and manufactured products. The 4.5-foot channel was soon viewed as inadequate, and legislation enabled the USACE to begin to deepen the river to a 6-foot-deep channel. Hundreds of wing dams and closing dams were built. Wing dams made of brush and stone extended from the riverbank at a 90 degree angle and directed the water flow toward the main channel. Closing dams used in conjunction with the wing dams blocked the flow of water between the main channel and side channels and backwaters. A few of these structures are still used on the upper Mississippi River to deepen the main channel.

It became apparent that even a 6-foot channel was insufficient to effectively permit commercial navigation to easily flow between St. Louis and Minneapolis, and the 9-foot channel navigation project was initiated (figure 21.1) in 1930. This legislation authorized the USACE to create a minimum 400-foot-wide and 9-foot-deep main channel by constructing 29 locks and dams on the upper Mississippi River (map 21.2). Most of the locks and dams are numbered in order from north to south. There is one missing lock and dam, number 23 (Missouri reach), and three additional lock and dams referred to as Upper St. Anthony Falls (currently closed; map 21.2 and figure 21.3), Lower St. Anthony Falls, and Lock and Dam 5A in Minnesota; which were not assigned a primary number. This system created what is commonly called a "stairway of water" as the Mississippi falls 420 feet from the Falls of St. Anthony in Minnesota to Lock and Dam 27 in Granite City, Illinois.

The locks and dams are maintained by the USACE from Upper St. Anthony Falls (Upper and Lower St. Anthony Locks and Dams) to Chain of Rocks (Lock and Dam 27) downstream (map 21.2 and figure 21.4). Each lock and dam complex creates an upstream pool. The slack water pools behind the dams allow towboats and other river vessels to be raised and lowered as they proceed from one pool to the next [7]. The locks provide a collective 404 feet of lift. Figure 21.4 lists the pool number, locality, lock name, mile marker, and distance from the confluence at Cairo, Illinois.

This new lock and dam system with a nine-foot channel enables diesel powered river vessels to push steel barges with bigger freight loads the entire distance of 850 river miles from mile marker 858 in Minneapolis, Minnesota, downstream to Cairo, Illinois. Innovations in lock and dam design, especially the

FIGURE 21.3 The St. Anthony Lock and Dam (left) and the Falls of St. Anthony, Minneapolis, Minnesota, are easily viewed from the stone bridge east of the falls. The lock and dam was closed in June of 2015 for repairs.

Pool	Locality	Lock	Mile Marker
USAF Pool	Minneapolis, Minnesota	Upper St. Anthony Falls Lock	854
LSAF Pool	Minneapolis, Minnesota	Lower St. Anthony Falls Lock	853
Pool 1	Minneapolis, Minnesota	Lock 1	848
Pool 2	Hastings, Minnesota	Lock 2	815
Pool 3	Welch, Minnesota	Lock 3	797
Pool 4	Alma, Wisconsin	Lock 4	753
Pool 5	Minnesota City, Minnesota	Lock 5	738
Pool 5A	Fountain City, Wisconsin	Lock 5A	728
Pool 6	Trempealeau, Wisconsin	Lock 6	714
Pool 7	La Crescent, Minnesota	Lock 7	703
Pool 8	Genoa, Wisconsin	Lock 8	679
Pool 9	Lynxville, Wisconsin	Lock 9	648
Pool 10	Guttenberg, Iowa	Lock 10	615
Pool 11	Dubuque, Iowa	Lock 11	583
Pool 12	Bellevue, Iowa	Lock 12	557
Pool 13	Clinton, Iowa	Lock 13	522
Pool 14	LeClaire, Iowa	Lock 14	493
Pool 15	Rock Island, Illinois	Lock 15	483
Pool 16	Illinois City, Illinois	Lock 16	457
Pool 17	New Boston, Illinois	Lock 17	437
Pool 18	Gladstone, Illinois	Lock 18	410
Pool 19	Keokuk, Iowa	Lock 19	364
Pool 20	Canton, Missouri	Lock 20	343
Pool 21	Quincy, Illinois	Lock 21	325
Pool 22	New London, Missouri	Lock 22	301
Pool 24	Clarksville, Missouri	Lock 24	273
Pool 25	Winfield, Missouri	Lock 25	241
Mel Price Pool	East Alton, Illinois	Melvin Price Lock	201
Pool 27	Granite City, Illinois	Lock 27 (Chain of Rocks)	185

FIGURE 21.4 List of pools and locks on the upper Mississippi River.

roller-gate dam, improved the efficiency and reliability of the lock and dam system to manage river water levels. This integrated transportation system has provided manufacturing and agriculture a reliable and inexpensive way to move fertilizers and other production inputs as well as raw commodities and finished products within the United States and for export at the Port of New Orleans. The lock and dam infrastructure across the entire Mississippi and Ohio river system is 60 to 80 years old and in need of twenty-first-century updates. However, environmental groups are concerned about changes in river and floodplain habitats, and budget restraint advocates caution that proposed multibillion dollar river projects lack economic, social, and environmental justification. Ecological management of the river system and allocation of scarce resources is a current and future challenge for this inland waterway.

Growth of the Inland Waterway as a Commercial Highway

United States rivers have been the roads that moved settlers west, connected them to river and coastal ports, and enabled agriculture and other interior industries to expand their markets beyond local users to foreign export [8]. Prominent river ports, such as Chicago, Illinois (see chapter 23), and St. Louis, Missouri (figure 21.5), were economic, cultural, and social hubs that helped to unify the United States as a nation. Today, barges loaded with fertilizers, grain, gravel and sand, coal, petroleum, and manufactured products are still pushed by towboats up and down the Mississippi River and its tributaries to loading and unloading facilities at river ports.

A large variety of riverboats and vessels are used to transport people and products from one place to another. The tow, which is one towboat pulling/pushing one or more barges (figure 21.6), is the most common way products are moved on the river. Different types of barges are used to move freight depending on whether the cargo needs protection from the weather. Bulky solid cargo such as dry cement, fertilizers, corn, wheat, soybeans, and other farm products are carried in covered barges to keep them dry. Bulk products including wood chips, scrap metal, coal, stone aggregates, and large finished products such as steel and wire coils and wind turbine blades do not need protection from the weather and are carried in open hopper barges. Liquid cargo (tank) barges carry chemicals, petroleum, oil, molasses and other liquid products. The deck barge carries equipment and materials that can be tied down and don't need weather protection.

Managing Upper Mississippi River Tributaries
The Hennepin Canal

The completion of the Hennepin Canal in 1907 reduced the distance from Chicago, Illinois, to Rock Island, Illinois, on the Mississippi by 419 miles. The canal was intended to connect the Illinois and Mississippi rivers (figure 21.7) and increase passenger and commercial exchanges throughout the region. Land and river surveys authorized in 1871 provided data for siting the canal and dredging alluvial soils in the channel of the ancient Mississippi. During the construction of the Hennepin Canal, separate legislation gave the USACE authorization to deepen the channel and widen the locks on the Illinois and Mississippi rivers. This increased river traf-

FIGURE 21.5 Barge traffic on the upper Mississippi River near St. Louis can be viewed from the St. Louis Arch. The port at St. Louis ranked 18 in total tonnage at US ports in 2011.

FIGURE 21.6 Barge traffic from Chicago and the Great Lakes is transported down the Illinois River to its confluence with the Mississippi River at Grafton, Illinois.

FIGURE 21.7 The Hennepin Canal walking path at Rock Island, Illinois, offers areas for picnics, fishing, and scenic views.

fic from Grafton, Illinois, at the confluence of the Mississippi and Illinois rivers upstream to Chicago and had the effect of making the Hennepin Canal obsolete before it was fully operational. As a result, the 75.2-mile-long Hennepin Canal with a 29.3-mile feeder canal was used primarily for recreation by the late 1930s (figure 21.7). Concrete instead of cut stone was used to build the lock chambers, and similar to the Panama Canal, a feeder canal from a human-made lake provided water for the canal [9]. The Hennepin Canal had 33 locks [10] and Marshall gates, which were specially made for the system. It originally had nine aqueducts that carried water and vessels when the canal crossed larger rivers and streams [10]. Today a paved trail along the Hennepin Canal is used for walking, jogging, biking and snowmobiling depending on the season.

The Missouri River

The Missouri River, a great river in its own right, is a major tributary of the upper Mississippi River. Its confluence with the upper Mississippi is located at Spanish Lake, just north of St. Louis, Missouri (figure 21.1). The Missouri River is the longest river in North America (see map 1.1) and flows from the Rocky Mountains of western Montana to the east and south for 2,341 miles before merging with the Mississippi River. It carries water from a semiarid region encompassing more than 0.5 million square miles or 320 million acres (map 21.1; figure 21.8).

Ten major groups of Native Americans historically used the Missouri River and its tributaries in Montana, the Dakotas, and Missouri for daily living and as trade and transportation routes. The river is part of their culture, providing spiritual, social, physical, and economic resources. With the arrival of the Spanish and French, the river became a route for adventure and trade and opened the west to settlers in search of new lives and opportunities. In the 1803 Louisiana Purchase, the United States gained the lands through which the Missouri River flowed (see chapter 1), and exploration began in earnest to find a river route connecting the Atlantic and Pacific oceans. Lewis and Clark traveled the entire length of the Missouri River in search of the Northwest Passage (1804 to 1806), and with great disappointment concluded there was no river connecting the two oceans across the United States.

Like the upper Mississippi and Ohio regions, the Missouri River basin in the twentieth century was managed to control flooding, improve navigation, and generate hydroelectric power. The main stem of the Missouri River has 15 dams and reservoirs with hundreds more on its tributaries. The main stem river's

FIGURE 21.8 The Little Missouri River, a tributary of the Missouri River, runs through the buttes and mesas of Theodore Roosevelt National Park, North Dakota.

length has been decreased by almost 200 miles with the removal of meanders and channelization to deepen the river for navigation. Much of the basin is located in a dry climate. Thus, the river and its tributaries are important sources of water for agricultural irrigation and industrial and community growth.

Public Lands, River Ecology, and Recreation

The upper Mississippi River topography, vegetation, and water surfaces are particularly scenic, and several states have set aside public lands for preservation, ecological protection, and recreational uses. Limestone bluffs and gorges carved from the rapids and waterfalls from St. Anthony Falls (Minneapolis, Minnesota) downstream to St. Paul, Minnesota, are visible from park overlooks, walking trails, and a wonderfully preserved stone bridge that crosses the Mississippi River just below St. Anthony Lock and Dam. Below downtown St. Paul the river passes through a wide preglacial valley which extends southward for many miles. Minnesota, Wisconsin, Illinois, Missouri, and Iowa have wildlife refuges and three National Park Service sites (figure 21.9). One park, the Mississippi National River and Recreation Area in Minnesota encompasses 54,000 acres and 72 river miles. A second National Park Service site at Harper's Ferry, Iowa, is the Effigy Mounds National Monument with a visitor's center and trails to observe the animal shaped mounds. Lastly, the Jefferson National Expansion Memorial offers a birds-eye view of the Mississippi River from the Gateway Arch in downtown St. Louis, Missouri.

The locks and dams on the upper Mississippi River have created lakes and extensive marshes, swamps, open sloughs, and backwater sloughs that provide natural habitats for a wide variety of wildlife (figure 21.10). More than 125 species of fish and 30 species of freshwater mussels live in the reaches of the upper

FIGURE 21.9 Lookout Point located above the sandstone bluffs that confine the Mississippi River near Mississippi Palisades State Park offers a panoramic view of the upper Mississippi River landscape.

FIGURE 21.10 The backwaters of the upper Mississippi River south of Lake Pepin on the Illinois side form an extensive wetland complex that protects the river ecology and provides recreational hunting, fishing, and boating opportunities.

Mississippi River. The Upper Mississippi River National Wildlife and Fish Refuge running from Alma, Wisconsin, downstream to Rock Island, Illinois (figure 21.9), is part of the Mississippi Flyway. Sandstone bluffs high above the river overlook backwaters, marshes, bottomland forests, sloughs, and forested islands.

Private and public marinas and recreational areas offer facilities for boat launching, camping, swimming, picnicking, and bird watching. Year-round fishing, waterfowl hunting, water sports, and island camping present unique opportunities to observe waterfowl nesting and hatching. The riparian floodplain forest at the National Audubon Field Station above Melvin Price Lock and Dam (number 26) across the river from Alton, Illinois, is also a good site to view local and migratory birds.

The health of the river and its water quality continue to be a deep concern along the entire length of the river. Runoff, soil erosion and river sedimentation, and off-field and off-farm nitrogen and phosphorous losses from cultivated crops as well as agricultural and industrial chemicals are threats to the river ecosystem. Pharmaceuticals and endocrine-disrupting chemicals in river water are new sources of concern. Minnesota, Wisconsin, Iowa, Illinois, and Missouri are working together to find solutions to water impairments that impact local waters and downstream Gulf of Mexico hypoxia conditions. Lake Pepin, Minnesota, a large natural lake that is part of Pool 4, is experiencing eutrophication from agricultural runoff. Nutrient impairment is occurring not just in the main stem river but also in off-channel streams suggesting that the entire upper Mississippi River basin land uses and practices need to be reexamined. These nutrients accelerate the growth of algae and duckweed and reduce light penetration to underwater aquatic vegetation, a food source for fish and aquatic life including waterfowl.

Managing for Navigation

Management of water resources and navigation on the upper Mississippi River today continues to provide vital social, economic, and environmental benefits to the people of this region and the United States. Unlike the lower Mississippi, much of the upper river is a series of pools created by a system of 29 locks and dams (map

21.2). A primary reason for these locks and dams on the river is to facilitate barge transportation, which moves raw agricultural commodities, fertilizer, forest products, petroleum, sand, stone, food, and manufactured products from the upper Midwest to ports south for domestic and export markets. The dams regulate water levels for the upper Mississippi River and also play a major part in regulating levels on the lower Mississippi. The effective management of commercial navigation on the Mississippi River has had important economic impacts on the port cities of the upper and lower Mississippi and tributaries. The Port of South Louisiana at the Gulf of Mexico was the lead United States port in 2011, carrying 246.5 million tons of domestic (125.7 million tons) and foreign cargo (120.8 million tons) [11]. This was a 4.3% increase from the prior year. That same year, the port at St. Louis, Missouri and Illinois, was ranked 18 with a total of 36.5 million tons, an 18.6% increase from 2010. Ports on the upper Mississippi from Minneapolis, Minnesota, to the mouth of the Missouri River recorded a 1.8% increase in ton-miles, representing the transport of 61.2 million tons along 663 river miles in 2011. During that same period, there was a 3.9% increase in ton-miles of cargo moving from the mouth of the Missouri River to the mouth of the Ohio River, totaling 106.6 million tons.

The Upper Mississippi River Nine-Foot Channel Navigation Project, authorized by the River and Harbor Act of 1930, created 244 miles of nine-foot channel navigation from Minneapolis, Minnesota, to Guttenberg, Iowa [12]. Much of the infrastructure in this lock and dam navigation system is between 50 to 70 years old. In June of 2015 the Upper St. Anthony Lock and Dam adjacent to St. Anthony Falls was closed for a variety of reasons including the need for maintenance and repairs. There is need for systematic investments in routine maintenance and repairs as well as close monitoring and assessment of locks and dams to preempt future failure. Close monitoring; channel dredging; maintenance of channel control structures such as wing dams, closing dams, and bank revetments; snag removal; and accurate channel marking are essential to keep the system operating at peak efficiency. These expenses, shared by the USACE, private shipping companies, and local port authorities, require substantial public investments and congressional authorization.

---

[1] Olson, K.R., and F. Christensen. 2014. How waterways, glacial melt waters, and earthquakes re-aligned ancient rivers and changed Illinois borders. Journal of Earth Sciences and Engineering 4(7):389-399.

[2] O'Bright, J.Y. 1989. The Perpetual March. An Administrative History of Effigy Mounds National Monument. Omaha, NE: National Park Service, Midwest Regional Office.

[3] Indorante, S.J., and A. Leeper. 2000. Soil survey of St. Clair County, Illinois. Washington, DC: USDA Natural Resources Conservation Service.

[4] Olson, K.R., R.L. Jones, A.N. Gennadiyev, S. Chernyanskii, W.I. Woods, and J.M. Lang. 2002. Accelerated soil erosion of a Mississippian mound at Cahokia site in Illinois. Soil Science Society of America Journal 66:1911-1921.

[5] Fowler, M.L. 1997. The Cahokia Atlas, a historical atlas of Cahokia archaeology. Revised Edition. Champaign, IL: University of Illinois at Urbana-Champaign.

[6] Saint Anthony Falls Heritage Trail. 2015. Interpretative plaques. Minneapolis, MN: Saint Anthony Falls Heritage Trail.

[7] Nelson, S.B. 1983. Water Engineering. In Standard Handbook for Civil Engineers, ed. F.S. Merritt. New York: McGraw-Hill.

[8] US Army Corps of Engineers. 2008. Gateway to Commerce: The US Army Corps of Engineers' 9-Foot Channel Project on the Upper Mississippi River. http://npshistory.com/series/archeology/rmr/2/intro.htm.

[9] Illinois Department of Natural Resources. 2016. Hennepin Canal Parkway State Park. Springfield, IL: Illinois Department of Natural Resources. http://www.dnr.illinois.gov/Parks/Pages/HennepinCanal.aspx.

[10] Illinois Department of Natural Resources. 2016. Hennepin Canal http://www.dnr.illinois.gov/recreation/greenwaysandtrails/Pages/HennepinCanal.aspx.

[11] US Army Corps of Engineers. 2012. The US Waterway System. Transportation Facts and Information. Navigation and Civil Works. Decision Support Center.

[12] US Army Corps of Engineers. 2015. Mississippi River 9-foot Project Channel Maintenance. http://www.mvp.usace.army.mil/Home/Projects/tabid/18156/Article/571029/upper-mississippi-river-9-foot-project-channel-maintenance.aspx.

Dredging of the Fractured Bedrock-Lined Mississippi River Channel at Thebes, Illinois

The usually abundant, slow soaking rain systems and evening thunderstorms that characterize the Great Plains climate from May through August [1] were absent in 2012. As a result, the Ohio and Mississippi rivers dropped to near record levels from July of 2012 through January of 2013, and the US Army Corps of Engineers (USACE) faced a new challenge to their ability to control the Mississippi River. The 2012 drought reduced the channel depths on the upper Mississippi River between Cairo, Illinois, and St. Louis, Missouri, to only one to six feet above the nine-foot-deep navigation shipping channel created by the USACE in response to the 1930 Rivers and Harbors Act. Of greatest concern was the bedrock-lined river shipping channel near Thebes, Illinois, which threatened to ground barge traffic transporting critical agricultural supplies, including fertilizers and grain.

The USACE systematically surveys the river bottom and routinely dredges sand accumulation within the Mississippi River to maintain the shipping channel. However, the Thebes section of the river posed a more difficult engineering situation. Ice Age glaciers and more recent seismic activity created the Thebes Gap in the upland bedrock ridge and rerouted the ancient Mississippi River through, rather than around, the upland bedrock ridge (map 22.1) of the former southern Illinois land bridge (see map 2.6). Throughout the summer of 2012, as the drought deepened and river levels fell, the USACE increased the removal of sand and other unconsolidated sediments along the upper Mississippi navigation channel. The six-mile fractured bedrock-lined channel, starting just south of Gale, Illinois, and extending past Thebes, Illinois, to Commerce, Missouri, and the underlying river bottom materials required substantive excavating of rock as the narrow bedrock channel under drought conditions became shallow with hidden and exposed rock (figure 22.1), a dangerous obstacle to barge and other boat traffic.

Historic Location of the Mississippi River Channel

Historically, the ancient Mississippi River turned southwest just south of Cape Girardeau (map 22.1) into the current state of Missouri and traveled more than 30 miles to the west before turning south. Then it flowed east and back toward Benton, Missouri, where

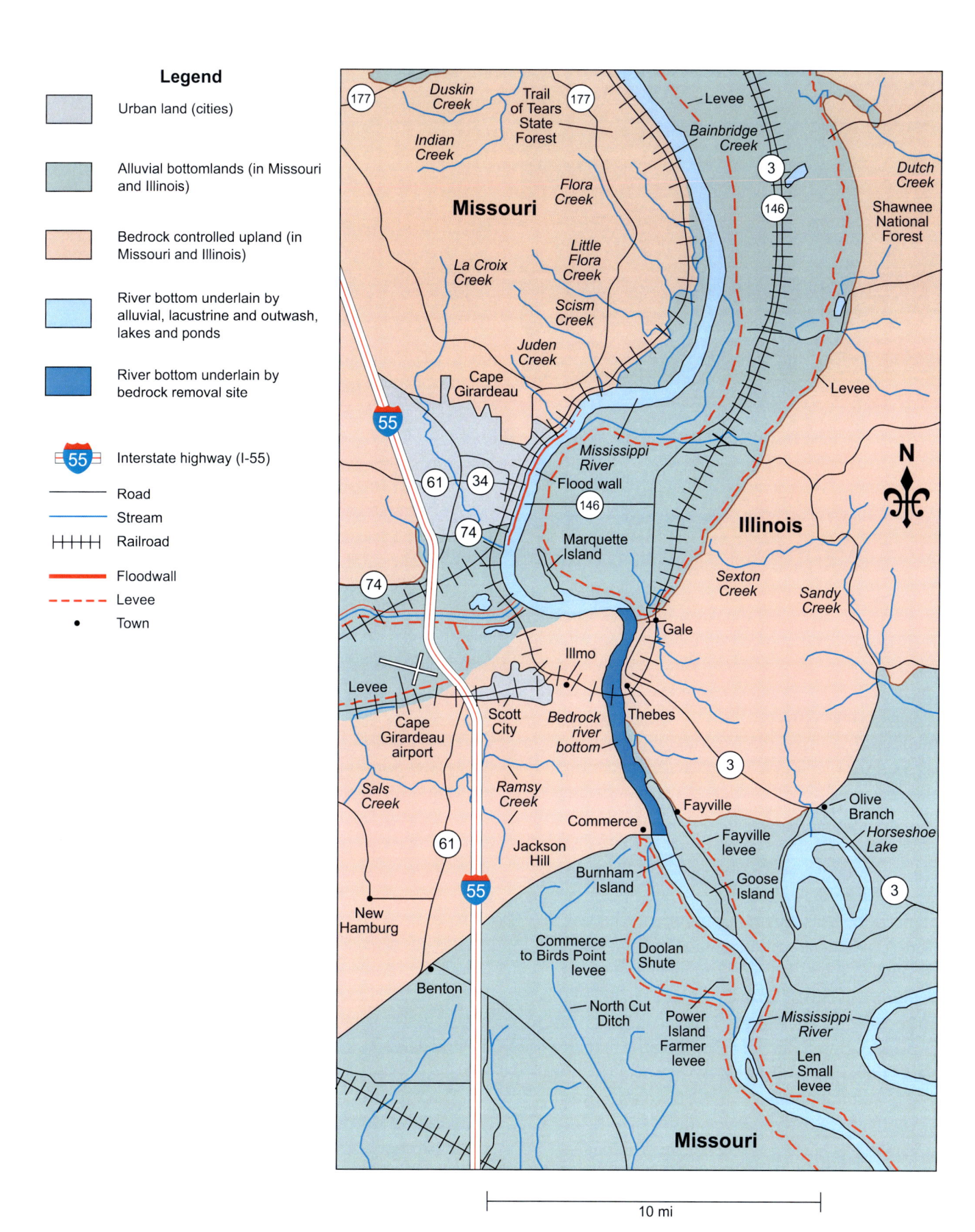

MAP 22.1 Thebes, Illinois, is located south of Cape Girardeau and Commerce, Missouri, on a narrow, six-mile bedrock-controlled upland stretch of the upper Mississippi River. This location made it an ideal ferry and railroad crossing for commerce between Illinois and Missouri.

it joined with the ancient Ohio River waters draining through the Cache River valley. The old riverbed from when the ancient Mississippi River flowed around the bedrock-controlled upland ridge is now alluvial bottomlands. The historic confluence was most likely west of Horseshoe Lake State Fish and Wildlife Area, which is

FIGURE 22.1 Bedrock was exposed and a threat to navigation on the Mississippi River near the Thebes railroad bridge on December 21, 2012, when the river reached a low of seven feet.

30 miles north of the current confluence of the Mississippi and Ohio rivers. The upland area west of Thebes, Illinois, and currently in Missouri would have been the southwesternmost point in Illinois had the Mississippi River course not changed. The ancient Mississippi River was rerouted at the end of the Great Ice Age, and east central Missouri and southern Illinois were engulfed in a shallow sea until the end of the Pennsylvanian period when the waters receded and regional elevation rose. Four glacier stages covered most of Illinois, including the Nebraskan, Kansan, Illinoian, and Wisconsinan (see map 2.3). After the last glacier advance, the melting ice flooded and altered the course of many channels and streams, including the Mississippi and Ohio rivers. Some geologists believe heavy seismic activity along the Commerce lineament about 10,000 to 12,000 years ago created a fault and helped the Mississippi River cut through the bedrock upland to create the Thebes Gap and a new confluence at Cairo, Illinois. The river then switched from a braided river to a meandering river through rock cuts that form the current state boundary between Missouri and Illinois. The Mississippi River in older days migrated rapidly by eroding the outside of a river bend and depositing on the inside of the river bend. Abundant oxbow lakes mark old positions of the channel that have been abandoned.

Early Holocene, late Wisconsin liquefaction features (where solid land turned into a liquid as a result of seismic activity) in western lowlands are thought to have been induced locally, possibly by the Commerce Fault as a result of earthquake upheaval along the Commerce geophysical lineament running from central Indiana to Arkansas [2]. The New Madrid area has been the center of seismic activity for thousands of years. This seismic activity affected the Mississippi River and perhaps the Ohio River by rerouting the waters and causing uplift of surrounding land masses by as much as 13 feet in 1,000 years. The last significant seismic activity in the form of quakes was in AD 1450 to 1470 and AD 1811 to 1812.

Floodwaters of the Mississippi River did not initially pass through this rather narrow channel and valley but instead were routed by the bedrock uplands near Scott City, Missouri, through an opening in the upland ridge 30 miles to the southwest. Then the river turned back to the east near Benton, Missouri, and merged with the ancient Ohio River southeast of Commence, Missouri (map 22.1). Over time, floodwaters of the ancient Mississippi River (from north) and ancient Ohio River (from the south) cut a valley trench along the Commerce Fault and through the bedrock-controlled upland west of Thebes, shortening the distance the Mississippi had to travel from 45 miles to 6 miles. The two historic rivers joined south of Commerce, Missouri, and Olive Branch, Illinois, and west of Horseshoe Lake creating a constantly changing confluence of these two mighty rivers [3]. It appears the bedrock upland was worn away by both rivers after seismic activity, and the creation of the Commerce Fault contributed to the opening of the bedrock-controlled channel following the last glacial advance approximately 10,000 to 12,000 years ago.

Central Great Plains 2012 Drought

The lack of rain throughout 2012 created the most severe summertime seasonal drought over the central Great Plains in the last 117 years [1] with major impacts on Mississippi River commerce due to reduced water flows. This unpredicted drought reduced corn yields 26% below the average regional 166 bushels per acre yield and soybean yields 10% below projected 44 bushels per acre as estimated by USDA. The National Oceanic and Atmospheric Administration Drought Task Force and the National Integrated Drought Information System assessment report of the central Great Plains in 2012 reveals a number of unusual aspects of this surprise drought [1].

The 2012 drought followed an upward trend of increased summertime Great Plains rainfall since the early twentieth century; the last major drought was in 1988. Droughts in the Great Plains occur when atmospheric moisture, both absolute and relative, is deficient and are often linked to the absence of processes that normally produce rain [1]. These processes include springtime low pressure systems with warm and cold fronts that lift air masses to produce rain and frequent summer thunderstorms that provide the bulk of July and August precipitation. The principal source of summer water vapor in this region, the Gulf of Mexico, had an appreciable reduction in northward meridian winds and a 10% reduction in climatological water vapor in 2012, creating in the Great Plains the greatest cumulative rainfall deficit since record keeping began in 1895 [1]. According to the National Oceanic and Atmospheric Administration Task Force, the immediate causes of the drought were meteorological, and the underlying causes for these conditions were assessed as unrelated to ocean surface temperatures or to changes in greenhouse gases.

Dredging of the Rock-Lined Channel near Thebes, Illinois

The USACE Mississippi River lock and dam system on the upper Mississippi River maintains a navigation channel of a minimum of nine feet of water with the last lock and dam at Granite City, Illinois. After the Missouri River joins the Mississippi at St. Louis, Missouri, the combined flow of the two rivers has historically been sufficient to maintain the nine-foot channel without locks and dams. Specialized barges pushed by towboats carry a wide variety of products for domestic and foreign markets, including fertilizer, grain (corn, wheat, and soybean), sand, coal, chemicals, petroleum, oil, molasses, and equipment.

Funds were appropriated in 2012 to begin dredging of the bedrock-lined Mississippi River channel near Thebes, Illinois. The USACE usually dredges alluvial and outwash sediment with specialized equipment, such a hydraulic dredge, a crane, or a backhoe, to maintain a 300-foot wide and 9-foot deep shipping channel. However, there was great concern as to whether these techniques would work in the Thebes Gap section of the Mississippi River channel. The six mile section of the Mississippi River channel, two miles to the north of the 1905 Thebes railroad bridge and four miles to the south, is underlain with fractured bedrock with the distance between the Illinois and Missouri bedrock-controlled escarpment only 4,000 feet and the river channel 2,000 feet wide. There is very little bottomland since most of the area is occupied by the Mississippi River between the two bedrock uplands. With bedrock exposed on both sides of the channel and underlying the river (figure 22.1), dredging of rock is difficult (figure 22.2).

The consolidated, rocky bedrock bottoms, with pinnacles of rock sticking up in the shipping channel, can be hit by heavy barges when the Mississippi River is low. In 2012, rock protrusions in the channel outside the shipping lane destroyed many propellers on boats used by local fishermen. Excavators and a dragline were loaded on barges and moved out into the channel to two separate locations on December 18, 2012, to begin the 30- to 45-day dredging process (figure 22.2). It was anticipated that explosives would be required to loosen some of the attached rock prior to removal. However, giant excavators proved to effectively loosen and remove massive amounts of bedrock and rock materials without the need for explosives. The rocks were removed using spud barges and a hydrohammer (a huge aquatic jackhammer) to break up bigger chunks of rock for removal by the giant excavators [4]. This technique was much faster than expected, and 75% of the project was completed by February 1, 2013. Excavation occurred during the daytime, and the Mississippi River remained open each night for barge traffic. By February, the river began to rise from increased runoff in the upper Mississippi, and the excavators could no longer reach and remove rock at the bottom of the shipping channel. The barges with these excavators were then moved 31 river miles north to remove additional rock from the shipping channel near Grand Tower, Illinois.

FIGURE 22.2 River bottom bedrock is dredged using an excavator to increase the navigation depth.

Little Egypt

The location of Thebes, Illinois, was determined by a number of geologic and cultural events which made the area unique. Thebes, a Mississippi River town, would more likely have been settled 30 miles to the west if the Mississippi and Ohio rivers had not cut a channel through the upland between Gale, Illinois, and Commerce, Missouri. The earliest recorded settlement was by the Sparhawk brothers prior to the 1830s, and the town was called Spar Hawk's Landing. Poplar (*Linodendron tulipifera* L.) and other tree logs from the surrounding area were hauled here for transport downriver to ship builders in New Orleans, Louisiana. The upland Stookey and Alford soils [5] were timbered and of little value for agricultural use. Haymond, Birds, and Wakeland soils on the narrow bottomlands were subject to frequent flooding and were not drained or farmed, and agriculture has had little impact on the town.

The southern seven Illinois counties became known as the "land of corn," and the name Little Egypt appears to date back to 1831. Local history reports that on September 18, 1831, there was a corn killing frost that affected all of the northern Illinois counties, and these farmers turned to southern Illinois to supply their grain needs. Most of the northern soils used for corn production at that time were well-drained timber soils along the rivers and streams. In the 1800s, the prairie soils were too wet to farm and were not used to grow corn until after the Land Drainage Act of 1879. When the corn crop was killed by early frost in 1831, the northern farmers paid the southern farmers a high price to get the corn they needed. These farmers shipped the corn to northern Illinois using the Ohio, Mississippi, and Illinois rivers in the winter and spring of 1832. This exporting of corn gave the northern farmers the perception that the region with the fertile, black, alluvial soils was "the land of corn," and they started to use the name Little Egypt to describe the Cache, Ohio, and Mississippi valley areas where the corn was grown. The town of Thebes was established by President Andrew Jackson in October of 1835. The historic courthouse was built by 1848 and still stands today (figure 22.3). The railroads from Chicago in the 1850s began to extend into Little Egypt but were limited by their ability to cross the Mississippi River. Many railroads converged on Thebes, and a ferry service developed to get the trains and materials across this narrow stretch of river.

Thebes Railroad Bridge

Thebes is the only place on the Mississippi with a bedrock channel and very narrow valley for a railroad bridge. The Mississippi River channel is about 2,000 feet

FIGURE 22.3 The historic Thebes courthouse built in 1848 commands a high view overlooking the Mississippi River.

FIGURE 22.4 The reinforced concrete, two track railroad bridge was built in 1905 to connect Illinois to Missouri and enabled east-west commerce throughout the region.

wide at Thebes (map 22.1), and the distance between the bedrock-controlled uplands and ridgetops is less than 4,000 feet. This was noted by the local railroads who initially had to use ferry service to get the trains across the Mississippi River. Thanks to the presence of the bedrock upland with Stookey and Alford soils [5] and the bedrock underlying the bottomland soils and Mississippi River, the Thebes location was the perfect place to construct a solid, reinforced concrete, two track railroad bridge. This bridge was engineered to withstand the pressure of two heavily loaded trains at the same time. In 1905, the railroad bridge (figure 22.4) was built to replace the ferry service, which could not keep up with the demand and had become a choke point for southbound trains out of Chicago, Illinois. The bridge took entire trains across the Mississippi River and eliminated the need for the ferry service.

Five local railroads pooled their resources to build a permanent bridge. The bridge was designed by Ralph Modjeski, a famous bridge builder. The original design called for two railroad tracks that could be used at the same time and for an auto deck to be added at some point in time. The deck was never added, but the bridge was engineered to handle the extra weight. Due to this extra strength and solid bedrock foundation, this bridge was long known as the strongest bridge to span the Mississippi River. Bridge abutments were made out of reinforced concrete and anchored into the bedrock escarpment (figure 22.4) on both valley walls (figure 22.5) and at the river bottom. The bridge structure is located at river mile marker 42.7. The normal river elevation is 308 feet at Thebes, and the bridge is 104 feet above the river. The total length of the bridge is 4,000 feet, with the longest span across the shipping channel of 651 feet. Unfortunately, the 1905 bridge streamlined the flow of rail traffic, and the trains no longer had a reason to stop at Thebes. As a consequence, Thebes experienced hard times, and the population declined rapidly. The two track bridge stands today (figure 22.1) and still handles 35 trains per day after more than 110 years of service.

Drought and River Navigation

The 2012 central Great Plains drought eclipsed the driest summers of 1934 and 1936 at the height of the Dust Bowl, substantively reduced the water flows of river systems, and severely curtailed commerce on the upper Mississippi River [1]. Following early 2011 snowmelt, heavy rains, extreme flooding, and levee breaching along the Mississippi River, the rapid onset of drought in 2012 was unexpected and challenged the USACE to maintain a safe river depth above the 9-foot navigation channel for barge traffic. The USACE successfully dredged (figure 22.2) other parts of the Mississippi River to keep the shipping channel open, but those channels were underlain with unconsolidated sediments (sands and alluvial materials) that could be removed with equipment routinely used to maintain the river depths and widths for navigation. However the dredging of the six-mile, narrow, bedrock-lined channel near the town of Thebes, Illinois, required large excavators capable of breaking loose the consolidated river bottom in order to deepen the channel. Without the dredging work by the USACE, the shipping on the Mississippi River would have stopped, possibly for months. The 9-foot deep and 300-foot wide Mississippi River channel was dredged at a time when the excavators could easily reach the bottom of the shipping lane and were able to restore and maintain the shipping lane for barge traffic as water levels dropped during the drought of 2012 to 2013.

FIGURE 22.5 Exposed valley wall bedrock at Thebes, Illinois, provided an anchor to the bridge abutment on the east side of the Mississippi River.

---

[1] Hoerling, M., S. Schubert, K. Mo, A. Kouchak, H. Berbery, J. Dong, A. Kumar, V. Lakshmi, R. Leung, J. Li, X. Liang, L. Luo, B. Lyon, D. Miskus, X. Quan, R. Seager, S. Sorooshian, H. Wang, Y. Xia, and N. Zeng. 2013. An interpretation of the origins of the 2012 central Great Plains drought. Assessment Report. National Oceanic and Atmospheric Administration Drought Task Force Narrative Team. http://www.drought.gov/drought/content/resources/reports.

[2] Vaughn, J.D. 1994. Paleoseismological studies in the Western Lowlands of southeastern Missouri. Final Technical Report to US Geological Survey. Reston, VA: US Geological Survey.

[3] Olson, K.R., and L.W. Morton. 2013. Impact of 2011 Len Small levee breach on private and public Illinois lands. Journal of Soil and Water Conservation 68(4):89A-95A, doi:10.2489/jswc.68.4.89A.

[4] Plume, K. 2012. Rock clearing begins on drought-hit Mississippi River. Reuters. December 18, 2012.

[5] Parks, W.D., and J.B. Fehrenbacher. 1968. Soil Survey of Pulaski and Alexander counties, Illinois. Washington DC: USDA Natural Resource Conservation Service.

The Illinois Waterway Connecting the Mississippi River and Great Lakes

In the seventeenth century, the French built trading forts between the Illinois River and Lake Michigan in the Illinois Territory. The first known Europeans to travel through the area were Father Jacques Marquette and Louis Joliet, who used the Chicago Portage in their travels. The Chicago Portage was a wet, swampy, frozen, or dry area depending on the season (known locally as Mud Lake) near the western tip of Lake Michigan. It connected the Great Lakes via the Chicago River to the Des Plaines River, a tributary of the Illinois River. Joliet is credited with remarking that a canal would remove the need to portage around the swampland and enable the French to expand their empire from Montreal to New Orleans. The canal was never built by the French. At the end of the French and Indian War in 1763, the area was ceded to the British, and was then awarded to the new United States by the Treaty of Paris (1783). However, these explorers understood the economic, political, and social importance of a water crossing over the St. Lawrence Continental Divide, which separated the Great Lakes basin from the Mississippi River basin (map 23.1). The construction of the Illinois and Michigan Canal in 1848 was one of the first efforts to realize the vision of Joliet and Marquette. Later, the Chicago Sanitary and Ship Canal was built in the 1900s and today serves as a main route for commercial shipping (figure 23.1). This established Chicago as a transportation hub connecting the Great Lakes waterways to the Mississippi River and the Gulf of Mexico via the Illinois Waterway.

The Illinois Waterway drops from 578 feet above sea level at Lake Michigan to 419 feet at the mouth of the Illinois River as it flows southwest into the Mississippi River. A system of eight locks on the Illinois River managed by the US Army Corps of Engineers (USACE) controls water flow along the 336-mile system to assure a 9-foot deep navigation channel (map 23.2). Extreme rain events in recent years have made managing navigation a little more difficult. In the spring of 2013, the highest floodwater levels in the last 70 years were recorded on the Illinois Waterway [1]. Many towns along the Illinois River are leveed to protect against high water. However, the river overflowed into the town of Marseilles, Illinois, damaging homes, a local school, and downtown properties. The unexpected high water unmoored several barges tied up in the pool behind the

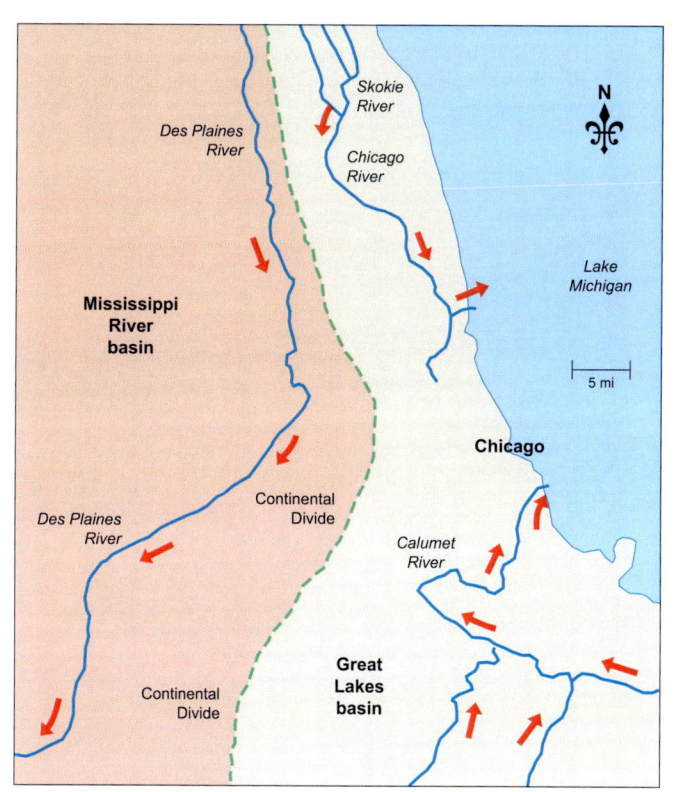

MAP 23.1 The St. Lawrence Continental Divide prior to the 1900s separated the Great Lakes and the Mississippi River basins. Water from the Chicago River flowed north into Lake Michigan, and water from the Des Plaines River flowed south into the Illinois River, a tributary of the Mississippi River.

Marseilles Dam. River traffic along the waterway was halted when the dam Tainter gates were damaged by the barges and were not able to be fully closed to manage the pool level behind the dam [2].

The 2013 Marseilles experience is a reminder of the system-wide interdependence of the port cities along navigable waterways. The Illinois Waterway, running from the Calumet River in Chicago to Grafton, Illinois, where the Illinois River flows into the Mississippi River (see figure 21.4), is a major tributary of the Mississippi and Ohio rivers inland navigation system. It consists of seven water systems: Illinois River, Des Plaines River, Chicago Sanitary and Shipping Canal, South Branch Chicago River, Calumet-Sag Channel, Little Calumet River, and the Calumet River. The Illinois River, the primary water body of the waterway, runs west and then south through a portion of the ancient Mississippi River valley (map 23.2) as it flows toward St. Louis, Missouri. It originates southwest of modern-day Joliet where the Des Plaines River and the Kankakee River converge.

Illinois and Michigan Canal

The Illinois and Des Plaines rivers and the construction of canals, towpaths, and locks to increase navigation were key factors in the early settlement and economic growth of Chicago and Illinois. The soils of central Illinois were fertile, but agriculture was primarily subsistence due to lack of transportation connections to population and trade centers. The Illinois and Michigan Canal completed in 1848 became an important link in opening new markets by connecting the Illinois River to Lake Michigan. A survey of the region authorized by Samuel D. Lockwood in 1824 influenced the location of the canal and Illinois's northern border (see chapter 3). The construction of the Illinois and Michigan Canal required cutting a channel through a ridge about 12 miles from the lakeshore. This ridge, known locally as the Chicago Portage, separates the Mississippi River basin from the Great Lakes drainage basin (map 23.1). Historically used by Native Americans, the ridge provided an accessible pathway through the swampy lowlands between the Chicago River (in the Great Lakes basin) and the Des Plaines River (in the Mississippi River basin). These two drainage basins were connected when the Illinois and Michigan Canal was cut across this continental divide.

The capacity to commercially ship grains and obtain fertilizers by way of the Great Lakes and the Mississippi River made agriculture profitable. The 96-mile-long Illinois and Michigan Canal drops 140 feet from the Chicago River at Bridgeport to LaSalle-Peru on the Illinois River and required 17 locks and four aqueducts for navigation. Towpaths along the 60-foot-wide and 6-foot-deep canal were used by mules who "towed" or pulled barges up and down the canal. Most of the canal work was done by Irish immigrants who had previously worked on the Erie Canal [3]. Canal building was physically demanding and dangerous, and many workers lost their lives, although there are no official records that document the number who died. From 1848 to 1852 the canal was a highly traveled passenger route. However, passenger service abruptly ended in 1853 when the Chicago, Rock Island, and Pacific Railroad, which ran parallel to the canal, was opened. The canal continued to be profitable for transporting bulk commodities and had its peak shipping year in 1882; it remained in use until 1933. The wider and shorter Chicago Sanitary and Ship Canal constructed in the 1900s replaced much of the transportation functions of the Illinois and Michigan Canal (figure 23.2). In 1964 the canal and towpath were designated a National Historic Landmark, ensuring that many of the canal's engineering structures and segments between Lockport and LaSalle-Peru remain accessible for viewing and recreational uses.

The Chicago River as Waste Disposal

Chicago quickly became an industrial boom city with the growth of river traffic and railroads bringing agricultural products, cattle, hogs, and sheep to the shores of Lake

FIGURE 23.1 Barges are pushed by a tugboat downstream on the Chicago Sanitary and Ship Canal north of Lockport Lock and Dam.

Michigan. The Union Stockyard opened in December of 1865 and supplanted many of the city's small stockyards. The refrigerator railcar invented in 1878 by Chicago packer Gustavus Swift allowed Chicago to set prices low enough to capture eastern markets and centralized the meat slaughter, packing, and shipping industry [4]. The Chicago River was important to the industrial growth of the city not only for navigation but for waste disposal. Stockyard, local industry, and home and farm wastes were dumped into the river and washed into Lake Michigan. The waste load in the river increased even more after the Great Chicago Fire of 1871 as Chicago rebuilt along the shores of the Chicago River. Although the Illinois and Michigan Canal was deepened to improve sewage disposal by moving it more quickly into the lake, the canal and river continued to be badly polluted.

Unrestricted dumping into the canal from fast-growing industries, the stockyards, and city sewers soon affected the safety of Chicago's drinking water supply drawn from Lake Michigan. The drinking water issue came to a head in 1885 when a large rainstorm carried refuse and heavily polluted waters from Bubbly Creek into the canal and the Chicago River and then more than two miles out into the lake where city water intakes were located. Surprisingly no epidemics occurred, but this was a wake-up call to city leaders and businesses that water conditions needed to be addressed. Four years later, the Illinois legislature created the Chicago Sanitary District, today called the Metropolitan Water Reclamation District, to redirect the polluted waters away from Lake Michigan and into the Des Plaines and Illinois rivers. The plan to dilute the contaminated waters downstream was operationalized by reversing part of the Chicago River and the canal so water flowed south (map 23.1). However, it was not until the Chicago Sanitary and Ship Canal was opened in 1900 (map 23.3) that the river's flow was fully reversed.

Chicago Sanitary and Ship Canal

Attempts to deepen the Illinois and Michigan Canal and reverse the river's flow were not long lasting or effective. A new plan to redirect the city's waste water away from Lake Michigan emerged with the formation of the Sanitary District of Chicago. Isham Randolph, the chief engineer for this newly created district, oversaw the design and construction of the Chicago Sanitary and Ship Canal, which opened the Illinois Waterway to large vessels transporting freight between Lake Michigan and the Gulf of Mexico (figure 23.1). Many of the engineers who worked on this huge earth-moving operation later applied their skills and experiences to the construction of the Panama Canal. The canal is 28 miles long, 202 feet

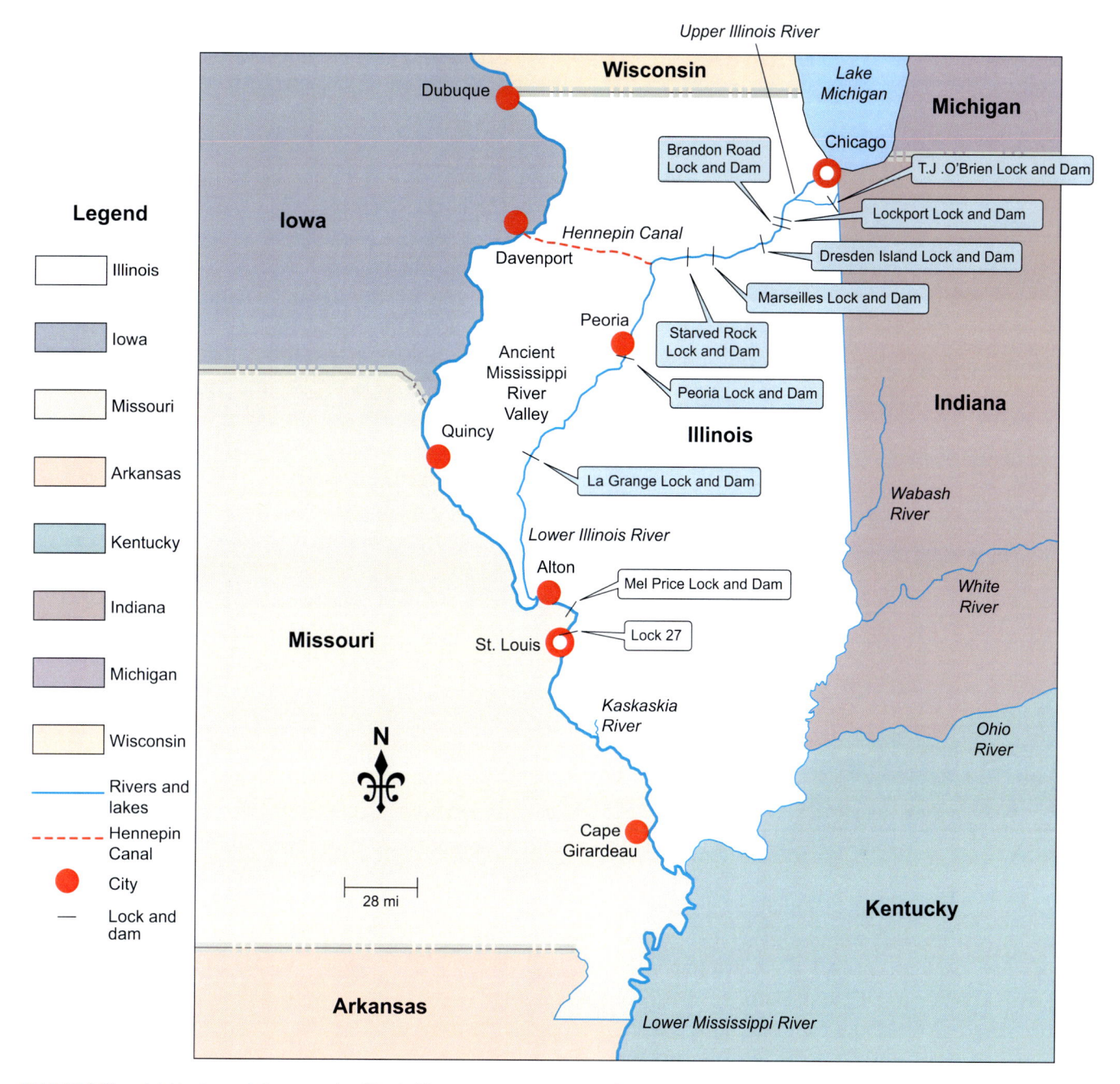

MAP 23.2 The eight locks and dams on the Illinois Waterway ensure a nine-foot navigation channel for commercial shipping and recreational boating from Lake Michigan to the Mississippi River.

wide, and 24 feet deep and the only shipping connection between the Great Lakes Waterway (specifically Lake Michigan by way of either the Chicago River or the Calumet-Saganashkee Channel) and the Mississippi River system, by way of the Illinois and Des Plaines rivers. It linked the south branch of the Chicago River to the Des Plaines River at Lockport and reversed the flow of the Chicago River in January of 1900.

The water flow direction before and after the construction of the Chicago Sanitary and Ship Canal is shown in maps 23.1 and 23.3, respectively. Note that map 23.1 does not include the flow path of the Chicago Sanitary and Ship Canal. Additional construction from 1903 to 1907 extended the canal to Joliet and totally replaced the Illinois and Michigan Canal with the wider and deeper Chicago Sanitary and Ship Canal (figure 23.2). A few years later, the Chicago Waterway system was expanded by the construction of two additional canals, the North Shore Channel in 1910 and the Calumet-Saganashkee Channel in 1922 (map 23.3).

The Chicago Sanitary and Ship Canal is designed to take water from Lake Michigan and discharge it into the Mississippi River watershed (map 23.3). A specific quantity of water was authorized to be diverted away from Lake Michigan under provisions of the US Rivers and Harbors Acts. The reverse flow and increased volume of water has effectively flushed untreated sewage away from Lake Michigan. However, water limits have

185

FIGURE 23.2 The old Illinois and Michigan Canal (right) flows under a railroad bridge as it merges with the wider and deeper Chicago Sanitary and Ship Canal (left) just above the Des Plaines River.

FIGURE 23.3 Flying Asian carp jump out of the water to travel upstream on the Illinois River at the Marseilles Dam in June of 2015.

not been honored or well regulated over the years with impacts on water levels to the US-Canadian managed Great Lakes and St. Lawrence River. Litigation began as early as 1907 when a court suit, Sanitary District of Chicago vs. United States, was taken to the Supreme Court. States downstream of the canal sided with the Chicago Sanitary District; a few years later in Wisconsin vs. Illinois the issue continued to be litigated. By 1930, management of the canal was turned over to the USACE, and the flow of water into the canal was reduced while retaining the important navigation function. Court decisions have pushed the sanitary district to invest in treating the city's raw sewage. Today, an international treaty with Canada and the governors of the Great Lakes states jointly monitors and regulates water diversions from the Great Lakes system through the International Joint Commission.

Bighead and Silver Carp Invade the Chicago Canals and Threaten the Great Lakes

An unexpected consequence of the nine-foot-deep Illinois Waterway is the potential transfer of invasive fish species from the Mississippi River to the Great Lakes. Silver carp (*Hypophthalmichthys molitrix*) from Asia were first introduced with the approval of the US Environmental Protection Agency (USEPA) in the 1970s to help remove algae from catfish farms in Arkansas. A number of Asian carp, noted for jumping or flying above the water (figure 23.3), escaped the fish farms and migrated up the Mississippi River. Carp are a fast-growing, aggressive, and adaptable fish that out-compete indigenous fish species for food and habitat. According to the National Wildlife Federation, Asian carp consume up to 20% of their bodyweight per day in plankton, and some species can grow to over 100 pounds [5]. They now threaten to enter the Great Lakes through the constructed canals connecting the Great Lakes to the Mississippi River watershed.

There are several kinds of carp that have spread (or are spreading) throughout North America. According to freshwater aquatic ecologist Cory David Suski at the University of Illinois, the two species that have garnered the most attention at present are the silver carp and bighead carp (*Hypophthalmichthys nobilis*). There are other carp (black carp [*Mylopharyngodon piceus*] and grass carp [*Ctenopharyngodon idella*]) that are less prominent, but silver and bighead are the two that are the biggest concern. There are about 18 different paths by which carp could pass from the Mississippi River basin into the Great Lakes. Most of these only have water at certain times of the year. The canals and rivers of the Chicago Waterway are the primary concern largely because they have continuously flowing water. Suski finds that the carp don't displace native fishes, and he is not aware of any displacement to date. It appears that the carp out-compete the native fish for food, and several species—particularly filter feeding fishes—have experienced reduced numbers and are not thriving.

Three electric fish barriers have been built by the USACE to prevent Asian carp from entering Lake Michigan (figure 23.4). The Chicago Sanitary and Ship Canal was temporarily closed on December 2, 2009, after the USACE disclosed on November 20, 2009, that a single sample of Asian carp DNA had been found above the electric barrier (map 23.3). The USEPA and the Illinois Department of Natural Resources applied a fish poison, rotenone, in the water to kill any fish that had escaped north of the Lockport electric barrier. Inspection of the fish kill and two months of intensive commercial fishing and electrofishing did not find any Asian carp. Alarmed by the potential disaster of Asian carp in Lake Michigan, Michigan State Attorney General Mike Cox filed a lawsuit on December 21, 2009, with the

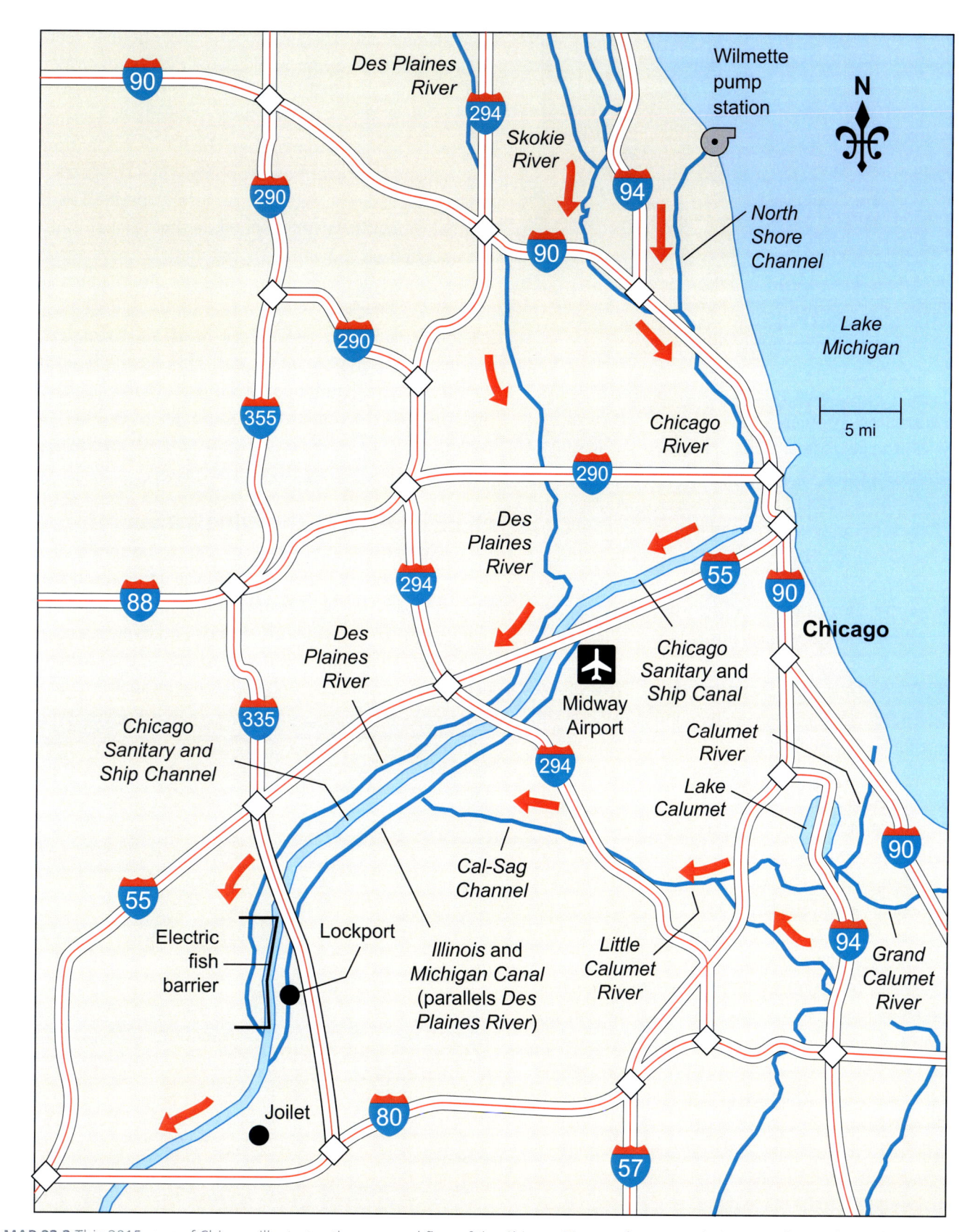

MAP 23.3 This 2015 map of Chicago illustrates the reversed flow of the Chicago River so that water drains away from Lake Michigan and into the Des Plaines River via the Chicago Sanitary and Ship Canal.

US Supreme Court seeking the immediate closure of the Chicago Sanitary and Ship Canal [3]. Co-defendants named in the lawsuit were the State of Illinois and the USACE, who constructed the canal. The main arguments presented against closing the canal were economic with estimates of more than $1.5 billion a year in lost

FIGURE 23.4 Electric fish barriers constructed in the Chicago Sanitary and Ship Canal are designed to prevent invasive species of Asian carp from passing from the canal into Lake Michigan.

revenues from millions of tons of iron ore, coal, grain, and other cargo shipments not able to use the waterway as well as job losses. Great Lakes states of Michigan, Minnesota, and Ohio rebuttal claimed that the sport and commercial fishery and tourism of the entire Great Lakes region was at risk, an annual economic loss valued at $7 billion. The US Supreme Court rejected the request for a preliminary injunction closing the canal on January 19, 2010, and the ruling was upheld by the United States Court of Appeals in 2011.

Marseilles Lock and Dam on the Illinois River

Downriver from the Chicago Sanitary and Ship Canal and the Lockport Lock and Dam are six locks and dams on the Illinois River (map 23.2). The locks and dams on the Illinois Waterway were designed by the USACE to function as a unit to maintain a nine-foot navigation channel. The Marseilles Lock and Dam system is located upstream about eight river miles from Starved Rock Lock and Dam. The Marseilles Lock at mile marker 244.6 is 2.4 miles west and downstream from the Marseilles Dam located at the town of Marseilles (map 23.4). The dam at mile marker 247 was constructed in 1933 to maintain the navigation pool between Marseilles and Dresden Island locks. It lies adjacent to the upstream end of the Marseilles Canal, which was created as a bypass to the rapids used to generate hydroelectric power [6]. The Illinois River at Marseilles splits into three channels (map 23.4): Marseilles Canal, which flows directly into the Marseilles lock; the rapids with water flow controlled by the dam; and the human-constructed channel on the Marseilles side of the river that runs through the former hydroelectric-powered mill. Note the location shown in map 23.4 of the abandoned Illinois-Michigan Canal that runs through the city of Marseilles.

The dam at Marseilles is a gated structure with eight 60-foot-wide submersible Tainter gates (figure 23.5), which cover a total width of 552 feet, and a 46.5-foot section containing an abandoned ice chute [6]. The main dam has a normal head of about 13 feet and maintains an upper pool at an elevation of 483.2 feet [6]. The submersible Tainter gates (16 feet high with a radius of 25 feet) are used as a spillway and are frequently adjusted to maintain the 9-foot navigation channel and to prevent overtopping when the gates are fully closed. The gates are remotely operated (with manual capability) using a schedule that (1) maintains a flat pool behind the dam, (2) prevents excessive scouring (e.g., one gate wide open while remaining gates are closed), (3) varies to reduce vulnerability to floating ice and debris, (4) minimizes out-draft, and (5) assures approximate equal use of the gates [6]. During flooding conditions, the gate schedule attempts to manage the high discharge created from wide-open gates and the considerable turbulence below the spillway, which has high potential for downstream scouring.

Flood of 2013 on the Illinois River

Record flooding on the Illinois River in the spring of 2013 raised the pool levels behind the dams along the Illinois River. Heavy rain and runoff from tributaries along with strong winds created river currents and conditions that made it difficult to secure the barges being moved in the shipping channel or anchored along

FIGURE 23.5 Seven partially sunken barges at the Marseilles Dam blocked the flow of water and prevented the gates from fully opening to release floodwater in April of 2013. Photo credit: Major General John W. Peabody, commander of the Mississippi Valley Division, US Army Corps of Engineers, and president of the Mississippi River Commission.

FIGURE 23.6 A partially sunken barge in April of 2013 blocked the flow of floodwaters and damaged the Tainter gates that control the level of water in the pool behind the Marseilles Dam.

the channel. On April 19, 2013, the currents and winds on the Marseilles pool above the dam caused seven barges to break free and crash into the Marseilles Dam. The unmoored barges were caught in the currents of the Marseilles pool and struck and damaged five of the eight Tainter gates at the dam leaving two gates with 16- to 20-foot holes [7]. Four of the seven barges partially sank in front of the southern gates (figure 23.5), and the other three barges blocked the water flow through three middle gates (figure 23.6). As a result, the blocked gates were unable to fully open to release the additional floodwaters. Water backed up into the Marseilles pool, topped riverbanks with and without a low levee protection, and flooded the bottomland alluvial soils (Sawmill, Millesdale, DuPage, Lawson, and Benton soils) in the town of Marseilles (map 23.4). Most of these poorly drained soils were developed under prairie and in alluvium over outwash or limestone. The floodwaters also flowed into the third channel running through Marseilles, the old diversion previously used to generate hydroelectric power, and then back into the Illinois River west of the Marseilles Dam.

Approximately 1,500 residents were evacuated from the low-lying areas, and over three feet of floodwater surrounded 200 homes and building structures and destroyed at least 24 homes. Many other buildings required substantial repairs, and a large number of appliances and household furnishings had to be discarded. When the rains stopped and upstream runoff slowed, floodwaters in Marseilles drained out through a diversion channel previously used by a former hydroelectric plant. There was an immediate and substantial federal disaster relief response including the Federal Emergency Management Agency, Lutheran Early Response, and Team Rubicon workers. Damages were in the millions of dollars, and many residents and relief workers were

FIGURE 23.7 Fast-moving floodwaters in the spring of 2013 deeply scoured the riverbank at Illini State Park.

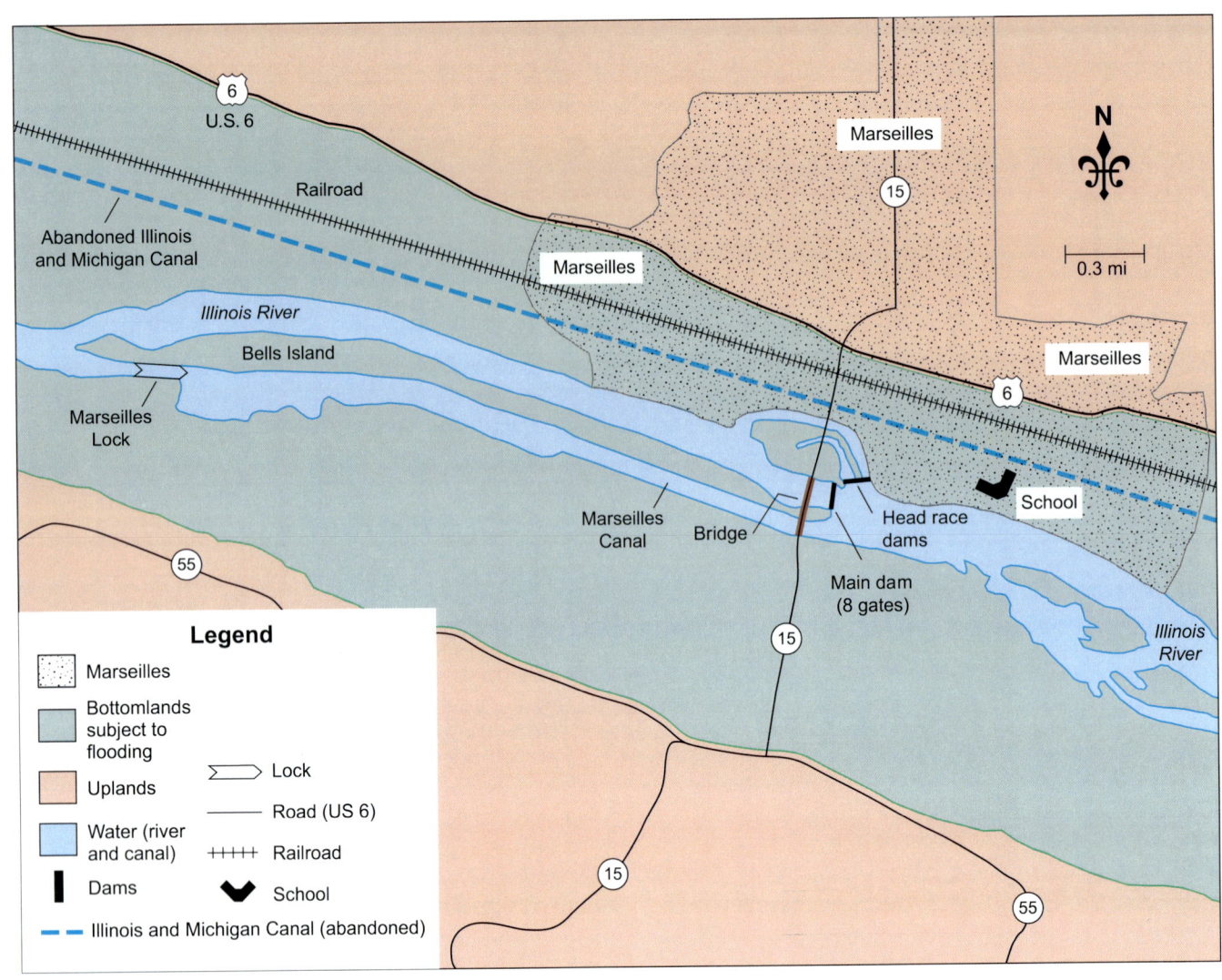

MAP 23.4 A portion of the town of Marseilles is located on bottomlands vulnerable to flooding. The lock and dam system on the Illinois River at Marseilles has a navigation canal on the south side that allows river traffic to bypass the rapids on the north side of the river.

FIGURE 23.8 Relief workers deal with mold and remove damaged items from flooded homes located in the Illinois River bottomlands in the summer of 2013.

housed in area hotels for weeks. The barge accident and 2013 flooding of Marseilles did some damage to parklands (figure 23.7), sidewalks, and roads. The greatest agricultural impact was the suspension of the shipping of fertilizers and grains. Damages to the dam structure reduced the capacity to manage the Marseilles water pool for navigation on the Illinois Waterway and to limit flooding in the town of Marseilles.

Homes, a school, and public buildings were flood damaged. Floodwaters reached depths of 3.3 feet in many buildings, and the lower one-third of the walls had to be removed and replaced. In the weeks immediately after the barge accident, hundreds of relief workers (figure 23.8) helped remove the damaged appliances and household items, or inspected the properties to assess the extent of the damage. The response teams helped clear out flood debris, damaged appliances, and furniture; removed walls and cabinetry; and treated the houses for mold. Dumpsters and portable bathrooms were brought in for use by residents and relief work-

FIGURE 23.9 Relief workers load trucks to haul away damaged household items in the town of Marseilles, Illinois, in the summer of 2013.

ers. The damaged appliances were placed out on the streets, put in dumpsters, or hauled to a public parking lot where relief workers sorted damaged property into categories and hauled the damaged household items away (figure 23.9). Long-time residents—including some that had lived there for 45 to 65 years—reported they had never experienced this kind of flooding before.

The Marseilles Elementary District 150, which serves 632 students, sued the Ingram Barge Company for $6.4 million for flood damages [8]. By June 30, 2013, 40 Marseilles property owners also filed claims against Ingram with more claims expected. The school claim was filed in Chicago federal court, and lawyers argued that the Nashville-based company "breached its duty" to safely secure the barges and was at fault for massive flooding of the city. The damage to the school was mostly covered by insurance from the Illinois School District Agency, through a self-insured pool of several school districts, except for an $834,000 deductible. To hold Ingram legally and financially responsible, it was necessary for the school district to file the claim. The lawsuit claims Ingram "had a duty to operate its vessels safely and to secure its barges under tow to prevent against a break away." School district lawyers claimed that because of the breakaway and resulting crash, "the flow of the Illinois River was impeded and altered such that floodwater flowed ashore, causing extensive flood damages to the Marseilles Elementary School" [8].

In anticipation of damage suits, Ingram filed for protection in federal court in May of 2013, asking it be freed from liability for the flood damage or at least be restricted to the value of the seven barges and towboat, or $4.2 million. Ingram Senior Vice President Dan Mechlenborg said the company does not believe it was responsible for the flooding of Marseilles and the school district. The record flooding could have resulted in the flooding of Marseilles on or after April 18, and evacuation notices were issued before the Marseilles Dam accident occurred. When the record high flood peak was predicted to occur on the evening of April 18, police began notifying and evacuating 1,500 residents from low-lying sections of Marseilles (map 23.4). Some residents were disabled and bedridden and had to be moved along with medical equipment. The USACE is investigating to determine if all or part of the flooding was as a result of the seven barges slamming into the Marseilles Dam.

Damages to Marseilles Lock and Dam and Barge Removal by the US Army Corps of Engineers

Within days of the accident, three of the barges in front of the dam were removed. The Marseilles pool behind the dam was lowered to make dam repairs and remove the partially sunken barges. However, it took many weeks after the barge accident for USACE cranes to offload cargo onto other barges so the damaged barges could be refloated and removed.

The dam and riverbed underwent rigorous inspection and monitoring to determine potential threats to the structure and to public safety. Riverbed scouring can undercut and weaken the dam's foundation and lead to future catastrophic failure [9]. Another storm system in late May and early June of 2013 flooded the watershed and raised the discharge rate through the gates and complicated the dam repair. One significant concern was the potential for increased scouring under the dam and on the north side of the riverbank since only four of the Tainter gates on the north side of the river could be fully opened. Dam failure could damage

FIGURE 23.10 The US Army Corps of Engineers constructed a dike and pond to temporarily manage the water pool behind the dam after the gates were damaged in April of 2013 at Marseilles.

commercial vessels and recreational craft if the pool of water behind the dam dropped.

Shutdown of Shipping and Boating on the Illinois River

The USACE created a temporary rock dike dam (figure 23.10) during the four weeks after the accident to permit repairs to the three most severely damaged gates. The temporary U-shaped dike downstream from the three gates included a series of culverts to hold or release water from the pond as needed, substituting for the controlled spillway effect under normal conditions. The temporary dike was able to hold enough water to elevate the navigation pool by late May 15, 2013, in order to restore boat traffic. The Marseilles Phase 1 repair costs were about $10 million. The current cost estimate to complete Phase II repairs ranges from $30 to $50 million for a total of $40 to $60 million. The money comes from the emergency response or disaster funds available to the USACE.

The decision to lower the Marseilles pool for dam repairs resulted in suspension of all use of the pool for shipping and travel. This temporarily blocked navigation use of the Illinois Waterway for barges and boats passing up and down river. This portion of the Illinois River and north is used for shipping bulk commodities downstream to the Gulf of Mexico ports. The Coast Guard declared the Illinois River unnavigable and any use of the river required their permission. This effectively reduced river traffic between the Marseilles Lock and Dam and Seneca, Illinois, for many weeks while the damaged Marseilles Dam was being repaired. The navigational channel between Marseilles and the Dresden dams remained open for local traffic between April 27 and May 7, 2013.

The Illinois Waterway

The Illinois Waterway and feeder canals continue to be an essential transportation corridor for moving goods from the Great Lakes to St. Louis, Missouri, and the port of New Orleans, Louisiana. The eight locks and dams on the Illinois River enable the USACE to manage the river levels for commercial shipping and other types of water traffic. The construction in 1900 of the Chicago Sanitary and Ship Canal opened this waterway to international markets from the Gulf of Mexico to the Great Lakes via the St. Lawrence River to the Atlantic Ocean. The reversal of the Chicago River to flow out of rather than into Lake Michigan improved the lake water quality and highlighted the urgency for Chicago to address residential and industrial sewage treatment and disposal.

Unpredictable weather and extreme rain events, such as in 2013, can cause hard-to-control flooding and river currents, which can damage river vessels, waterway control structures, and communities along the river. Damage to the Marseilles Dam when unsecured barges crashed into the dam and damaged the Tainter gates affected navigation on the Illinois Waterway, caused economic harm to Marseilles port facilities, and put at risk the adjacent community. The USACE is still assessing whether the barge accident caused or enhanced the flooding in Marseilles and damaged the elementary school and the local residences. The Illinois Waterway is one of the most used in the nation, and local ports along it as well as the USACE recognize the need to continually monitor climate and weather and the condition of lock and dam infrastructure to prepare for unexpected hazards and risks.

---

[1] National Oceanic Atmosphere Administration. 2013. Historic crests. Morris, IL: National Weather Service, Advanced Hydrologic Prediction Service.

[2] Olson, K.R., and L.W. Morton. 2014. Runaway barges damage Marseilles Lock and Dam during 2013 flood on the Illinois River. Journal Soil and Water Conservation 69(4):104A-109A, doi:10.2489/jswc.69.4.104A.

[3] Illinois Department of Natural Resources. 2016. Illinois and Michigan Canal. Springfield, IL: Illinois Department of Natural Resources. http://www.dnr.illinois.gov/recreation/greenwaysandtrails/Pages/IMCanal.aspx.

[4] Pacyga, D.A. 2015. Slaughterhouse: Chicago's Union Stock Yard and the World It Made. Chicago IL: University of Chicago Press.

[5] National Wildlife Federation. 2015. Asian Carp Threat to the Great Lakes. Merrifield, VA: National Wildlife Federation. http://www.nwf.org/Wildlife/Threats-to-Wildlife/Invasive-Species/Asian-Carp.aspx.

[6] Cooper, D.R., T.L. Fagerburg, T.N. Waller, S.W. Guy, and A. Tuthill. 2001. Monitoring of Marseilles Dam submersible gates, Illinois River, Illinois. Engineer Research and Development Center ERDC TR-01-15. Washington, DC: US Army Corps of Engineers.

[7] Plume, K. 2013. Damaged Illinois River lock may hinder barges for weeks. Reuters, April 25, 2013.

[8] Stout, S. 2013. Marseilles Elementary seeks millions from barge company. Ottawa Daily Times, June 25, 2013.

[9] US Army Corps of Engineers. 2013. Salvage operations on remaining barges continue. April 27, 2013. Washington, DC: US Army Corps of Engineers.

Soil Degradation and Flooding Risk Decision Making in Leveed Agricultural Landscapes

Levee-protected agricultural lands are some of the most fertile and productive soils in the world. These lands, which are part of the global food security network, are highly vulnerable and continually at risk of river flooding and levee breaching. Most types of river flooding have repetitive behaviors presenting known risks. However, highly variable weather and a shifting climate can change the frequency, seasonality, and severity of flood events, often in random ways, creating unpredictable risks [1, 2]. The uncertainty and nonlinear second and third order effects of the global climate system can amplify the nonuniform distribution of precipitation and threaten the integrity of dams, levees, and other structures designed to protect land uses adjacent to rivers [3].

More than 75% of the disasters that have occurred globally over the past decade have been triggered by weather- and climate-related hazards such as floods, storms, and drought. In the United States much of the 1993 flooding was associated with sand boils and structural failure of levees (rather than overtopping) due to prolonged high flood stages and unusually large runoff in systems that were cut off from historical floodplains [2, 4]. Flooding of agricultural lands, particularly those adjacent to rivers and alluvial river plains, can have high impact and persistent effects on soil erosion and degradation; crop productivity; and economic, social, and ecological conditions. The 2008 Intergovernmental Panel on Climate Change Report concluded that current water management practices may not be sufficient to cope with impacts of a changing climate and draws specific attention to flooding risk in agricultural and ecological systems. For example, the Mississippi River basin experienced major flooding and levee breaching in 1993 and 2011 with damage in the billions of dollars to levees, agriculture, livestock, fields, farm buildings, and equipment [2, 3, 4].

A new generation of engineers is calling for risk management engineering that extends beyond risk minimization and strengthening of physical infrastructures [5]. They promote a system approach that uses information feedback loops to minimize the consequences of failure and increase the flexibility of engineered, natural, and social systems to better respond to unstable and unpredictable conditions. This kind of management integrates structural solutions with adaptive management strategies by continuous monitoring and assessments of land use changes, soil and water damages from flooding and levee breaching, economic and social conditions, and stakeholder perceptions and concerns. [5, 6]. These assessments provide valuable feedback that improves capacity to develop solutions that accomplish societal goals. In this chapter, river bottomland flooding and vulnerability to levee breaching in the United States are discussed using southeast Missouri (Bootheel) leveed agricultural lands (see chapters 10 through 13) for illustration. Historical land use patterns of leveed lands and the great flood of 2011 on the Mississippi River reveal the impacts of flooding and levee breaching on soil conditions and agricultural productivity as well as public tensions associated with recovery and reconstruction. The linking of scientific knowledge and social values and concerns is central to effectively managing leveed agricultural land under changing conditions to address risks and future uncertainty.

Leveed River Bottomlands and Levee Breaching

Leveed river bottomlands are designed to protect human populations and various land uses, including agriculture, from flooding. When a levee fails, the damage caused by floodwaters and contamination of water and land is significant [7]. Water-borne sediments often cover plants and soils and fill in road ditches, drainage ditches, and waterways, or re-enter water in rivers, streams, and lakes. Frequently crater lakes are created by floodwaters either topping or pouring through the levee breach, and substantive gullies develop [8]. These gullies and land scour areas can extend into the floodplain several miles beyond the breach into fields or along ridges. As the water slows, the coarse sediments, such as sand, are deposited first on the alluvial soils followed by silt and clay.

Sediment is the primary water pollutant on a mass basis, and the sediment often carries with it other nutrients and pollutants including pathogens, hydrocarbons, and pesticides. Once fields dry out, thin sediment deposits may be incorporated into the soil with tillage. The effects on soil productivity and crop production are thought to be minimal. However, thick sediment deposits, such as sand deltas, require piling up and removal to restore agricultural functionality. The land scouring and erosional processes remove topsoil and create eroded phases and depositional phases on a soil and sometimes subsoil. The result is a less productive soil, even if land is reshaped and reclaimed [9, 10]. In addition, the sediment can block highway ditches and drainage ditches. This makes it difficult to remove excess water from the poorly drained soils and return the land to agricultural production.

The soil types; hydrogeologic features; volume of flow; time of year; and agricultural use of fertilizers, pesticides, and other chemicals affect the extent of land scouring and sedimentation. These factors and upstream point sources such as sewage treatment plants, storm sewer drainage, and other urban land uses influence the fine-scale remediation needed. Floodwater can also damage surface and subsurface water and impact water tables within the watershed. The productivity of these soils, including their capacity to hold moisture under future drought conditions compared to the original soils, is not measured. Thus, effects of sediment deposition and land scouring on soil profiles and productivity are often unknown. This makes it difficult for agency technical staff, local leadership, and farmers to have sufficient information to effectively restore soil productivity and put in place strategies and infrastructure to prepare for future flood events.

Most research related to the impact of flooding on floodplain soils has focused on natural, seasonal flood events where the inundation and subsequent drainage of the land occurs as relatively slow, low-energy processes. In contrast, levee breaches result in a very fast, high-energy release of large quantities of water onto the floodplain. A closer examination of the New Madrid Floodway, Missouri, and the US Army Corps of Engineers (USACE) induced breaching during the 2011 Ohio and Mississippi rivers flood offers an opportunity to synthesize lessons learned about river flood conditions, impacts of levee breaching on agricultural lands, and the social tensions associated with managing leveed landscapes.

New Madrid Floodway, Missouri
Historical Land Uses

The New Madrid Floodway, located immediately southwest of the confluence of the Mississippi and the Ohio rivers at Cairo, Illinois (see map 10.1), at 279 feet above sea level, was designed by the USACE in the aftermath of the deadly 1927 flood. The original frontline levee, which forms the eastern boundary of the floodway, was intended to protect land uses within the floodway until the Mississippi River reached the 55-foot stage, at which time the floodwater could naturally overtop the frontline levee. The USACE obtained easements between 1928 and 1932 from the landowners giving the right to pass floodwater into and through the New Madrid Floodway. The Flood Control Act of 1965 authorized modification of the New Madrid Floodway opera-

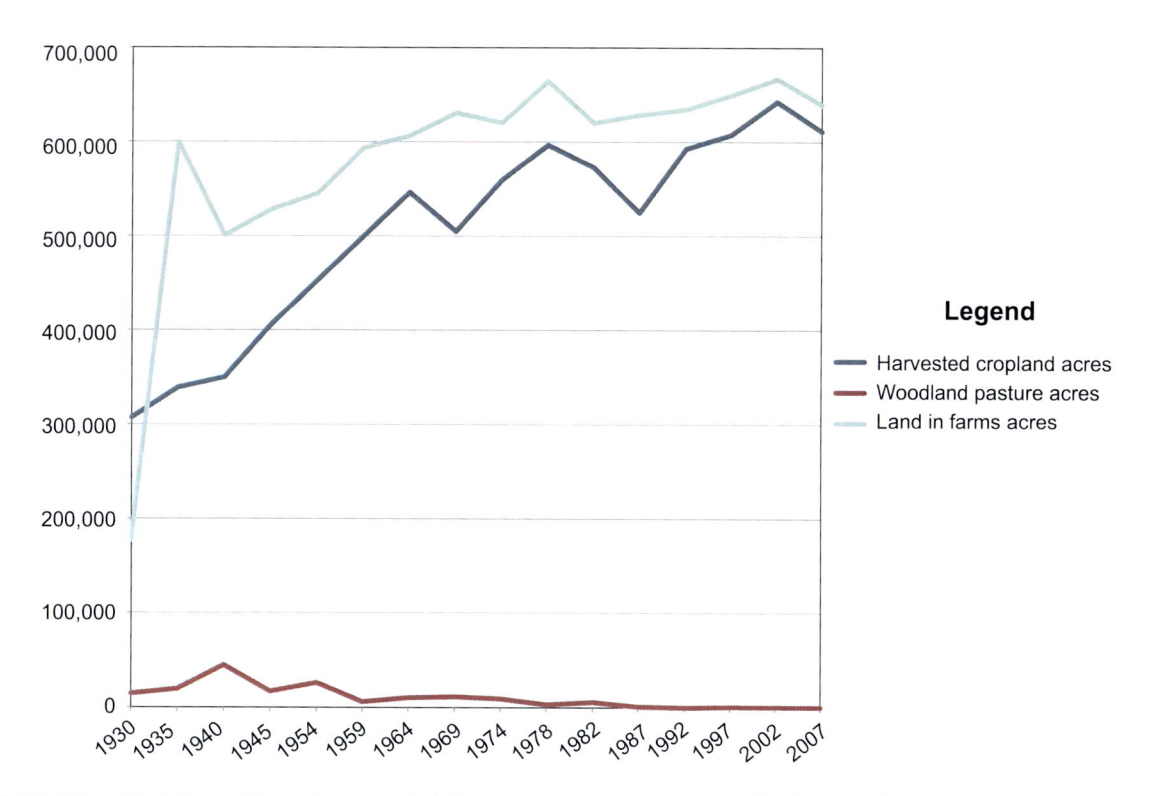

FIGURE 24.1 New Madrid and Mississippi counties' (Missouri) land uses from the USDA Census of Agriculture, 1930 to 2007.

tional plan; levees were raised and new easements were obtained. When the weather forecast predicts a 60-foot or higher Ohio River peak on the Cairo gage, the USACE must make a decision about the deliberate breaching of the New Madrid Floodway frontline levee fuse plugs to reduce pressure on the Cairo levee system and protect downstream cities and levees.

A look at the historical land use patterns of Mississippi River bottomlands places in perspective the implications and impacts of induced and natural breaching on levee-protected lands. Prior to settlement, the Missouri Bootheel contained more than half of all the state of Missouri's original 4.8 million acres of wetlands. Over time, almost all of these acres of wetlands and forested bottomlands were cleared, drained, and leveed for production agriculture, leaving 800,000 acres of wetlands in Missouri in 2013. The evolution of the New Madrid Floodway from forested bottomlands to productive agricultural lands is reflected in the land use change patterns between 1930 and 2007 of New Madrid and Mississippi counties (figure 24.1), a portion of which are levee-protected lands within the New Madrid Floodway. The US Census of Agriculture farmer-reported data for New Madrid and Mississippi counties show 6,510 farms with 338,988 acres in harvested cropland and 19,513 acres of woodland pasture in 1935. Seventy-two years later, in 2007, there were 578 farmers of record, harvested cropland acres had almost doubled to 610,979 acres, and woodland pasture substantively decreased to 139 acres. Although corn, soybean, wheat, cotton, and rice are the main cultivated crops in this region, an intensification of soybean production can be observed from 1945 to 2007 (figure 24.2). This likely reflects farmer adaptive management responses to seasonal wetness and flooding in these bottomlands as the soybean can be planted in early summer after saturated and flooded soils have drained.

Soil Functional Uses and Productivity

The characteristics of different soil series affect the functional uses and ecosystem services that the soil provides [11]. The flooding process can alter these functional uses when land is eroded by water and recreated as new soil where silt is deposited when sediment laden water slows down. Flooding can have beneficial effects: replenishing agricultural soils with new nutrients (when the water is not contaminated) and transporting sediment downstream to maintain delta and coastal areas [2]. However, flooding can also leave behind infertile sand and degraded soils, thus changing the soil functionality to a less than optimal state as soil organic matter is lost. Alterations in soil functionality can change its ability to sustain biological activity and productivity. Changes also affect how well soil regulates water, filters nutrients, buffers and detoxifies organic and inorganic materials, and stores and cycles nutri-

FIGURE 24.3 Wetlands and ponds in the deep gullies of O'Bryan Ridge (October of 2013) replace a productive soybean field after the May of 2011 levee breach and flooding.

a democracy obtains information about stakeholder beliefs, concerns, and opinions but alone are insufficient in guiding effective management. Full consensus is difficult and often not achievable or even desirable as there is frequently a strong preference for self-interest, the status quo, and a lack of knowledge about the floodplain as an ecological system [3, 15]. Public deliberative processes provide space to communicate the (a) problem of uncertainty, (b) facts associated with managing river ecosystems under changing current and future conditions, and (c) diverse values and concerns of stakeholders.

Local public agencies and private stakeholders with intermediary land use and water management responsibilities (e.g., levee districts, planning commissions, and soil and water conservation districts) can be barriers or enablers in facilitating how scientific facts and social values are linked. These leaders are key conduits of information exchange among local landowners and residents; federal and state agencies; and nonlocal publics with specific, larger societal interests. They play central roles in assessing the social, economic, and biogeophysical situations after disaster events. They can communicate known science about soil, hydrology, wetlands, and agricultural landscapes, and propose a variety of solutions to reduce future vulnerability and risk. They can also facilitate trust among sectors and between citizens and government agencies so resources can be mobilized. Social distrust of government is a major barrier to developing resilient, diversified river landscapes with complementary wetland and agricultural uses [15]. Trust is essential if adaptive management policies are to effectively combine engineering solutions with resilience-based management that reduces risk and vulnerability of levee-protected agroecosystems.

Generating New Solutions

Purposeful stakeholder engagement not only offers an information forum but can also generate new solutions. One public testimony to the 2013 USACE mitigation proposal noted that the agency environmental report

> ...did not contain an agronomic section where these details would be discussed...the economic opportunity cost of not providing the option of using a corn-soybean; corn-soybean-soybean; or corn-wheat-soybean rotation should be factored...it would be reasonable to figure the cost of potential crop productivity losses from increased crop pests when a single crop is used over the years.

Further the testimony asserted, "...this is an important oversight, as demonstrated in the report's economic section..." The economic report referenced notes, "key assumptions are missing," notably evidence of current agricultural production.

A main concern underlying this testimony is the need for data and assessments that can guide adaptive management in the context of reconstruction after flooding and the reevaluation of land uses for increased resilience to future disruptions. Adaptive management entails social, economic, and biogeophysical adjustments based on past events such as flooding disasters, or adjustments in anticipation of future hazards and risks. Planning that accomplishes adaptive management integrates engineering risk and broader landscape resilience approaches. This includes comprehensive assessments before and after flood events, such as assessment of soil characteristics and degradation, hydrology, wetland habitats, and social and economic conditions [5, 6, 16]. The 2011 flood event and the New Madrid Floodway levee breaching and reconstruction provide important lessons in developing public policies that are responsive to the complexity of the coupled human-natural system at local, regional, and national scales.

Managing land and living in a floodplain means farmers, residents, industries, and supporting institutions as well as public and private levee districts must always assume there will be another flood event. They need short- and long-term strategies as well as public policies to (a) sustain their systems of levees, (b) address breaching events and reclaim agricultural lands, and (c) put in place plans that anticipate future events. Levees are complex engineered systems linked to river systems, wetland and agricultural systems, and social systems. Due to incomplete knowledge of these dynamic systems and how they interact, future levee redesigns must not only account for risks to the engineered system but also risks and uncertainty associated with land use and social, economic, soil, and hydrologic conditions.

Soil Assessment

Resilience analysis and engineering is premised on the unknown risks that can't be planned for. Management focuses on preparing for emergent and unexpected events by continuously gathering new information and using these data as feedback loops to adjust as conditions change. In agricultural-leveed landscapes new information about soil damage and agronomic impacts from breaching and flooding is needed each time a levee fails. Soil condition assessments as part of the resilience analysis would offer (a) improved delineation of eroded and depositional soils associated with levee breaching, (b) better measurements of soil deposition and land scouring, and (c) finer resolution mapping of key hydrogeologic features. These assessments would increase the capacity of the USACE, local USDA Natural Resource Conservation Service technical specialists, Extension agronomists, soil and water conservation district commissioners, and levee district leadership to address short-term structural repairs. They would also enable strategic landscape level redesign to balance production agriculture and wetland ecosystem services needed to improve the resilience of the floodplain system.

New spatial technologies such as geographic information systems (GIS), light detection and ranging (LiDAR), and remote sensing are tools for assessing disasters and building a hazard information database to guide decision making for preparedness, response, and recovery. GIS utilizes spatially referenced data, integrating these data into electronic digital maps. Remote sensing data are obtained from sensors on fixed wing aircraft and satellite links and provide earth surface imagery. New unmanned aerial vehicles (UAV) offer huge potential for gathering site-specific and landscape-level data to better track real-time change. These technologies hold great potential to assess current conditions and develop models for scenarios to guide future flooding and levee breaching disaster preparedness and remediation.

However, these technologies are dependent upon accurate soil survey data obtained from field measurements. Many of the published US county soil survey maps are one-time surveys and are 1 to 30 years old. At best, they reflect eroded conditions, deposition, and degradation at the time the soil survey was made. Changes that have occurred from land use practices (e.g., cultivation of marginal lands, drainage of wetlands, or poor agricultural management practices) or from subsequent flood events are not reflected on the published soil maps. Levee breaching and flooding and their impacts on soil and soil productivity need to be documented in updated soil surveys. Restoration plans can be developed based on these updated soil surveys and would include locations of permanent soil productivity losses; damaged or abandoned levees; crater lakes, gullies, and thick sand deposits; sediment-filled drainage and road ditches; and land scouring. Soil degradation may be so severe in some locations, as in the case of the gully field on O'Bryan Ridge, that the land use has to change from agricultural use to wetlands (figure 24.3) with a loss in soil productivity and agricultural production (see chapter 13). Any flooding-related damages to the soils can result in changes in soil series on the maps, result in new reconstructed soils, or a change the erosion or depositional phases of existing soils. Thus, accurate soil surveys and maps are a critical basis for developing soil and water conservation plans.

Updating of the national soil survey after every levee breach is congruent with the Committee on Increasing National Resilience to Hazards and Disasters recommendations to establish a disaster-related database to better quantify risk models and structural vulnerability [17]. This recommendation could be implemented by an agreement between the USACE and the USDA Natural Resource Conservation Service to ensure a rapid federal response after levee breach and flooding. This could be part of the federal government's response to a disaster, along with emergency funds for restoration work including drainage ditch opening, levee repairs, crater lake filling, gully repairs, and sand deposit removal.

Assessments of Stakeholder Values, Perceptions, and Social and Economic Conditions

Adaptive management that includes deliberative processes beyond public hearings for gathering information about stakeholder concerns and social and economic conditions can increase decision making capacities. In managing the larger floodplain system, science and technologies must be linked to social values if social learning and behavior changes are to occur [14]. There are a variety of social science tools for assessing economic conditions, stakeholder values, willingness to participate in incentive programs, impacts of rules and regulation, perceptions of threats to physical safety, vulnerability of livelihoods to increased weather uncertainty, and evaluation of agency-proposed technical solutions [13]. For example, a 2013 survey of landowners [15] reported that program design and delivery of voluntary conservation programs influenced willingness to participate in adding biodiversity to land management plans. This kind of information could be particularly valuable in developing policies and programs that combine agricultural production with wetland management that reconnects the floodplain to the hydrology of the river.

Citizen assessments and participation in public decision making often reveal current and emerging divergent opinions that can lead to polarized positions as well as bring to light areas of agreement and common ground. Stakeholder consensus on levee and floodplain ecosystem management is highly unlikely in many instances. However, understanding the heterogeneity of fears and motivations for how land is managed acknowledges the variety of preferences, attitudes, and cultures and can lead to creative collaborative solutions and compromises. This information can help guide agency and public decision makers in negotiating solutions congruent with local values and increase policymakers' understanding of stakeholder fears and concerns associated with threats to safety and livelihoods as well as conflicting interests associated with restoration of river habitats and agricultural land uses.

Assessment tools such as surveys and listening sessions are particularly effective when findings are shared with stakeholders and presented in combination with biogeophysical and ecosystem data and the problems associated with managing the floodplain. Providing various stakeholders access to information including factual science about climate and weather patterns, river hydrology, soil and agronomic factors, levee structures, and bottomland ecosystems increases local knowledge and understanding of the landscape-level problem. Further, public forums offer stakeholders opportunities to contribute their experiential knowledge and engage in dialogues about what the problem is, impacts on their livelihoods, and strategies for addressing and adapting to changing conditions.

Stakeholder assessment and engagement can encompass use of websites and social media to make factual, accurate data accessible and gather feedback in a timely manner. However, this forum of exchange is not a substitute for creating and strengthening local and regional relationships and networks. Workshops, public meetings, goal-oriented committees, and public spaces for informal discussions can build trust; offer venues for exploring and negotiating solutions among divergent, competing values and interests; and meet multifunctional goals.

Focus on Flooding Solutions

Every watershed on the Mississippi and Ohio rivers has to deal with potential flooding during the rainy season, with or without levee breach issues. Knowledge gained from past episodic disasters can break down barriers to change and become a source of new information used to reframe future decisions as public agencies, private organizations, and citizens work to prepare for future disruptions [18]. Levees serve as valuable infrastructure in protecting the productivity of agricultural bottomlands. However, they may be inadequate if the distribution, seasonality, and intensity of precipitation patterns change. Restoration of large-river floodplains utilizing the natural ecosystem to mitigate flood hazard and risks associated with extreme precipitation events and changing climate is part of the solution [19]. Returning all leveed river bottomlands to their original wetland state has political, social, and economic barriers that make this change in land use highly unlikely and, in many cases, undesirable under current conditions. However, as government agencies, technical advisors, and society better understand the ecological functions of the river floodplain and the roles that hydrology, wetlands, and soils play in filtering, absorbing, and storing floodwater, there may be an increased willingness to adapt and live with floods. Social-ecological systems are dynamic and continually adapting (and mal-adapting) in unpredictable ways. While focus on risks to levee design may meet goals of efficiency and temporarily hold equilibrium, additional agroecosystem strategies that balance social, economic, and ecosystem vulnerabilities are needed to build resilience. Taken together, assessments of stakeholder values, knowledge, and willingness to adapt and assessments of changing soil conditions and other ecosystem functions are essential feedback information to the scientific analytics and deliberative processes necessary to guide planning and adaptive management for future uncertainties.

---

[1] Coumou, D., and S. Rahmstorf. 2012. A decade of weather extremes. Nature Climate Change 2:491-496.

[2] Wisner, B., P. Blaikie, T. Cannon, and I. Davis. 2004. At Risk: Natural Hazards, People's Vulnerability and Disasters. 2nd edition. New York: Routledge.

[3] Mileti, D.S. 1999. Disasters by Design: A Reassessment of Natural Hazards in the United States. Washington, DC: Joseph Henry Press.

[4] Olson, K.R., and L.W. Morton. 2012. The impacts of 2011 induced levee breaches on agricultural lands of the Mississippi River Valley. Journal of Soil and Water Conservation 67(1):5A-10A, doi:10.2489/jswc.67.1.5A.

[5] Park, J.T., P. Seager, S.C. Rao, N. Convertino, and I. Linkov. 2012. Integrating risk and resilience approaches to catastrophe management in engineering systems. Risk Analysis, doi: 10.1111/j.1539-6924.2012.01885.x.

[6] Morton, L.W., and K.R. Olson. 2013. Birds Point–New Madrid Floodway: Redesign, reconstruction, and restoration. Journal of Soil and Water Conservation 68(2):35A-40A, doi:10.2489/jswc.68.2.35A.

[7] US Army Corps of Engineers. 2010. National Report: Responding to National Water Resources Challenges. Washington, DC: USACE Civil Works Directorate. http://www.building-collaboration-for-water.org/Documents/nationalreport_final.pdf.

[8] Londono, A.C., and M.L. Hart. 2013. Landscape response to the international use of the Birds Point New Madrid Floodway on May 3, 2011. Journal of Hydrology 489:135-147.

[9] Olson, K.R., J.M. Lang, J.D. Garcia-Paredes, R.N. Majchrzak, C.I. Hadley, M.E. Woolery, and R.M. Rejesus. 2000. Average crop, pasture and forestry productivity ratings for Illinois soils. Bulletin 810. Urbana-Champaign, IL: University of Illinois, College of Agriculture, Consumer, and Environmental Sciences.

[10] Olson, K.R., and J.M. Lang. 2000. Optimum crop productivity ratings for Illinois soil average crop, pasture and forestry productivity ratings for Illinois soils. Bulletin 811. Urbana-Champaign, IL: University of Illinois, College of Agriculture, Consumer, and Environmental Sciences.

[11] Hatfield, J., and L.W. Morton. 2013. Marginality Principle. *In* Principles of Sustainable Soil Management in Agroecosystems, eds. R. Lal and B.A. Stewart, 19-55. Boca Raton, FL: CRC Press.

[12] Bendix, J., and C.R. Hupp. 2000. Hydrological and geomorphological impacts on riparian plant communities. Hydrological Processes 14:2977-2990.

[13] Morton, L.W., and S.S. Brown. 2011. Pathways for Getting to Better Water Quality: The Citizen Effect. New York: Springer Science and Business.

[14] Dietz, T. 2013. Bringing values and deliberation to science communication. Proceedings of the National Academy of Sciences 110 (suppl.3):14081-14087.

[15] Sorice, M.G., D.O. Oh, T. Gartner, M. Snieckus, R. Johnson, and C.J. Donlan. 2013. Increasing participation in incentive programs for biodiversity conservation. Ecological Applications 23(5):1146-1155.

[16] McLaughlin, D., and M.J. Cohen. 2013. Realizing ecosystem services: Wetland hydrologic function along a gradient of ecosystem condition. Ecological Applications 23(7):1619-1631.

[17] National Academies. 2012. Disaster Resilience. Committee on Increasing National Resilience to Hazards and Disasters. Committee on Science, Engineering, and Public Policy. Washington, DC: The National Academies Press.

[18] Sidle, R.C., W.H. Benson, J.F. Carriger, and T. Kamai. 2013. Broader perspective on ecosystem sustainability: Consequences for decision making. Proceedings of the National Academy of Sciences 110(23):9201-9208.

[19] Goodwell, A.E., Z. Zhu, D. Dutta, J.A. Greenberg, P. Kumar, M.H. Garcia, B.L. Rhoads, R.R. Holmes, G. Parker, D.P. Berretta, and R.B. Jacobson. 2014. Assessment of floodplain vulnerability during extreme Mississippi River flood 2011. Environmental Science and Technology, doi: 10.1021/es404760t.

Managing Ohio and Mississippi River Landscapes for the Future

25

Moving water is a powerful force that humans have attempted to tame and harness since civilization began. Although levee construction on the Mississippi River occurred as early as 1717 to protect the low-lying port of New Orleans, serious river engineering to manage the river and its tributaries did not begin until 1824 [1]. The Swamp Land Acts of 1849 through 1860 accelerated private construction of levees and ditches to drain river bottomlands and manage internal and river flooding. However, it took the catastrophic flood of 1927 for the US government to fully invest in a system of levees, floodwalls, diversion ditches and dredged channels, floodways, and upland reservoirs. Management of the Ohio and Mississippi river system has evolved over the last hundred years as we have learned more about relationships among uplands, headwater streams, floodplains, and rivers. Many of the lessons were learned the hard way. Unexpected heavy winter rains in December of 1936 and a prolonged four-week January storm in 1937 over the Ohio River valley generated a 1,000-year flood from Pittsburgh, Pennsylvania, to Paducah, Kentucky, that destroyed river cities without floodwalls and severely damaged even those river cities with floodwalls. Fifty-four people died and thousands more were displaced, agricultural bottomlands were flooded, telegraph service was lost, and washed out railroad beds halted freight transportation for many months.

The completion of the Kentucky Dam on the Tennessee River after the 1937 flood increased upland water storage and improved capacity to manage floodwater downstream from the confluence of the Ohio and Tennessee rivers. In the intervening years many upland reservoirs have been built to control the volume and timing of water runoff as it flows into main stem rivers toward the Gulf of Mexico. In addition to flood control, these lock and dam reservoirs have increased dry season stream flows and made year around commercial navigation possible. Early snowmelt and prolonged heavy rainfall in early spring of 2011 over the Ohio and Mississippi river valleys created another record flood, with a river crest at the Cairo confluence (figure 25.1) that exceeded the 1937 flood. In 2011 public and private levee systems breached (map 25.1), and the US Army Corps of Engineers (USACE) activated the New Madrid Floodway. Agricultural crops were lost, and substantial

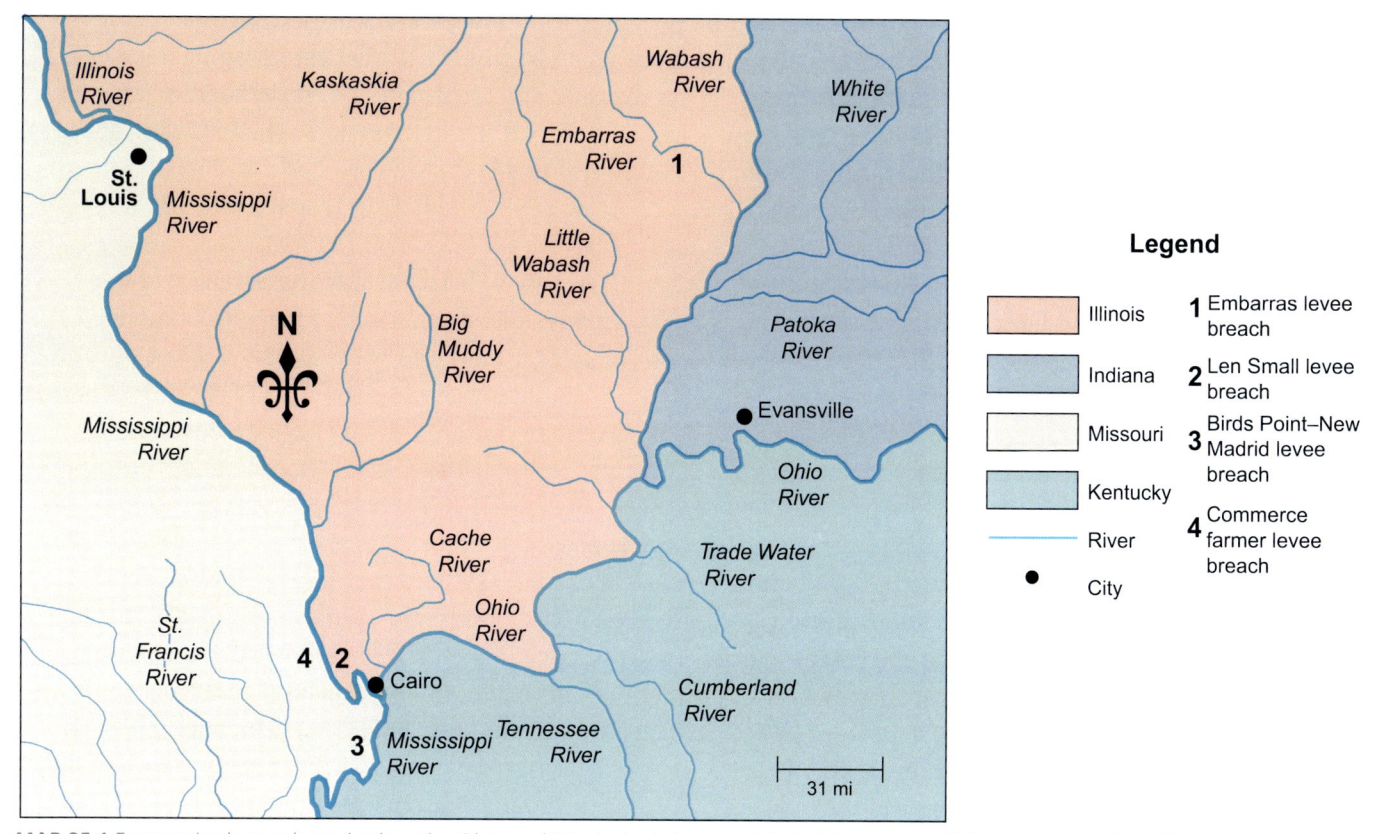

MAP 25.1 Four major levees breached on the Ohio and Mississippi rivers and their tributaries and damaged agricultural lands during the flood of 2011.

soil degradation occurred, including thick sediment deposits over fields, craters, and severe gully erosion; however, no lives were lost and there was limited property damage. Reclamation of agricultural lands, rebuilding of levees, sediment removal from road and drainage ditches, and repair of homes and agricultural structures have been costly to taxpayers and private landowners.

Despite the risks of living along the Mississippi and Ohio rivers and their tributaries, these waterways have been and continue to be sources of ecological diversity and abundance and economic prosperity fostering cultural and social centers. These rivers form the backbone transportation system for the central United States, a region rich in natural resources, fertile soils, and abundant water, making it one of the world's largest producers of corn, soybeans, and other agricultural products. The construction of lock and dam systems on the upper Mississippi and the Ohio rivers created the largest navigable inland waterway in the world, which today transports more than 60% of all US grain shipments for domestic uses and export.

This inland waterway not only has a vibrant and productive past and a vital economic present, it also has enormous potential to meet many of society's needs in the twenty-first century. Pressures to achieve global food security will only increase as world population, currently 7.3 billion, is projected to continue to grow during the next century. The agricultural productivity of central United States is key to meeting this need. Water security is one of the greatest challenges the United States and the world face now and in the coming decades. Water scarcity and quality for drinking, agricultural irrigation, industrial processes, and ecosystem services are limiting factors in assuring food security, human health, and well-being [2]. The Mississippi and Ohio river system has water in abundance, but it should not be taken for granted. The soil, vegetation, and ecosystem resources that filter and replenish these waters are at risk and must be protected to ensure quality and abundance. Climate and weather variability across the Ohio and Mississippi river landscapes affect the water cycle. The Third National Climate Assessment released in 2014 observes the modern-day landscape has experienced increases in annual precipitation and river-flow in the Midwest and Northeast over the last 50 years [3]. The 37% increase in very heavy precipitation events from 1958 to 2012 is projected to continue to increase into the future, intensifying flooding and intraseasonal droughts. This has implications for not only flood risk and crop production but also maintenance of water depth for navigation during extreme dry periods. Further, land use and land management practices within the river system have increased rates of upland erosion and discharge of sediments [1]. These increases in sediment discharge require more frequent channel and river port dredging to maintain adequate navigation depth and port viability. In addition to increased flood risks, higher air and water temperatures and more intense precipitation and runoff are decreasing lake and river water quality with increased transport of sediments, nitrogen, and other pollutant loads.

Extreme flooding events along the Mississippi and Ohio rivers and their tributaries well illustrate the continuing challenges of public agencies (e.g., Mississippi River Commission [MRC], USACE, National Weather Service [NWS], National Oceanic and Atmospheric Administration [NOAA], and Federal Emergency Management Agency [FEMA]), river municipalities, and private levee districts to anticipate risk and manage emergency and evolving natural disasters associated with downstream flooding and increased pressure on levee-protected landscapes [4]. Of particular concern is the vulnerability of low-lying deltaic environments (river bottomlands), which are levee protected. The direct impacts of levee breaching on soil erosion, land scouring, sediment contamination, and sediment distribution, and the indirect impacts on social and economic activities, particularly agriculture, of flooded areas are extensive. We don't know when or where the next catastrophic flood event on the Ohio and Mississippi rivers will occur. The only certainty is that it will happen. So the following questions must be asked:

1. How can we be better prepared for the next flood event?
2. How can we better realize the potential of this unique inland waterway to meet navigation, ecosystem services, water supply, recreation, and other quality-of-life goals?
3. What are the adaptive management strategies needed to manage this river system for the future?

Throughout this book, we have illustrated many aspects of river systems, the up- and downstream connectivity of lands and waters, and the direct and indirect cascading effects of engineering and management in one locale spilling over into other locales. Public testimonies at MRC and USACE low and high water public hearings well reflect this connectivity when port authorities request annual dredging to remove silt and sediment deposited from upstream waters and discuss

FIGURE 25.1 The confluence of the Mississippi and Ohio rivers south of Cairo, Illinois, during the flood of 2011 was more than five miles wide and inundated bottomlands in Kentucky, Missouri, and Illinois.

the need for flood easements with adjacent districts to store and redirect floodwaters when the river reaches specific elevations at their port. Managers of upland reservoirs evaluate how much water to release by monitoring downstream navigation channels depths to ensure barges and river traffic can keep moving. Local levee districts closely watch upstream rainfall events and tributary flooding to anticipate downstream risks to their levee system. Farmers anxiously track upland rainfall as it drains into and fills local drainage ditches to determine field conditions to make planting decisions and nitrogen applications, and under worst case scenarios, when to begin sandbagging.

Managing complex river systems to ensure healthy, resilient landscapes in the near and distant future requires recognition that knowledge about these systems is incomplete. Public agencies, private organizations, landowners, and residents that live in floodplains must plan for the unexpected and be prepared to adapt when the unexpected occurs [5]. The Mississippi and Ohio river system is a huge, sprawling landscape that needs an extensive coordination and communication network of public-private partners and multiple funding sources. Many kinds of experts and local knowledge are needed to engineer and manage river watersheds to achieve resilience [5, 6]. These partnerships must be committed to continuous monitoring and assessment of the river and its landscape. They must be willing to learn from the past and able to integrate new science and technologies in managing the routine and anticipating the unexpected, adjusting and adapting as conditions and situations change. The USACE provides critical engineering services in managing and protecting river resources. Their public mission is diverse, encompassing navigation, flood risk management, river ecosystem protection and restoration, regulatory oversight, water supplies, and hydropower production. Central to accomplishing this mission is the administrative and mission-vision leadership they provide in coordinating, communicating, and creating spaces for iterative dialogues among the many stakeholders that use, manage, and value river landscapes. Local partnerships with levee districts, river port authorities, agriculture and associated enterprises, and community leaders are necessary for this mission to be accomplished.

In this concluding chapter we learn from the past and look to the future to make recommendations that we think will increase the resilience of river systems. By resilience we mean the capacity to absorb shocks and disturbances and yet retain the ecological, social, and economic structure and functionality of the system. Despite good engineering and planning efforts of humans, infrastructure system failures occur for a variety of reasons. Snow melts. Then it rains and rains. Roads and fields flood. Rivers exceed flood stage and push beyond the capacity of leveed infrastructures to keep the river out. Levees breach. Barges at high water become unmoored and damage floodgates that control water flow. Levee districts lack funds to repair and maintain their infrastructure. Citizen leaders, local organizations, and state and national agencies with different priorities and resources miscommunicate or worse don't communicate at all. Engineers, soil scientists, land use managers, port authorities, and technical support staff lack sufficient data to calculate accurately river conditions, to estimate infrastructure needs, or model out 500- or 1,000-year flood events.

Two kinds of observations and recommendations are presented: postflood assessment and management, and ongoing investments in physical and social infrastructures to improve future adaptive responses. Our list is not intended to be comprehensive but is derived from our expertise in soil science and human-social sciences. It is grounded in our knowledge of agriculture and natural resource systems and observations of leveed and unleveed landscapes; synthesized from listening to stakeholders who own and manage land in the alluvial bottomlands, levee district leaders, and upstream and downstream rural and urban stakeholders; and refined by spirited dialogues with other scientists, technical staff, leadership in public agencies and private organizations, and private landowners.

Postflood Assessment and Agricultural Lands Management

Almost 10% of the Mississippi and Ohio river watershed is alluvial bottomland and is used primarily for crop production. Between 10% (Illinois) and 30% (Missouri) of individual state crop, agricultural, and food production comes from bottomland soils. Millions of acres of agricultural floodplain lands are drained, levee-protected, and irrigated. Today, the Mississippi levee system has over 3,500 miles of public and privately managed levees from Cape Girardeau, Missouri, to New Orleans, Louisiana. Many more miles of levees are found in the Missouri River subwatershed [7] and adjacent to the upper Mississippi and Illinois rivers; the Ohio, the Tennessee, the Wabash, and the Cumberland rivers; and their tributaries. Levee breaches in 1927 flooded 27,000 square miles to a depth of 30 feet, including thousands of acres of fertile Mississippi river agricultural bottomland, and effectively ended the plantation cotton system [8].

The Flood Control Act of 1928 built more levees; made existing ones higher and stronger; and created three floodways, including the New Madrid Floodway, to divert floodwaters and reduce downstream water pressure on levees. The Flood Control Act of 1936 made flood control a federal policy and officially recognized the USACE as the major federal flood control agency. In the 1940s the Tennessee Valley Authority (TVA) built the Kentucky Dam (see chapter 19) on the Tennessee River to better control the fast rise of the Ohio River during spring rains and slow its rush to the Cairo confluence. Despite public and private investments and extensive engineering efforts, once-in-a-lifetime flood events in 1945, 1975, 1993, 1997, 2008, and 2011 on the upper Mississippi, Missouri, and Ohio rivers continued to result in flooding, levee damage, destruction of property, and devastation of soil resources. These floods led to record erosion levels on both bottomland and upland soils. Further, natural and induced levee breaches on Ohio and Mississippi rivers resulted in short- and long-term soil contamination and agricultural crop damages.

Addressing Soil Erosion and Degradation

Soil erosion caused by these floods brings into question the adequacy of current soil conservation practices and their implementation (or not) by landowners. Soil conservation for the most part is a social learning process whereby experience from past events informs changes in practices to prevent resource degradation during future events. The floods of 1993 and 1997 provided excellent opportunities for Midwest conservationists to improve upon conservation practices in preparation for future events, such as the 2008 flooding in the Mississippi and Missouri river basins. The impact on alluvial soils in these river basins was partially addressed by raising and strengthening some of the levees (see chapters 11, 12, and 17). In other areas the land use was converted from agricultural use to a conservation use. However, on the upland Midwest soils the flooding lessons were not learned. If conservationists and landowners had learned from the past, the 2008 and 2011 floods would not have had as much land scouring and soil erosion-related destruction as we see on both the upland and bottomland soils.

The question remains: what have we learned from the 1993, 2008, and 2011 floods, and will we implement practices to protect against future floods? These floods in the upper Mississippi river basin caused considerable devastation with extensive property loss. The estimates of financial loss because of structural damages can be readily assessed when each property has a known market value. However, we do not have this kind of cost analysis or market value data for the soil and water degradation damage. Further, even if the data existed, the eroded soil cannot be easily replaced. Thus, we need a plan and commitment to save our soil.

Conservationists and soil scientists have not recommended changing the current tolerable soil loss (T) values for soils (as high as 5 tons per acre) on the uplands and bottomlands of the Midwest. The T values are set based on how fast topsoil and subsoil are formed in specific parent materials. The problem is two-fold. The acceptable T value loss metrics are likely set too high. And second, cultivated soils are subject to more intense rainfall events before and during planting when soils are not protected by vegetation. As a result greater soil erosion, water runoff, and sedimentation occur than our equations would predict. Perhaps one starting point is to assume a more intense rainfall factor (based on different weather and climate scenarios) than currently used to calculate soil loss (for the Universal Soil Loss Equation [USLE], the Revised Universal Soil Loss Equation 2 [RUSLE 2], and the Water Erosion Prediction Project [WEPP]). A 2013 study of an upstream Iowa agricultural landscape with intense row crop cultivation suggests that land use changes could reduce flood events, decreasing both the number and frequency of severe flooding [9]. Several scenarios were modeled, and the greatest flood risk reduction was found to be associated with conversion of all cropland to perennial vegetation. While this is not practical from an economic nor food security point of view, a second scenario of converting half of the land to perennial vegetation or extended rotations could have major effects on reducing downstream flooding and reducing soil erosion.

This serves as more evidence of the importance of increasing infiltration and reducing soil loss from tillage, erosion, and water runoff when soils on the uplands are used for row crops. While many soil and water conservation management and cropping practices (such as terraces, grassed waterways, strip cropping, fewer row crops in the rotation, and conservation tillage including no-till) reduce soil erosion when utilized, there is a need to expand the use of filter strips; utilize cover crops on sloping and eroding soils; increase the use of conservation tillage; add small grains and forages into the crop rotations; construct more temporary water storage dams, check dams, or retention ponds on the uplands; and take highly eroded lands out of row crop production and replace with perennial forages or timber (see chapter 9). If more runoff water and sediment can be retained on the uplands for a longer period of time, infiltration will increase, and crop production losses will be reduced, resulting in less degradation of the bottomland soils and less sediment in the surface waters.

It is also critical to assess the impacts of flooding on agricultural lands post–levee breaching to guide adaptations that prepare for future flood risks. Assessment of land scouring and deposition effects on soil productivity and long-term agricultural production is key to understanding the impacts of flooding on soils and profitability of future crops. Levees protect public and private lands from the consequences of periodic flooding. However, when they fail naturally or as a result of human induced breaching, the consequences are disastrous and can take different forms. The damages include crop loss; levee damage; crater lakes; gullies; thick sand deltaic deposits; scoured land; irrigation equipment destruction; soil and water degradation; building structure and farmstead damage; blockage of drainage and road ditches; road deterioration; and ecological damage to forests, parklands, and wetlands. The effects of levee breaches and flooding on soils and soil productivity are seldom determined since updated soil surveys are not routinely made in response to levee breaching and flooding. In the case of the O'Bryan Ridge gully field (see chapter 13) following the opening of the New Madrid Floodway, the damage to soils after restoration attempts included the permanent loss of 30% of the agricultural productive capacity as result of land use conversion, land scouring, water erosion, and gully field formation with little deposition of sediments since the rushing floodwaters drained quickly and transported the sediments from the field.

Resurvey and Assess Soil Conditions

There is a need to resurvey and assess soil conditions following natural and human induced levee breaches to (*a*) improve characterization and measurement of eroded soils and distribution of sediment contaminants after breaching, (*b*) assess contamination effects on soil productivity and long-term agricultural production, and (*c*) reassess current levee location and design in response to expected future increase in extreme weather patterns (flooding and drought) and changing climate conditions. Better data and assessment of soil conditions postflooding can provide valuable guidance in the restoration of craters, gullies, land scoured areas, and contaminated sediment depositional sites and thereby improve remedial effectiveness, future risk analysis, and levee management decision making. This information can increase the capacity of public and private levee districts to evaluate and restore sediment contamination sites created after a levee is breached and increase the resilience of the agricultural landscape to manage future high water and flood events. Reiterating our recommendation from chapter 24, an agreement between the USACE, MRC, and the USDA Natural Resource Conservation Service to conduct a land scouring and deposition surveys after every levee breach and to update the soil survey maps would ensure more effective responses.

A pattern of intensive resource use—human, equipment, energy, financial, and social—emerges from levee breach events and reconstruction investments. These encompass levee repair, return of land to productivity, and creation of a landscape that is less vulnerable to future flooding and levee breaching stress. Resilience analysis utilizes continuous monitoring and assessment but assumes that there will always be unidentified or emergent factors that cannot be accounted for. Expectations of the unpredictable lead to the development of more flexible engineering and management that better respond to uncertainty and "surprise" conditions. Engineers, soil scientists, farmers, agricultural production specialists, and rural community leaders in levee-protected regions should consider alternative designs that incorporate natural wetlands and bottomlands into the levee system to increase capacity to deal with the unpredictable. Designs that integrate natural wetlands can reduce water pressure on levee systems; increase water storage capacity; absorb and transform excess nutrients that degrade water quality; reduce social, biophysical, and economic impacts of soil degradation and contamination; and improve the overall resilience of agricultural productivity in deltaic environments [9, 10, 11, 12, 13].

Maintenance and Modernization Investments in Physical and Social Infrastructure

The full potential of the Mississippi and Ohio river system has not yet been realized. While it has a glorious and colorful past, the nation and its leadership have not yet captured and reproduced a compelling vision for the future of this unique inland waterway. A unified vision and purposeful investments in physical and social infrastructures are necessary to create a world-class river system that achieves the multifunctional goals of a national economic engine that relies on and protects ecological resources and is a source of technological, social, and cultural vibrancy. Currently, it has pockets of prosperity and poverty; cities, ports, and levee districts that compete for scarce federal dollars; and fragmented priorities and investments that benefit some locales and disadvantage others.

We first observe that this inland waterway is a complex human-natural system, which is geographically distributed, subject to high levels of local variability, and yet a unified whole. Three areas of action are proposed to prepare, guide, and adapt this amazing resource for the future. The first is the physical infrastructure. There is an urgent need to operate and maintain navigation as an inland system and invest in repair and modernization of aging lock and dam structures. Much like the national highway system, this inland river system, encompassing more than 40% of the United States, needs substantive systematic infrastructure investments. Failure and closure of one lock and dam because routine maintenance and repairs have not been kept up to date harms the entire system. Silting in of one port along the river removes a node in the transportation system that affects river traffic and reverberates throughout the economies of that port and the whole system. Second, managing complex systems requires data about individual as well as integrated components of the entire system in order to monitor, evaluate conditions, and adaptively manage. Standardized metrics are essential to a system approach of management. Data must be spatially comparable across the system, able to be aggregated, and accurately modeled when primary data are not easily assessable. These scientific data must be readily available and accessible to the many local, regional, and national partners that evaluate technologies and best practices, and make a myriad of daily decisions associated with river management. Last, the river system is deeply intertwined with human and social systems. Human perspectives and goals, social relationships, and actions are key factors that reflect how the river system is valued and cared for as well as influence how it is managed.

Inland Navigation System

Prior to the 1820s, periods of drought often affected river navigation on the Ohio (see chapter 18), Tennessee (see chapter 19), Cumberland (see chapter 20), Upper Mississippi (see chapters 21 and 22), and Illinois (see chapter 23) rivers. When Lewis and Clark headed down the Ohio River in 1803, there were no locks and dams. In dry years the water depth was very low, and navigation was often delayed until high water. The major physical hurdle on the Ohio River was the Falls of the Ohio River near Louisville, Kentucky, which steamboats could only pass over when the river was high. The USACE, in 1825, began building locks and dams on rivers to permit year-round navigation and shipping. Today we have a lock and dam system that assures a nine-foot navigation depth for the entire length of the Ohio River and the upper Mississippi from Cairo, Illinois, to Minneapolis, Minnesota. After the construction of the Kentucky (1940s) and Barkley (1960s) reservoirs, it was possible to release sufficient water for weeks or even months to maintain an additional four feet of water in the lower portion of the Ohio River. These early wicket dams and lock chambers have been systematically replaced (see figures 18.4 and 18.5). The Olmsted Lock and Dam on the Ohio River just north of the Cairo confluence is slated for completion in 2020. This modern, state-of-the-art infrastructure (see figures 18.2 and 18.9) is being built at a cost of $3 billion and replaces these last two aging wicket dams and locks on the Ohio River. However, many locks and dams throughout the upper Mississippi River and Illinois River systems are aging, with resources limited for even routine maintenance and repairs. These systems also need to be modernized.

Locks and dams on the upper Mississippi and Ohio rivers and their tributaries increased the volume and weight capacities of barge and river traffic and boosted the economies of river cities. Many river ports are at the intersection of rail lines making the port city a transportation hub for agriculture, mining, and other commodities for domestic use and export. The current and future viability of these river ports and harbors throughout the system is dependent on retaining and increasing commerce through active routine maintenance and modernization of the navigation system. America's Watershed Initiative Report Card for the Mississippi River [14], published in 2015, gives the Mississippi River infrastructure maintenance a D+. The USACE operation and management budget covers only routine maintenance and consistently underestimates the amount needed to keep inland navigation infrastructure operating. Deferred maintenance and repairs are increasing and increase the risk of unscheduled delays. High water events deposit silt and sediment loads in these ports and over time can reduce a 25-foot harbor to 8 or 9 feet, limiting the tonnage barges can carry. Small ports and harbors often need to be dredged annually, especially after multiple high water events. However, they currently lack a dependable strategy for funding dredging operations, which require annual congressional appropriations. Not only must local leadership continue to lead the way in maintaining and modernizing their own port, they must also build and strengthen partnerships with other ports up and downstream. Cooperative agreements, knowledge exchanges, resource sharing, and a cohesive vision for the Mississippi and Ohio rivers as an inland waterway with global reach can bring visibility and additional investments to the system.

Increased Science and Data Availability for Improved Decision Making

Scientists track the frequency, duration, magnitude, timing, and rate of change of water, material, and biotic fluxes to downstream waters [15] to measure the degree of connectivity of land and water within a watershed. These factors have cumulative effects across the entire watershed and influence the variety of functions that rivers and their floodplains provide. Streams, wetlands, and rivers can serve a number of functions simultaneously and affect the structure and function of downstream waters [15]. These functions include (*a*) export of downstream water, soil, nutrients, and organisms; (*b*) removal and storage of sediment, contaminants, and water; (*c*) provision of habitat for organisms; (*d*) transformation of nutrients and chemical contaminants into different physical or chemical forms that make them less harmful; and (*e*) regulation and delay of the release of floodwater, sediment, and concentrated contaminants.

The structure and shape of rivers and floodplains and their relationship with each other is always changing and continually evolving with changes in land use, climate, and human activities. Mainline levees block river flooding. Interior drainage ditches and large pumps drain surface and groundwater seepage to protect agricultural and urban land uses. Floodplain lakes and backwaters are scoured during high flows and accumulate fine grain sediments during low water periods. Navigation pools behind locks and dams have changed sedimentation and shoreline erosion processes. Clearing the river of woody debris, construction of channel training structures such as chevrons and wing dams, dredging, and redistribution of dredged material are modifications that have changed the geometry of river channels and floodplains [1]. These modifications have stabilized the main channel, reduced the width, and deepened the river as intended. Dams have increased water levels, slowed current velocities, and flooded low-lying floodplains within the navigation pool [1].

Over time, wind-driven and boat-generated waves in impounded areas of navigation pools have eroded shorelines resuspending and redistributing sediments. Sedimentation is among the most critical problems in the river and a major concern to natural resource managers (ecological impacts), river port authorities (dredging is costly), and the USACE (maintaining the navigation channel). The USACE analyzes data on channel geometry, river contours, sediment delivery to the river, hydrologic records, and river engineering structures to track changes and evaluate their cumulative effects. However, the capacity to forecast accurately changes in the geometry of river channels and floodplains is limited by insufficient past and current condition data. Data such as floodplain topography; sediment delivery rates from tributaries; and quantitative measures of geomorphic responses to impoundment, river regulation, and channelization are dynamic and continually in flux and need documentation [1].

Data limitations are also a challenge in preparing river and flood forecasts. Two types of data are fundamental for the NWS to issue river forecasts and flood warnings [16] necessary for the USACE, levee districts, and landowners to prepare for potential flood conditions. The first is the river stage or the water depth, usually measured in feet. The second is the total volume of water that flows past a point on the river for some period of time (flow or discharge), measured in cubic feet per second or gallons per minute. River stage and river discharge are measured at a specific location on the river called a stream-gaging station. The US Geological Survey (USGS) operates and maintains a network of 7,292 stations throughout the United States, almost 4,000 of which are used to forecast river depth and flow conditions [16]. Most of these gaging stations are automated with sensors that continuously monitor and report river stages to one-eighth of an inch. Battery-powered stage recorders with satellite radios transmit data to USGS and NWS computers even when high waters and strong winds disrupt normal communication systems. This is essential, especially at remote sites, for tracking how quickly water is rising or falling. River discharge is usually estimated from preestablished rating curves that represent the relationship between river stage and discharge. USGS field personnel periodically measure river discharge in person to detect and track changes in discharge and assure the rating curves reflect real-time conditions as accurately as possible. Flood conditions can effect scouring and deposition of sediment as well as in-stream bed and bank roughness. These in turn can change the river stage and discharge relationship resulting in the need to develop a new stage/discharge rating.

Flood stage metrics are based on the impact to people in a specific location, that is, the water level at which the river threatens lives, property, or navigation. Flood stage on the river gage is commonly measured at the level of the water surface above an established zero point at a given location. The zero references a point within 10 feet of the bottom of the channel, which is also usually the mean sea level. Flood stage is only calculated for bodies of water that affect communities. For example, the Cairo gage at flood stage is 40 feet in the Ohio River channel with a sea level elevation of 310 feet. The peak 2011 flood levels measured 62 feet on the Cairo river gage (a sea level elevation of 332 feet), or 22 feet above flood stage. There are five levels of flooding which are used to communicate flood risk and potential impacts to human settlements [17]. The first is an "action stage" where the water surface is near or slightly above its banks with water overflowing into parkland or wetlands but not human-made structures. "Minor flood stage" is slightly above flood stage with minor flooding of low-lying farmland or roads. "Moderate flood stage" begins to inundate buildings, close roads where low-lying areas are cut off, and cause some evacuations. "Major flood stage" is significant, life-threatening flooding with low-lying areas completely covered, buildings submerged, and large-scale evacuations [18]. A "record flood stage" is the highest level that a river has reached since flood measurements were historically recorded on that particular river gage. However, a record flood does not necessarily have to be a major flood, but is simply the highest level ever recorded on that community's river gage.

It should be readily apparent to the reader that river gage measurements and definitions of major and minor flood stages at Cairo, Illinois, are quite different numbers with different meanings than those at Paducah, Kentucky; Keokuk, Iowa; or Cincinnati, Ohio. River gage data are location-specific. Local residents know what the numbers on their gage mean in relationship to potential for flood damage to crops and local infrastructure, and the need to evacuate. River gages are not a standardized metric that can be used to monitor and assess changes in the river system. Inconsistencies in river stage data make local decision making difficult. Decisions are made based on river stage forecast. Historic river stages are used as analogs for when water will cover certain local roads; whether to sandbag around homes and buildings; whether to evacuate; and the urgency, timing, and speed to take action. Further complicating system-wide monitoring and assessment is the fact that current river elevation data are not reported in the same way for dam or project structure elevations. There is a need to standardize reference river system metrics using sea level elevations so they have system-wide meaning. Locals already know the interpretation and meaning of readings on their own gage but often do not know the implications of upriver or downriver gage readings. This hampers downriver decision making as leaders track upstream conditions in order to make timely, appropriate decisions.

Related to this issue are data metrics and limitations noted by levee districts and their engineers who need to make flood easements and agreements up- and down-river with other districts. These agreements are needed to redirect floodwater and make floodwater storage arrangements. When agreements are not in place regarding when to accept water from another district at a specified elevation, sequencing of the pools and storage as river elevations change is quite difficult. There is a need for standardization of river level metrics and a lot more data on tributaries and calculations for storing and holding upland water in order to manage main stem river elevations. This will help set appropriate elevations based on engineering science and local knowledge as triggers for activating agreements. Currently local gages and flood records used as the baseline for measuring changes in the river height do not translate well across the system. This makes it difficult to maintain infrastructure and adapt to changing conditions when science-based changes in the river profile are not up to date or easily compared across the system.

Human Perspectives, Social Relationships, and Actions
It is well recognized that managing river landscapes involves a great deal of engineering as well as the physical and natural sciences [19]. Often overlooked is the human factor—the patterns of civilization, the human and social decisions and actions that underlie the remaking of the natural environment to reflect human values and aspirations [6, 20]. The Mississippi and Ohio river system is a multiple-use resource shared by many. This "public commons" presents huge issues of how to manage to meet complementary and competing goals within resource constraints. The USACE is charged by Congress to engineer this resource to ensure navigation, mitigate flood risk, protect the river ecosystem, and provide regulatory oversight. However, engineering science is silent on how to select project locations and chose from a variety of possible designs to select those most socially acceptable. Further, legislation, policies, regulations, and planning documents do not provide adequate guidance for prioritizing projects, evaluating engineering designs, or assuring local or regional support for engineered projects.

People have diverse and conflicting beliefs, attitudes, and opinions about the value of the river system and how it should be managed. The uses of this resource involve public and private lands, agricultural practices and policies, natural resource rights, public water supplies and disposal, flood risk perceptions and expectations, lifestyle and consumption of nature behaviors, and allocations of moral and financial responsibilities [6]. Individual and local self-interests often compete for "winning" their preferred project and resources to construct it. These self-interests lead to fragmented solutions with unintended downstream or upstream consequences. Self-interests can also polarize cross-sectoral interests and block capacity to manage the river as a whole system. For this river system to become a world-class system, the people of the region need to view it as a shared, public commons worthy of investing time, energy, and financial resources that are of benefit to the whole region. They must have a vision of it as a unique inland waterway and develop a shared normative understanding about its economic, social, and ecological importance. They must be willing to place the public good over personal self-interest and accept the rights and obligations of living, working, and owning land in this region.

How is a shared vision constructed? How do we create communities of cooperation that don't ignore or belittle the diverse self-interests and sector-specific economic, environmental, or social concerns but listen and learn from each other in order to find shared solutions to common problems? Social science [6, 20, 21] suggests four key elements are foundational to constructing a civic structure capable of realizing system level goals: (*a*) a common vision; (*b*) iterative exchanges of knowledge and perspectives; (*c*) public and private collaborative partnerships in the public interest; and (*d*) processes and mechanisms that integrate and utilize scientific and nonscientific knowledge in priority setting and mobilization of resources to accomplish the shared vision.

The USACE is well positioned to provide the mission-vision leadership and develop mechanisms and processes for integration of scientific and nonscientific knowledge. River management requires communication, cooperation, coordination, and joint investments across many federal and state public agencies (e.g., FEMA, NOAA, US Environmental Protection Agency, and USDA NRCS), local municipalities, levee and soil and water districts, private organizations, and individual landowners and managers. Thus, as a public agency they cannot single-handedly develop the vision nor carry it alone. However, the federally mandated annual high and low water public hearings conducted by the USACE and MRC are critical forums that provide neutral space for public dialogue, learning, and listening exchanges on river issues. These iterative exchanges of knowledge and perspectives among landowners and managers, stakeholders, public agencies, not-for-profit organizations, citizen leaders, and taxpayers provide opportunities for the construction of shared concerns and initiation of collaborative efforts to find and implement solutions in the public interest. The USACE creates a respectful and orderly process for listening and information exchange. They use the public hearing forum to convey that citizen voices are heard and are part of the public record. These hearings enable public exchanges that communicate engineering challenges and progress. They are a place where sectoral organizations and individuals can publicly voice frustrations and concerns, recommend resource allocations, suggest technologies, and bring scientific knowledge to problem identification and potential solutions. Equally importantly, these hearings are opportunities for stakeholders to express gratitude for and acknowledge the value of public and private projects that have met community's needs.

Effective management that reflects citizen public interests depends on building cross-sectoral and geographically diverse partnerships. There are abundant examples of public and private co-joint partnerships throughout the river system. Leadership for these partnerships has developed historically and continues to emerge along the entire spatial and temporal scale, including levee districts and local port authorities. The Sny Island Levee and Drainage District in Illinois (see chapter 4) and Little River Drainage District in Missouri (see chapters 5 and 6) are examples of such partnerships. Collaborative partnerships are built from social relationships and sectoral networks of trust and mutual respect that share common goals. For the inland waterway vision and profile to be raised to a national level, these effective local partnerships need to extend their geographic and sectoral relationships to encompass a larger network. Another effort, the America's Watershed Initiative (http://americaswatershed.org/), a public-private-sector collaborative has begun working to find solutions to the challenges of managing the Mississippi River [14]. Their steering committee represents a diversity of sectors including conservation, navigation, agriculture, flood control and risk reduction, industry, academic, basin associations, local and state government, and the USACE/MRC.

These partnerships foster the passion and energy necessary to continually reinforce the shared vision of a world-class inland waterway and public norms of civic cooperation. Lastly, public agencies and public-private partnerships have a variety of processes and mechanisms they can use to bring scientific information to bear on management decisions. They also have roles that help ensure that scientific and nonscientific knowledge are integrated into priority setting, decision making, and mobilization of resources. These processes include formal and informal public gatherings ranging from in-person and virtual meetings; uses of print and visual media, websites, chat rooms, and Twitter; river festivals; and community and river-wide celebrations. Regular established and ad hoc workgroup meetings around specific action items, sectoral and civic organizational meetings, and cross-geographic groups all have potential to communicate and reaffirm a region-wide vision and provide opportunities for groups, agencies, and individuals to act on that vision.

Final Observation

Much can be learned by observing and studying the human and natural systems of river landscapes. We framed this book as a series of short case studies about leveed agricultural lands, river navigation, upland reservoirs, and landscape management for flood risks. Together these stories reveal that change is the only certainty in river systems. Many factors influence and affect change. The connectivity between soil and water creates vulnerability and opportunity. People differ greatly in their vision for and functional uses of river landscapes. Managing for resilience can best prepare us to adapt to future unknown risks and catastrophes.

---

[1] Theiling, C. 1999. River geomorphology and floodplain habitats. In Ecological Status and Trends of the Upper Mississippi River System 1998: A report of the Long Term Resource Monitoring Program. April 1999. LTRMP 99-T001. La Crosse, WI: US Geological Survey, Upper Midwest Environmental Sciences Center.

[2] Morton, L.W. 2014. Achieving water security in agriculture: The human factor. Greening the Agricultural Water System. Agronomy Journal 106:1-4, doi:10.2134/agronj14.0039.

[3] Melillo, J.M., T.C. Richmond, and G.W. Yohe (eds). 2014. Highlights of Climate Change Impacts in the United States: The Third National Climate Assessment. US Global Change Research Program. Washington DC: US Government Printing Office.

[4] Camillo, C.A. 2015. Protecting the Alluvial Empire: The Mississippi River and Tributaries Project. Vicksburg, MS: Mississippi River Commission.

[5] Park, J., T.P. Seager, P.S.C. Rao, M. Convertino, and I. Linkov. 2013 Integrating risk and resilience approaches to catastrophe management in engineering systems. Risk Analysis 33(3):356-367, doi:10.1111/j.1539-6924.2012.01885.x.

[6] Morton, L.W., and S. Brown. 2011. Pathways for Getting to Better Water Quality: The Citizen Effect. New York: Springer Science and Business.

[7] Gellman, E.S., and J. Roll. 2011. The Gospel of the Working Class. Urbana-Champaign, IL: University of Illinois Press.

[8] Barry, J.M. 1997. Rising Tide: The Great Mississippi Flood of 1927 and How It Changed America. New York: Simon and Schuster.

[9] Schilling, K.E., P.W. Gassman, C.L. Kling, T. Campbell, J.K. Jha, C.F. Wolter, and J.G. Arnold. 2013. The potential for agricultural land use change to reduce flood risk in a large watershed. Hydrological Processes, doi:10.1002/hyp.9865.

[10] Weber, W.L. 2015. On the economic value of wetlands in the St John's Bayou-New Madrid Floodway. Cape Girardeau, MO: Southeast Missouri State University, Department of Economics and Finance.

[11] Polasky, S.T., K. Johnson, B. Keeler, K. Kovacs, E. Nelson, D. Pennington, A.J. Plantinga, and J. Withey. 2012. Are investments to promote biodiversity conservation and ecosystem services aligned? Oxford Review of Economic Policy 28(1):139-163.

[12] Jenkins, W.A., B.C. Murray, R.A. Kramer, and S.P. Faulkner. 2010. Valuing ecosystem services from wetlands restoration in the Mississippi Alluvial Valley. Ecological Economics 69:1051-1061.

[13] Kozak, J., C. Lant, S. Shaikh, G. Wang. 2011. The geography of ecosystem service value: The case of the Des Plaines and Cache River wetlands. Illinois Applied Geography 31:303-311.

[14] America's Watershed Initiative. 2015. Methods report on data sources, calculations, additional discussion, America's Watershed Initiative Report Card for the Mississippi River. October 9. http://americaswater.wpengine.com/wp-content/uploads/2015/10/Mississippi-River-Report-Card-Methods-v10.1.pdf.

[15] US Environmental Protection Agency. 2015 Connectivity of streams and wetlands to downstream waters: A review and synthesis of the scientific evidence. EPA/600/R-14/475F. Washington, DC: Office of Research and Development US Environmental Protection Agency.

[16] Mason, R.R., and B.A. Weiger. Stream gaging and flood forecasting. US Geological Service. http://water.usgs.gov/wid/FS_209-95/mason-weiger.html.

[17] National Weather Service. Glossary. Flood stage; zero datum; mean sea level. http://w1.weather.gov/glossary/.

[18] Lowery, B., C. Cox, D. Lemke, P. Nowak, K.R. Olson, and J. Strock. 2009. The 2008 Midwest flooding impact on soil erosion and water quality: Implications for soil erosion control practices. Journal of Soil and Water Conservation 64(6):166A, doi:10.2489/jswc.64.6.166A.

[19] Camillo, C.A. 2012. Divine Providence; The 2011 flood in the Mississippi River and tributaries project. April 5. Protecting the Alluvial Empire: The Mississippi River and Tributaries Project. Vicksburg, MS: Mississippi River Commission.

[20] Morton, L.W. 2008. The role of civic structure in achieving performance-based watershed management. Society and Natural Resources 21(9):751-766.

[21] Morton, L.W., Y.C. Chen, and R. Morse. 2008. Small town civic structure and intergovernment collaboration for public services. City and Community 7(1):45-60.

AUTHORS

Kenneth R. Olson is professor of soil science in the Department of Natural Resources and Environmental Sciences, College of Agricultural, Consumer, and Environmental Sciences, University of Illinois, Urbana, Illinois. He is an internationally known soil scientist who has published extensively on soil conservation, soil management, soil erosion and degradation, soil productivity, soil characterization and interpretation, and methodologies for measuring change over time in soil organic carbon.

Lois Wright Morton is professor of sociology in the Department of Sociology, College of Agriculture and Life Sciences, Iowa State University, Ames, Iowa. Her research encompasses the social relations and civic structure of place associated with agriculture, performance-based agroecosystem management, locally led watershed management, impacts of long-term weather change on agricultural land use decision making, rural development, and natural resource management. She has published a number of books including *Pathways for Getting to Better Water Quality: The Citizen Effect*.

INDEX

Page numbers in **bold** denote maps and page numbers in *italics* denote photographs or illustrations.

A

adaptive management, 3, 68, 96, 193–195, 198–201
agriculture, 50, 61, 169, 195, 207. *see also* crops
    Big Swamp, 39, 40–42, 46, 48–51
    flooding, 61–68
    future management, 205–207
    levee breaches, 61–68, 100–104
    levee-protected, 6, 61, 75, 98, 106, 193–201
    New Madrid Floodway, 79–80
    swamp conversion, 32–41, 50, 57, 107
Alexander County, IL, 108–109, 114–117, 120–122
Allegheny Mountains, 143, 149
alluvial plains, 42, 50, 54, 90, 95–96, 109, 136, 193
Appalachians, 9, 10, 15–16, 44, 159–160
aqueducts, 6, 29, 172, 183
Arkansas, 23–24, 39–40, 49–50
Army Corps of Engineers. *see* United States Army Corps of Engineers (USACE)
artificial crevassing, 77–78

B

backflow, 38, 43, 52, 112, 116–117, 120, 153
backup. *see* backflow
backwater, 56–57, 173–174
barges
    damage from, 182–183, 188–192
    for dredging, 179
    navigation, 146, 148, 156, 170–171, 208
    specialized, 179
    towed, 183
Barkley Lake, 159, 161–162
Barkley Locks and Dam, 5, 148, 159, 161–162
bayou, 32, 35, 52. *see also* wetlands
bedrock, 5, 8, 21–22, 47, 109, 176–181
benefit tax assessment, *see* tax assessment
berms, 121, 135, 136, 139
Big Oak Tree State Park, 56–57, 79, 80, 83–85, 96
Big Swamp, 32–43, 46, 49–50, 110
Birds Point-New Madrid Floodway. *see also* New Madrid Floodway
    creation of, 48
    gap, 55–56
    height, 116
    induced breaching, 2, 76–82, 93
    repair, 83–89
    St. John's Bayou basin changes, 52–53
black families, 148
blankets (impervious), 136

blowout holes. *see* crater lakes
bootlegging, 50
borders (geographic), 17–24, 151–152
bottomlands, 114, 207. *see also* floodplain
    drainage, 42–51
    forested, 32
    leveed, 61–62, 65, 68, 106–107, 120–121, 194
    reclamation, 89
    unleveed, 95, 120–121, 130
bridges, 155, 182
    damage, 93
    Mississippi River, 130, 180–181
    Ohio River, 144–145, 147–148
buildings, 67

C

Cache River Drainage District, 109
Cache River valley, 11, 22, 108–113
Cahokia, IL, 167
Cairo, IL, 5, 108, 123–131, 132–140
Cairo City and Canal Company, 126
canals, 17, 145
    Barkley Dam, 161–162
    Chicago Sanitary and Ship Canal, 182, 184–186
    Hennepin Canal, 13, 171–172
    Illinois and Michigan Canal, 19, 182, 183, 184, 188
    Louisville and Portland Canal, 4
    Marseilles Canal, 188
    Panama Canal, 38, 46, 172, 184
Cape Girardeau, 13–15
carp, 186–188
Castor River, 38–39, 42
Center Hill Dam, 163
channels. *see names of specific rivers*
Cherokee, 151
Chicago, 19, 183, 192
Chicago Portage, 182
Chicago River, 183–184
Chicago Sanitary and Ship Canal, 182, 184–186
Chickamauga Reservoir, 155–156
Civil War, 4, 17, 19, 126–127, 152, 160
Clarksville, IN, 143
climate, 1, 10, 16, 193. *see also* weather
    ancient MS and OH rivers, 8–15
    continental, 44–45
    shifts, 62, 68, 96–97

Third National Climate Assessment, 203
variability, 60, 90, 94–95, 112, 149, 162–164, 193, 203
coal, 9, 145
Commerce farmer levee, 48, *62*, 116, 117
Commerce Farmer Levee and Drainage District, 55
Commerce Fault, 21, 178
confinement-dispersion, 3, 7, 124, 162
confluence, 8, 10
  flooding, **3**, 108, 117, 130–131
  historic OH and MS rivers, 15, 54, 110, 114, 177–178
  OH and MS rivers, 2, 16, 76–**77**, 126
conservation practices, 74–75, 104–105, 206–207
Consolidated Drainage District, 1, 86–87
continental divide, 142, 182, 183
contour farming, 74
corn, 40–41, 80, 180
cotton, 36–37, 40–41, 90
crater lakes, 61–64, 67, 71, 73, 80, 85
crevasse. *see* levee breaches
Crooked Creek, 38–39, 42
crop residue, 66
crop rotation, 74, 91
crops. *see also names of specific crops*
  flooding, 55, 61, 66, 69, 79, 85–86
  production, 37, 40–41, 46–47, 167, 195, 203
  yields, 120, 153, 178
Crowley's Ridge, 9, 15, 33–35
culverts, 74
Cumberland River
  Barkley Dam, 5, 161–162
  ecology, 157
  geologic history, 8, 11–12
  managing landscapes, 159–164
  navigation, 4, 164
curfews, 129
cutoffs, 136

D

dams
  Barkley Locks and Dam, 5, 148, 159, 161–162
  Center Hill Dam, 163
  Illinois River, 16
  Kentucky Dam, 5, 148, 150, 154–155, 202
  Lock and Dam 1, 161
  Lock and Dam A, 161
  Marseilles Lock and Dam, 183, 188–192
  McAlpine Locks and Dam, 146
  Miami Valley, 154–155
  Ohio River, 5, 16, 146, 148–149
  Olmsted Lock and Dam, 5, 142, 148–149, 208
  upland water storage, 74

upper MS River, 16, 46, **168**, 170–171
Watts Bar Lock and Dam, 155–156
Wolf Creek Dam, 161, 163
decision making, 209–210
deltaic deposits, 9, 67, 72, 80, 85, 98
Department of Housing and Urban Development, 94
detention basins, 38–39, 48
Development Block Grant, 94
Dickens, Charles, 145
dispersion, 2
dispersion risk management, 2
diversion ditches, 110–111
diversions, upland, 42–51
downwarping, 8–9
Drainage District Law, 37
drainage districts. *see also* Little River Drainage District
  Cache River Drainage District, 109
  Commerce Farmer Levee and Drainage District, 55
  Consolidated Drainage District 1, 86–87
  Farmer Levee and Drainage District, 114
  Mingo Drainage District, 39
  St. John's Bayou Drainage District, 52–60, 78
  Sny Island Levee Drainage District, 25–31
drainage ditches
  Big Swamp, 39, 40
  sediment in, 66–67, 72–73, 80, 86–87
drainage structures, 74
dredging, 5, 110, 148, 176, 179
Driftless Area, **11**, 13, 166–167
drought, 14, 203
Drought of 2012, 2, 5, 57, 139, 158, 163, 176, 178–179
Dust Bowl, 181

E

earthquakes, 9, 21, 22, 155, 178
easements
  New Madrid Floodway, 55–56, 59, 77, 78, 91, 194–195
  permanent, 96
  to redirect flow, 210
  through eminent domain, 78, 91
ecology
  New Madrid Floodway, 196–197
  Tennessee River, 156–157
  upper MS River, 171, 173–174
economics, 140, 149, 157, 196. *see also* crops
  assessment, 200
  Big Swamp drainage, 46–47
  Commerce Farmer Levee and Drainage District, 55
  New Madrid Floodway, 196–197
ecosystem, 60, 149, 150, 196
effigy mounds, 167, 173

electric fish barrier, 186
Embarras River, 70–73
Emergency Watershed Protection (EWP), 86
eminent domain, 78, 91
employment, 46–47, 87, 154
environmental advocacy, 2
Environmental Defense Fund, 56–57, 94
environmental impact statements, 57–59
environmentalists, 57
evacuations, 127, 210
    2008, 69
    Cairo, IL, 129, 136, 140
    conditions for, 210
    Marseilles, IL, 189–190
    New Madrid Floodway, 93

F

Falls of St. Anthony, 168–169, 170
Falls of the Ohio, 4, 143
Farmer Levee and Drainage District, 114
fault zones, 9, 21, 22, 178, 195
Federal Emergency Management Association (FEMA), 94, 204, 211
ferries, **53, 84, 92, 177,** 181
fires, 147, 184
fish, 149, 157, 173, 186–188
Flood Control Act of 1928, 47, 76
Flood Control Act of 1936, 5, 141–142, 206
Flood Control Act of 1938, 141–142, 159, 161
Flood Control Act of 1954, 55
Flood Control Act of 1965, 77, 78, 91
Flood of 1927
    Cairo, IL, 127
    Headwaters Diversion System, 47, 48
    lessons from, 2, 40
    levee breaches, 76, 90–91, 98
Flood of 1937. *see also* New Madrid Floodway
    Cairo, IL, 127
    Cumberland River, 159
    lessons from, 202
    Mississippi River, 48
    MS River height, *55*
    New Madrid levee, 78
    Ohio River, 141, 147–148, 150–151, 202
    Paducah, KY, 153–154
    sharecropping, 91
    Tennessee Valley, 153–154
Flood of 2011
    Alexander County, IL, 117–118
    Cache River valley, 111–113
    Cairo, IL, 5, 123–130, 132–140

    Commerce farmer levee, 55, *56*
    confluence of MS and OH rivers, **3**, 5
    Cumberland River, 162–163
    economic losses, 90
    Headwaters Diversion System, 48
    Len Small-Fayville levee, 114–122
    levee breaches, 2, 98
    New Madrid Floodway, 83
    Pinhook, MO, 57
flood pulse, 38
flood repair, 121, 138
floodgate, 111
floodplains, 2, 16
floods. *see also* Flood of 1927; Flood of 1937; Flood of 2011
    1800s, 124, 141
    1993, 26, 117, 193
    1997, *55*
    2008, 25, 55, 69–75
    2013, 188–192
    internal, 52–60
    O'Bryan Ridge, 98–107
    seasonal, 2–3, 50–51, 61, 194
floodwalls, 47, 48, 61, 62–63, 130
floodways, 162. *see also* New Madrid Floodway
flour mills, 169
fluvial deposit, 9
food security (insecurity), 56, 193, 203
Food Security Act, 29
forests. *see also* Big Oak Tree State Park; timber
    flood damage, 106, 207
    loss of, 54–55
    Minnesota, 169
    riparian, 90, 121, 130, 174, 195
French and Indian War, 17, 182
fuse plugs
    Birds Point, 98, 118–120, 123, 129
    Birds Point-New Madrid Floodway, 83–85
    New Madrid Floodway, 77–79

G

gaging station, 149. *see also* river gage
General Survey Act of 1824, 4
geographic information systems (GIS), 199
geology. *see also* glaciation
    ancient Cumberland River, 159–160
    ancient MS River, 8–15, 176–178
    ancient OH River, 8–15, 159–161
    Cache River valley, 109
    Cairo, IL, 126
    Cumberland River, 163–164
    fault zones, 9, 21

Missouri, 21
Ohio River, 142–143
Ozark Plateau, 43–44
St. John's Bayou basin, 53–54
Sny River channel, 27
Tennessee River, 163–164
timescape, 9
upper MS River, 165–167
geopolitical boundaries, 17–24, 151–152
GIS (geographic information systems), 199
glacial lakes, 13, 14, 165–166
glaciation, 8, 10–16, **11**
Cache River valley, 109
Illinois, 70
Illinois River, 19
Mississippi River valley, 13–15
Missouri, 19
Ohio River valley, 21–22
St. John's Bayou, 54
Wisconsin, 19
Grafton, IL, 13, 172
Great Depression, 47, 50, 91
The Great Lakes, 182, 186–188
Great Plains, 178–179
Gulf Coast Sea, 9
Gulf Coastal Plain, 8, 9, **10**, 15
Gulf of Mexico, 3, 8, 13, 142, 165, 175, 192
influence on weather, 44, 147, 153, 161, 179
gullies, 61, 62, 64, 71–72, 80, 110
gully fields, 64–65, 80–81, 87–89, 98–101

H

Hadley-McCraney Diversion, 28
Headwaters Diversion System, 38–39, 42–46, 47–51
Hennepin Canal, 13, 171–172
Hickman levee, 48, 95, 116, 117
highways. *see* roads
homes, 48, 67, 93, 121, 141, 164, 189
Horseshoe Lake, 114, **115**, 121
human-natural systems, 6–7
hydraulic jumping, 65, 87
hydraulic roughness, 106, 196
hydroelectric power, 146, 149, 154–157, 161–162, 169, 172, 188–189
hydrology, 57, 60, 95, 112, 131, 134, 196
hypoxia, 174

I

Ice Age. *see* glaciation
Illinois, **20**
2008 flooding, 69–75

Alexander County, 114–122
border locations, 17–24
Cache River valley, 108–113
Cairo, 5, 123–131, 132–140
Len Small-Fayville levee, 114–122
Little Egypt, 179–180
Marseilles, 182, 188–191
Sny Island Levee Drainage District, 25–31
statehood, 126
Thebes, 5, 179–181
Illinois and Michigan Canal, 19, 182, 183, 184, 188
Illinois Drainage Act, 25
Illinois River, 10, 13, 16, 182–183, 188–192
Illinois Waterway, 182–192
impounds. *see* reservoirs
Indian Removal Act, 152
Indiana, 17, **18**, **20**, 69–75
infrastructure, 207–209. *see also* bridges; roads; structures
Ingram Barge Company, 191
inland seas, 9, 16
inland waterways, 142–143, 149, 165, 169, 171, 203, 207–209
Inland Waterways Commission, 150
insurance, 47–48, 66, 67, 80, 86, 94, 120
Interior Lowlands, 9, **10**, 15, 16, 44
Inter-River Improvement District, 39
invasive species, 57, 186–188
Iowa, **20**
Iowa River, 10, 69

J, K

Jadwin, Edgar, 127
Jackson Purchase, 151–152
Karnak levee, 111, 113
Kentucky
border locations, 17, **18**, **20**, 23–24
Flood of 2011, 130
Hickman levee, 48, 95, 116, 117
Paducah, 150–153
Kentucky Dam, 5, 148, 150, 154–155, 202
Kiser Creek Diversion, 28
Kochtitzky Family, 36–38

L

Lake Agassiz, 13, 14, 165
Lake Itasca, MN, 165
Lake Michigan, 17, **18**, 19, 182–188
Lake Warren, 13
Land-Between-the-Lakes, 163, 164
land bridges, 8, 11, 15, 43–44
Land Drainage Act of 1879, 180
land use, 90, 107, 195

lawsuits
    Chicago Sanitary and Ship Canal, 186–188
    Environmental Defense Fund, 56–57, 94
    farmers, 88–89, 195, 196–197
    landowners, 39, 78
    Marseilles Elementary District 150, 191
    Missouri Attorney General, 120, 196, 197
    National Wildlife Federation, 56–57, 94
    Pittsburgh vs. Wheeling, 144
    US vs Sanitary District of Chicago, 186
    Wildlife Defense Fund, 195–196, 197
legislation
    1879 Illinois drainage law, 75
    Drainage District Law, 37
    Flood Control Act of 1928, 47, 76
    Flood Control Act of 1936, 5, 141–142, 206
    Flood Control Act of 1938, 141–142, 159, 161
    Flood Control Act of 1954, 55
    Flood Control Act of 1965, 77, 78, 91
    Food Security Act, 29
    General Survey Act of 1824, 4
    Illinois Drainage Act, 25
    Indian Removal Act, 152
    Land Drainage Act of 1879, 180
    National Swamp Lands Acts, 4, 32, 35–37, 202
    Ohio River Navigation Modernization Program, 5, 146
    Ransdell-Humphreys Flood Control Act, 39
    Rivers and Harbors Act, 5, 146, 175, 176, 185
    Tennessee Valley Authority (TVA), 5, 148, 154–155, 156, 157, 206
    Water Resources Development Act, 149
Len Small-Fayville farmer levee, 48, 49
Len Small-Fayville levee, 108, 114–122
levee breaches, 62, 78, 90, 124, 127, 204
    Dorena, MO, 127
    Embarras River, 71–74
    Flood of 2011, 55, 129–130
    fuse plug, 118, 129
    impacts of, 61–68
    induced, 2, 76–82
    Karnak levee, 111
    Len Small-Fayville levee, 114, 118–122
    natural breach, 14
    New Orleans, 4
    O'Bryan Ridge, 98–107
    soil damage, 68, 194, 207
levees, 3, 50. *see also* Headwaters Diversion System; levee breaches
    Big Swamp conversion, 32
    Birds Point, 116, 117
    Commerce, 116
    earthen, 61, 62–63
    failure, 133
    frontline, 52, 77, 95, 116, 194
    gap in levee, 56, 94, 197
    Hickman levee, 48, 95, 116, 117
    Karnak, 111
    Len Small-Fayville farmer levee, 48, 49
    Len Small-Fayville levee, 108, 114–122
    levees-only strategy, 124, 162
    natural levee, 128
    New Orleans, 3–5
    protecting, 139–140
    reconstruction, 83–85, 116, 121
    Reevesville levee, 110, 111
    repair, 67
    saturation, 63
    setback, 77
    Sny Island Levee Drainage District, 27, 28–29
    temporary, 73
    topping, 63, 117, 129
Lewis and Clark Expedition, 4, 143, 145, 172, 208
liquefaction, 178
Little Egypt, IL, 179–180
Little River, 42–43, 50
Little River Drainage District
    Big Swamp, 32, 37–38
    developments from, 46–51
    legacies, 47–49, 50–51
    management area, 42, 43
    St. Francis River basin, 40
Lock and Dam 1, 161
Lock and Dam A, 161
locks
    Barkley Locks and Dam, 5, 148, 159, 161–162
    Illinois River, 16
    Lock and Dam 1, 161
    Lock and Dam A, 161
    Marseilles Lock and Dam, 183, 188–192
    McAlpine Locks and Dam, 146
    Moline Lock, 170
    Ohio River, 5, 16, 146, 148–149
    Olmsted Lock and Dam, 5, 142, 148–149, 208
    upper MS River, 16, 46, **168**, 170–171
    Watts Bar Lock and Dam, 155–156
Lookout Mountain, 152
Louisiana, 3–4
Louisiana Purchase, 4, 19, 143, 172
Louisville and Portland Canal, 4
low-gradient water, 48, 50, 53, 58
lumber mills. *see* sawmills

M

Main Ditch, 52, 55, 93, 108, 110
Marseilles, IL, 182, 188–191
Marseilles Canal, 188
Marseilles Lock and Dam, 183, 188–192
McAlpine Locks and Dam, 146
meander, 8, 54, 71, 112, 173, 178
meltwaters, 8, 9, 11, 49, 109, 126, 160, 165
Miami Valley, 154–155
Michigan, **18**
Miller City, IL, 108, 122
mines, 44
Mingo Drainage District, 39
Minnesota, 168–169
Minnesota River, 13
Mississippi, 23–24
Mississippi Embayment, 9, 15, 44, 53–54
Mississippi Flyway, 16
Mississippi River. *see also* New Madrid Floodway
   2008 peak, 69
   basin, 1, **2**
   confluence with Ohio River, 123–131, 177–178
   dams, 16, 46, **168**, 170–171
   future management, 5–6, 202–212
   geologic history, 8, 10–15, 19–21, 23–24
   Illinois state border, 17–21
   landscape today, 15–16
   Len Small–Fayville levee, 114–122
   navigation, 169–175, 181
   past management, 3–5
   Sny Island Levee Drainage District, 25–31
   upper tributaries, 171–173
   watershed, 61
Mississippi River and Tributaries Commission, 116
Mississippi River Commission, 4–5, 78, 116–117, 127, 204, 211
Mississippi River Diversion, 110–111
Mississippians, 167
Missouri. *see also* Birds Point–New Madrid Floodway; New Madrid Floodway
   Big Swamp, 32–43, 46, 49–50
   Bootheel, 54–55, 87, 90, 93–96
   border locations, 19, **20**, 23–24
   Commerce farmer levee, 48, *62*, 116, 117
   Commerce Farmer Levee and Drainage District, 55
   Inter-River Improvement District, 39
   Mingo Drainage District, 39
   Ozark Plateau, 43–46
Missouri Compromise, 17, 19
Missouri River, 15, 172–173
Moline Lock, 170
The Mound Builders, 167–169

MRC (Mississippi River Commission), 4–5, 78, 116–117, 127, 204, 211

N

National Flood Insurance Levee Evaluation, 47–48
National Oceanic and Atmospheric Administration (NOAA), 178–179, 204, 211
National Park Service, 173
National Road, 143, 144
National Swamp Lands Acts, 4, 32, 35–37, 202
National Weather Service (NWS), 163
National Wildlife Federation, 56–57, 94, 186
Native Americans, 152, 167–169, 172, 183
Natural Resources Conservation Service (NRCS), 67, 80, 86–87, 88, 95–96, 211
navigation
   Cumberland River, 4, 164
   Mississippi River, 139, 169–175, 181
   Ohio River, 4, 139, 141–149
   reservoirs for control, 164
   Tennessee River, 155–156
New Deal, 47, 91, 148
New Madrid Floodway, 52–53. *see also* Birds Point–New Madrid Floodway
   activation, 48
   easements, 55–56, 77, 194–195
   future strategies, 96–97
   gap, 55–56
   induced breaching, 2, 76–82, 90, 202–203
   land use, 90–93, 95–96, 194–195
   levee locations, 95–96
   long-term weather patterns, 94–95
   O'Bryan Ridge, 98–107
   Pinhook, MO, 93–94
   realignment, 94–96
   repair, 83–89
   soil use and productivity, 195–196
New Madrid Seismic Zone, 9, 21, 22, 155
New Orleans, LA, 3–4, 202
NOAA (National Oceanic and Atmospheric Administration), 178–179, 204, 211
no-till systems, 74
NRCS (Natural Resources Conservation Service), 67, 80, 86–87, 88, 95–96
nuclear power, 155–156

O

O'Bryan Ridge, 87, 98–107
Ohio, 154–155
Ohio River
   basin, 1–2, **2**

confluence with Mississippi River, 123–131, 177–178
confluence with Tennessee River, 152–153
dams, 5, 16, 146, 148–149
falls, 4, 143
flood control, 148–149
flooding, 108–113, 147–148
future management, 5–6, 202–212
geologic history, 8, 10–13, 21–24
Illinois state border, 17
landscape today, 16
locks, 5, 16, 146, 148–149
navigation, 4, 139, 141–149
past management, 3–5
Ohio River Navigation Modernization Program, 5, 146
Olive Branch, IL, 108, 122
Olmsted Lock and Dam, 5, 142, 148–149, 208
oxbows, 54, 65, 71, 114, **115**, 178
Ozark Plateau, 43–46

P

Paducah, KY, 150–153, 202
Panama Canal, 38, 46, 172, 184
parks. *see* Big Oak Tree State Park
partnerships, 210–212
peak flow, 29, 48, 129, 131, 139
peak forecast, 120
Pennsylvania, 144, 145, 147
pervious materials, 138
Pinhook, MO, 57, 93–94
piping, 63, 71, 134–135, 136, 138
Pittsburgh, PA, 144, 145, 147
pollution, 81, 183–184, 194
port cities, 5, 141, 175, 183, 202, 205, 207, 208
population growth, 50
Post Creek Cutoff, 109–112
poverty, 50, 154, 207
productivity indices (PI), 28
public engagement, 58, 197–199
public hearing, 58, 197, 204
public policy, 2
public tensions, 194
pulse (of the river), 141

Q, R

quarries, 44
raceway, 169
railroads, 36–37, 46, 126, 155, 156, 180–181, 183
Ransdell-Humphreys Flood Control Act, 39
rapids, 156, 173
realignment, 1, 16, 95
Reconstruction Finance Corporation, 47

recreation, 45–46, 173
Reevesville levee, 110, 111
relief wells, 6, 117, 132, **134**, 135–140
reservoirs
    Barkley Lake, 159, 161–162
    Big Swamp conversion, 39
    Chickamauga Reservoir, 155–156
    Cumberland River, 161–162
    Kentucky Dam, 150, 158
resilience, 7, 60, 95, 205, 207, 212
resilience management, 3, 7
resistance-adaptation, 96–97
restoration, 87–89
retention ponds, 74, 206
Revised Universal Soil Loss Equation, 206
rice, 41
risk management, 2, 3, 193–201
river discharge, 209
river gage, 52, 59, 91, 153
    Cairo gage, 113, 120, 127, 132, 163, 210
river stage, 140, 147, 209
Rivers and Harbors Act, 5, 146, 175, 176, 185
roads
    flooding, 72–73, 74, 93, 105–106
    highways, 50
    National Road, 143–145
    Old Pole Road, 36
    sediment on, 66–67, 80
    severed, 87–88

S

St. Francis River
    flood control, 40
    flow, 39, 49–50
    watershed, 32, 33, **36**, 42–43
St. John's Bayou, 33, 52
St. John's Bayou Drainage District, 52–60, 78
sand berm, 121
sand boils, 61, 63–64, 71, 117, 123, 132–135, 193
sand deltas, **62**, 67, 72, 85, 194
sand deposits, 64, 66, 73–74, 80
sandbag dikes, 63, 120, 134
sandbagging, 108, 197
sawmills, 108, 169
scouring, 3, 50, 61, 64–66, 87–89, 98, 100, 105, 114, 124, 196
sediment
    deposition, 61, 62, 86–87, 106, 174, 194
    primary pollutant, 194
    removal, 66–67
    transport, 80–81
sediment basins, 31, 75

Big Swamp conversion, 39
Sny Island Levee Drainage District, 27–29, 30
seepage, 63–64, 123, 133–138
seismic activity, 9, 16, 32, 49, 160, 176, 178
settlements 6, 54, 90, 150, 170, 172, 179
settling basins. *see* sediment basins
Seven Island Conservation Area, 96
sewage, 81, 132, 184–186, 192, 194
sharecropping, 50, 90, 91, 93
siltation, 5
sinkholes, 132–136, 139–140, 163
Sixmile-Bay Creek Diversion, 27, 28
sloughs. *see* bayou; wetlands
slurry trenches, 136, 138–139
Smithfield Locks and Dams, 146, 148
Sny Island Levee Drainage District, 25–31
Sny River, 27
social tradeoffs, 95, 140, 207
social values, 2, 7, 196, 198–199, 200, 205, 210–212
soil erosion. *see also* gully fields
   conservation practices, 74–75, 206–207
   Flood of 2008, 69–70
   levee breaches, 67, 106
   timber soils, 124
soil surveys, 3, 27, 98, 199–200, 207
soils
   alluvial, 61, 70, 110, 114–116, 120, 171, 206
   assessment, 199–200, 207
   degradation, 153, 193–201, 206–207
   drainage, 74–75
   functions, 195
   geological legacies, 7
   New Madrid Floodway, 79–80
   productivity, 28, 65, 83, 96, 101, 193–196, 199, 207
   reclamation, 89
   replacement, 65, 67, 85
   tolerable loss (T), 29, 206
   shearing, 135
   soil organic carbon, 135
   Udifluvents, 101
sorghum, 41
soybeans, 40–41, 80, 82, 100–101, 103–105, 120
stakeholders, 197–199, 200
steamboats
   Cumberland River, 4, 160
   Mississippi River, 4, 169
   Ohio River, 141, 143, 145, 208
   Paducah, KY, 152
stockyards, 183–184
strip cropping, 74

structures, 67
sublevees, 136
Swamp Lands Acts, 4, 32, 35–37, 202
swamps, 8, 32–43, 46, 49–50, 182. *see also* wetlands; bayous

T

tailwater, 156, 157
tax assessment, 25, 37, 39, 47, 55, 149
Teays River, 8, 10–11
Ten Mile Pond Conservation Area, 93
Tennessee, 23–24
Tennessee Divide, 164
Tennessee River, 5, 8, 11–13, 148, 150–158, 163–164. *see also* Kentucky Dam
Tennessee Valley Authority (TVA), 5, 148, 154–155, 156, 157, 206
terraces, 74, 75, 104, 206
Thebes Gap, 14, 49, 110, **118**, 176, 178–179
Thebes, IL, 5, 179–181
tile drains, 74
tillage, 81, 206
timber, 37, 46, 93, 110, 160–161, 169, 179
tobacco, 153
topsoil, 64, 69, 103, 105
tow *or* towboat, 171, 179
towpaths, 183
Trail of Tears, 152
transportation corridors, 143–145
Treaty of Paris, 17
TVA. *see* Tennessee Valley Authority (TVA)

U, V

uncertainty, 198
Underground Railroad, 126
Union Stockyard, 184
Union troops, 127
United States Army Corps of Engineers (USACE), 4–5, 25, 56–60, 94, 159
   flood response, 90, 98, 117, 123, 138–139, 153, 169, 202
   leadership role, 204, 211
   river navigation, 142, 176, 182
United States Department of Agriculture Risk Management Agency, 86
Universal Soil Loss Equation, 206
uplands, 112, 206
USACE. *see* United States Army Corps of Engineers (USACE)
USGS (United States Geological Survey), 209
vegetation, 50, 196
Virginia, 144–145
vision, 211

W

Wabash River, 11, 17–21, 70–72, 123, 126, 147, 152, 160
waste disposal, 183–184
Water Erosion Prediction Project, 206
water quality, 50, 174, 192, 203
water resources, 7
Water Resources Development Act, 149
water reverse-flow, 108, 111, 185
water rights, 151–152
water scarcity, 203
water velocity, 50
waterways, 74, 75
Watts Bar Lock and Dam, 155–156
Watts Bar Nuclear Power Plant, 155–156
weather, 85, 141, 143, 154, 159, 184, 200. *see also* climate
   2008 flood, 69, **70**
   leading to drought, 178–179
   leading to flood, 54, 56, 90–91, 128, 147, 153, 159
   long-term patterns, 94–95
   records, 38, 159
   sequential storms, 147
wetlands, 7, 31, 58–60, 96. *see also* bayou; swamps
   drainage, 35, 42, 112, 195
   forested, 57, 110
   gully, 101
   inventories, 35
   unclaimed gullies, 65
   value of, 50, 173, 207–209
Wetlands Reserve Program, 95, 96
wheat, 40–41, 80, 82
Wheeling, WV, 143–145
Whitewater River, 38–39, 42
wicket dams, 5, 146, 148, 208
wildlife, 16, *33*, 40, 59, 84
wildlife refuges, 57, 173, 174
wing dam, 170
Wisconsin, 19, **20**
Wisconsin River, 13, 166
Wolf Creek Dam, 161, 163
Woodland dwellers, 167